Grassland Productivity and Ecosystem Services

MIX
Paper from
responsible sources
FSC® C018575
FSC
www.fsc.org

Grassland Productivity and Ecosystem Services

Edited by

Dr Gilles Lemaire

INRA-UP3F, France

Professor John Hodgson

Massey University, New Zealand

Dr Abad Chabbi

INRA-UEFE, France

www.cabi.org

CABI is a trading name of CAB International

CABI
Nosworthy Way
Wallingford
Oxfordshire, OX10 8DE
UK

Tel: +44 (0)1491 832111
Fax: +44 (0)1491 833508
E-mail: cabi@cabi.org
Website: www.cabi.org

CABI
875 Massachusetts Avenue
7th Floor
Cambridge, MA 02139
USA

Tel: +1 617 395 4056
Fax: +1 617 354 6875
E-mail: cabi-nao@cabi.org

A catalogue record for this book is available from the British Library, London, UK.

Library of Congress Cataloging-in-Publication Data

Grassland productivity and ecosystem services / edited by Gilles Lemaire, John Hodgson, Abad Chabbi.
p. ; cm.
Includes bibliographical references and index.
ISBN 978-1-84593-809-3 alk. paper)
1. Grassland ecology. 2. Grasslands. 3. Rangelands. I. Lemaire, Gilles, 1945- II. Hodgson, J. (John), 1937- III. Chabbi, Abad.

QH541.5.P7G7128 2011
577.4--dc22

2011010104

ISBN–13: 978 1 84593 809 3

Commissioning editor: Rachel Cutts
Editorial assistant: Alexandra Lainsbury
Production editors: Simon Hill and Holly Beaumont

Typeset by Columns Design XML Ltd, Reading, UK.
Printed and bound in the UK by MPG Books Group.

Front cover photo: Ricardo Weiler & Aníbal F. Parera/The Alliance for The Grasslands. Sedge Wren (*Cistothorus platensis*) on tall grass (*Andropogon lateralis*) in 'La Primavera', a cattle ranch in southern Brazil where The Alliance for The Grasslands is currently working with a group of meat producers who carefully graze their natural pastures to provide habitat to endangered and flagship birds from the imperiled Pampas.

The Alliance for The Grasslands (Alianza del Pastizal) is a non-profit initiative of BirdLife International and its partners, now involving dozens of organizations and rural producers towards sustainable management of the grasslands in Brazil, Paraguay, Argentina and Uruguay (www.pastizalesdelconosur.org).

Contents

Contributors

Abberton, Michael, IBERS, Aberystwyth University, Gogerddan, Ceredigion, Wales, SY23 3EB, UK; E-mail: mla@aber.ac.uk

Abdelguerfi, Aïssa, ENSA, El Harrach, Algeria; E-mail: aabdelguerfi@yahoo.fr

Ameziane El Hassani, Tayeb, IAV Hassan II, Rabat, Morocco; E-mail: amezianet@gmail.com

Askew, Andrew P., Department of Biology, Syracuse University, 107 College Place, Syracuse, NY, USA; E-mail: a.p.askew@sheffield.ac.uk

Attard, Eléonore, Université de Lyon, INRA (USC 1193), CNRS (UMR 5557), Laboratoire d'Ecologie Microbienne, Bât Mendel, 43 bd du 11 Novembre 1918, 69622 Villeurbanne, France

Badenhausser, Isabelle, INRA, Centre d'Etudes Biologiques de Chizé (UPR 1934), 79360, Beauvoir sur Niort, France; E-mail: badenh@cebc.cnrs.fr

Balent, Gérard, INRA, UMR 1201 DYNAFOR, F-31326 Castanet-Tolosan, France; E-mail: gerard.balent@toulouse.inra.fr

Bol, Roland, Rothamsted Research, North Wyke, Okehampton, EX20 2SB, UK; E-mail: roland.bol@bbsrc.ac.uk

Bretagnolle, Vincent, CNRS, Centre d'Etudes Biologiques de Chizé (UPR 1934), 79360, Beauvoir sur Niort, France; E-mail: breta@cebc.cnrs.fr

Buchmann, Nina, ETH Zurich, Universitätsstr. 10, 8092 Zurich, Switzerland; E-mail: nina.buchmann@ipw.agrl.ethz.ch

Chabbi, Abad, INRA-UEFE, 86600 Lusignan, France; E-mail: abad.chabbi@lusignan.inra.fr

Collins, Rosemary, IBERS, Aberystwyth University, Gogerddan, Ceredigion, Wales, SY23 3EB, UK; E-mail: rpc@aber.ac.uk

de Faccio Carvalho, Paulo C., University Federal of Rio Grande do Sul, Porto Allegre, Brazil; E-mail: paulocfc@ufrgs.br

Decaëns, Thibaud, Laboratoire d'Ecologie, EA 1293 ECODIV, FED SCALE, Bâtiment IRESE A, UFR Sciences et Techniques, Université de Rouen, F-76821 Mont Saint Aignan Cedex, France; E-mail: thibaud.decaens@univ-rouen.fr

Delagarde, Rémy, Institut National de la Recherche Agronomique, UMR1080, Unité Mixte de Recherches INRA Agrocampus Ouest Production du Lait, 35590 Saint-Gilles, France; E-mail: remy.delagarde@rennes.inra.fr

Dumont, Bertrand, INRA Unité de Recherches sur les Herbivores, Centre de Clermont-Fd/Theix, 63122 Saint Genès Champanelle, France; E-mail: bertrand.dumont@clermont.inra.fr

Dunn, Robert M., Rothamsted Research, North Wyke, Okehampton, EX20 2SB, UK
Ferrier, Robert C., The James Hutton Institute, Craigiebuckler, Aberdeen, Scotland, AB15 8QH, UK; E-mail: bob.ferrier@hutton.ac.uk
Franzluebbers, Alan J., USDA – Agricultural Research Service, 1420 Experiment Station Road, Watkinsville GA 30677, USA; E-mail: alan.franzluebbers@ars.usda.gov
Fridley, Jason D., Department of Biology, Syracuse University, 107 College Place, Syracuse, NY, USA; E-mail: fridley@syr.edu
Fritz, Hervé, CNRA UMR 5558, LBBE, Université de Lyon 1, 1918 69611 Villeurbanne France
Fryxell, John M., Department of Integrative biology, University of Guelph, Canada; E-mail: jfryxell@uoguelph.ca
Garnier, Eric, CNRS, Centre d'Ecologie Fonctionnelle et Evolutive (UMR 5175), 1919 route de Mende, 34293 Montpellier Cedex 5, France; E-mail: eric.garnier@cefe.cnrs.fr
Gauffre, Bertrand, INRA, Centre d'Etudes Biologiques de Chizé (UPR 1934), 79360, Beauvoir sur Niort, France; E-mail: gauffre@cebc.cnrs.fr
Gibon, Annick, INRA, UMR 1201 DYNAFOR, F-31326 Castanet-Tolosan, France; E-mail: annick.gibon@toulouse.inra.fr
Grime, J. Philip, Department of Animal & Plant Sciences, University of Sheffield, Sheffield, S10 2TN, UK; E-mail: j.p.grime@sheffield.ac.uk
Hodgson, John, Institute of Natural Resources, Massey University, Palmerston North, New Zealand; E-mail: johnruth@inspire.net.nz
Huguenin-Elie, Olivier, Aroscope Reckenholz-Tänikon Research Station ART, CH-8046 Zurich, Switzerland; E-mail: olivier.huguenin@art.admin.ch
Humphreys, Mike, IBERS, Aberystwyth University, Gogerddan, Ceredigion, Wales, SY23 3EB, UK; E-mail: mkh@aber.ac.uk
Illius, Andrew W., Institute of Evolutionary Biology, School of Biological Sciences, University of Edinburgh, EH9 3JT, UK; E-mail: a.illius@ed.ac.uk
Kemp, David R., Charles Sturt University, University of Sydney Agricultural & Wine Sciences, PO Box 883, Orange, NSW 2800, Australia; E-mail: dkemp@csu.edu.au
Kleinman, Peter J.A., United States Department of Agriculture — Agricultural Research Service, Pasture Systems and Watershed Management Research Unit, University Park, PA 16802, USA; E-mail: peter.kleinman@ars.usda.gov
Lambert, M. Greg, AgResearch, Grasslands Research Centre, Private Bag 11008, Palmerston North 4442, New Zealand; E-mail: greg.lambert@agresearch.co.nz
Lavelle, Patrick, Université Paris VI/IRD, UMR 211 BIOEMCO, Centro Internacional de Agricultura Tropical (CIAT), TSBF_LAC, Ap Aereo 6713 Cali, Colombia; E-mail: patrick.lavelle@ird.fr
Le Roux, Xavier, Université de Lyon, INRA (USC 1193), CNRS (UMR 5557), Laboratoire d'Ecologie Microbienne, Bât Mendel, 43 bd du 11 Novembre 1918, 69622 Villeurbanne, France; E-mail: xavier.leroux@fondationbiodiversite.fr
Lecomte, Philippe, CIRAD Es UPMR Systèmes d'Elevage Méditerranéens et Tropicaux, Campus International de Baillarguet, 34398 Montpellier Cedex 5, France; E-mail: philippe.lecomte@cirad.fr
Ledgard, Stewart F., AgResearch, Ruakura ResearchCentre, Private Bag 3123, Hamilton, New Zealand; E-mail: stewart.ledgard@agresearch.co.nz
Lemaire, Gilles, INRA-UP3F, 86600 Lusignan, France; E-mail: gilles.lemaire@lusignan.inra.fr
Luo, J., AgResearch Limited, Ruakura ResearchCentre, Private Bag 3123, Hamilton, New Zealand; E-mail: jiafa.luo@agresearch.co.nz
Lüscher, Andreas, Aroscope Reckenholz-Tänikon Research Station ART, CH-8046 Zurich, Switzerland; E-mail: andreas.luescher@art.admin.ch
McDowell, Richard W., AgResearch, Invermay Agricultural Centre, Private Bag 50034 Mosgiel 9053, New Zealand; E-mail: richard.mcdowell@agresearch.co.nz

McIvor, John G., CSIRO Ecosystem Sciences, GPO Box 2583, Brisbane, Queensland 4001, Australia; E-mail: John.Mcivor@csiro.au

Macleod, Christopher (Kit) J.A., The James Hutton Institute, Craigiebuckler, Aberdeen, Scotland, AB15 8QH, UK; E-mail: kit.macleod@hutton.ac.uk

Marshall, Athole, IBERS, Aberystwyth University, Gogerddan, Ceredigion, Wales, SY23 3EB, UK; E-mail: thm@aber.ac.uk

Meiss, Helmut, INRA, Centre d'Etudes Biologiques de Chizé (UPR 1934), 79360, Beauvoir sur Niort, France; E-mail: meiss@cebc.cnrs.fr

Michalk, David L., Orange Agricultural Institute, Orange, NSW Australia

Monaghan, R.M., AgResearch, Invernay Research Centre, P.O. Box 50034, Mosgiel, New Zealand

Moraes (de), Anibal, University Federal of Parana, Curitiba, Brazil; E-mail: anibalm@ufpr.br

Morris, Craig D., Agricultural Research Council – Animal Production Institute (ARC-API), c/o University of KwaZulu-Natal, Private Bag X01, Scottsville 3209, South Africa; E-mail: morris@ukzn.ac.za

Navas, Marie-Laure, Montpellier SupAgro, Centre d'Ecologie Fonctionnelle et Evolutive (UMR 5175), 1919 route de Mende, 34293 Montpellier Cedex 5, France

Newton, Paul, AgResearch, Palmerston North, New Zealand; E-mail: paul.newton@agresearch.co.nz

Parsons, Anthony, Institute of Natural Resources, Massey University, Palmerston North, New Zealand; E-mail: A.J.Parsons@massey.ac.nz

Peyraud, Jean-Louis, INRA, Agrocampus Ouest, UMR 1080 Production du Lait, F-35590 St Gilles, France; E-mail: jean-louis.peyraud@rennes.inra.fr

Pilgrim, Emma S., Rothamsted Research, North Wyke, Okehampton, EX20 2SB, UK

Poccard-Chapuis, René, CIRAD Es UMR Systèmes d'Elevage Méditerranéens et Tropicaux, Campus International de Baillarguet, 34398 Montpellier Cedex 5, France; E-mail: poccard@cirad.fr

Poppi, Dennis P., Schools of Animal Studies and Veterinary Science, University of Queensland, Gatton 4343, Queensland, Australia; E-mail: d.poppi@uq.edu.au

Prache, Sophie, Institut National de la Recherche Agronomique, Unité de Recherches sur les Herbivores, Centre de Clermont-Fd/Theix, 63122 Saint Genès Champanelle, France; E-mail: sophie.prache@clermont.inra.fr

Recous, Sylvie, INRA, UMR614 FARE, 2 esplanade Roland Garros, 51000 Reims, France; E-mail: sylvie.recous@reims.inra.fr

Rowarth, Jacqueline, Massey University, Palmerston North, New Zealand; E-mail: J.S.Rowarth@massey.ac.nz

Rumpel, Cornelia, UMR Université Paris VI et XI-CNRS-IRD-INRA-Agroparistech, Thiverval-Grignon, France; E-mail: cornelia.rumpel@grignon.inra.fr

Russelle, Michael P., USDA - Agricultural Research Service, 1991 Upper Buford Circle, Room 439, St Paul, MN 55108, USA; E-mail: michael.russelle@ars.usda.gov

Snow, V.O., AgResearch Lincoln Research Centre, Private Bag 4749, Christchurch 8140, New Zealand; E-mail: val.snow@agresearch.co.nz

Soussana, Jean-François, INRA, UR 874, UREP Grassland Ecosystem Research, 234 Avenue Du Brézet, Clermont-Ferrand, F-63100 France; E-mail: jean-francois.soussana@clermont.inra.fr

Sulc, R. Mark, The Ohio State University, 224 Kottman Hall, 2021 Coffey Road, Columbus, OH 43210, USA; E-mail: sulc.2@osu.edu

Tallowin, Jerry R.B., Rothamsted Research, North Wyke, Okehampton, EX20 2SB, UK; E-mail: jerry.tallowin@bbsrc.ac.uk

Thornley, John, 6 Makins Road, Henley on Thames, RG6 1PP, UK; E-mail: johnthornley@waitrose.com

Introduction: Food Security and Environmental Impacts – Challenge for Grassland Sciences

Gilles Lemaire, John Hodgson and Abad Chabbi

The 'grassland' biome, defined as permanent vegetation of herbaceous plant communities, covers approximately 25% of the terrestrial Earth surface, approximately equivalent to the area of forest lands or the area of cultivated lands. Among the different types of grassland ecosystems, it is important to distinguish: (i) the climatically determined grasslands, in areas where water availability is not enough to allow the development of forest ecosystems and where natural vegetation remains at the herbaceous succession stage in dynamic equilibrium with herbivores (Lauenroth, 1979); and (ii) the anthropogenic grasslands, mainly located within temperate climate regions, where woody vegetation is excluded and herbaceous plant communities are maintained by means of human management for domestic herbivore production. Among these anthropogenic grasslands it is also possible to distinguish long-term naturalized grasslands from cultivated grasslands with a wide range of variation in degree of intensification. All these grasslands play an important role in agronomic, economic and social activities in the exploitation of their resources by domestic herbivores for food and fibre production. The efficiency of exploitation of the vegetation resource and its renewal by means of more or less intensified management systems has been the main focus for 'grassland science' within the different regions of the world during the last century (see Lemaire *et al.*, 2005). Nowadays, facing the problem of sustainable development, it is necessary also to consider the important role of grassland ecosystems in the dynamics of atmosphere, hydrosphere and continental surface interactions that drive changes and environmental hazards at local, regional and global levels, and also how the functioning of the grassland ecosystems can adjust to these changes.

The overall objective of this book is to analyse the present knowledge and the future trends of research for combining, in more synergistic ways, the classical view of grasslands, as a resource for secure feeding of an increasing human population, with the more recent perspective of the contribution of grasslands to the mitigation of environmental impacts and biodiversity erosion as consequences of human society activities. For this purpose, we used the concept of 'ecosystem services'. This concept has been developed within the Millennium Ecosystem Assessment (2005). Human well-being and ecosystems are linked by 'ecosystem services' – the benefits human societies and individual people can obtain from ecosystem functioning. These ecosystem services are of different natures:

- **Provisioning services** such as food, drinking water, timber, fibre and fuel.

- **Regulating services** such as climate, floods, diseases, wastes and water quality.
- **Cultural services** such as recreation, aesthetic enjoyment and spiritual fulfilment.
- **Supporting services** such as soil formation and protection, primary production processes, biogeochemical cycles and dynamics of biodiversity – all functions necessary for providing other services.

The concept of 'ecosystem services' is encapsulated within the definition of 'sustainable development', which argues that the three components – environmental, economic and social – must be analysed together in a deliberately multi-disciplinary approach. In this sense, the title of the book could be considered as not corresponding well to this requirement because productivity of grassland seems to be opposed to the services this ecosystem could provide, even though food production must be considered as an ecosystem service among others. The clear identification of the grassland productivity function involved with other and not directly productive functions is an editorial choice, and it identifies the challenge we have nowadays for moving from a traditional vision of grasslands used exclusively for domestic herbivore production to a more integrated vision, where all ecosystem services and their multiple interactions have to be considered together.

The chapters of this book are organized within five sections dealing with the different functions and the main ecosystem services expected from grasslands: (i) domestic herbivore feeding and animal production; (ii) the regulation of biogeochemical cycles and its consequences for the environment; (iii) dynamics of biodiversity hosted by grasslands; (iv) integration of grasslands within sustainable animal production systems; and (v) interactions of grassland areas with other land-use systems at the landscape level. The chapters are conceived more as 'opinion papers' than as 'review papers', in order to focus more deeply on the research perspectives necessary for meeting the challenges of combining the necessary productivity of grasslands for human food production with the optimization of their intrinsic capacity to mitigate environmental deterioration and biodiversity erosion. So, even though each chapter is focused on a relatively well-defined process or function, owing to the necessity to mobilize and to prospect disciplinary advances, the overall objective of the book is to identify interactions and synergies among elementary processes and functions that are at the base of most of the ecosystem services that grasslands can provide to human societies. All the ecosystem services are not considered here. We concentrate our attention on provisioning, regulating and supporting services (see above), which are directly related to the domains of agronomy, ecology and environmental sciences. The cultural services are beyond the scope of this book, not because we consider these services as irrelevant and negligible for grasslands, but because of the difficulty, at this stage, to include directly economic, sociological, ethnological and political sciences. Nevertheless, we hope that this book will serve to open areas of discussion with these human science domains.

Four important issues are addressed across the different chapters and should serve as guidelines for the structure of the book.

Trade-off between Productivity and Environment Benefits of Grasslands

Grassland ecosystems can be considered as composed of permanent herbaceous vegetation interacting with soil abiotic and biotic components and with animal consumers through more or less important human management activities. As in the case of forest ecosystems, the permanency of soil–vegetation interactions leads to strong coupling between the different cycling elements, mainly carbon, nitrogen and phosphorus. These couplings operate at two levels: (i) in plants, where N and P are linked to C for organic matter synthesis and its accumulation and long-term sequestration in soils; and (ii) in soils, where microbes abundantly fed with C are able to recapture and recycle mineral N and P. In most circumstances, grasslands are able to sequester C, N and P for relatively long-term periods, so contributing to the atmospheric CO_2 sink, and to the reduction of emissions of N compounds to the hydrosphere and atmosphere and their associated environmental risks. So grasslands can be considered as a land-use system that is very favourable for preservation of the environment. Nevertheless, this idealistic view has to be moderated because herbivores deeply decouple C and N–P cycles, through their urine and faeces depositions and their CH_4 emissions, offsetting a part of the beneficial effect of grassland vegetation–soil interactions. It appears that, as intensification of grassland productivity is enhanced by large N and P inputs, and when a greater proportion of the primary production is harvested and consumed by herbivores through efficient grassland management, then the beneficial effects of grassland could progressively decrease and even disappear. So a trade-off probably exists between optimizing grassland productivity and optimizing beneficial environmental impacts from grassland areas, depending on agro-ecological and socio-economic conditions.

Functional and Conservation Role of Grassland Biodiversity

Grasslands are often considered as hot spots of biodiversity, reflecting the fact that they host many specialized plants, animals and microbes, all nested in a great variety of microhabitats. This is particularly true for semi-natural grasslands with species-rich meadows, but agricultural intensification, particularly in temperate regions through ploughing and reseeding mono-specific swards and adoption of silage instead of hay-making, has resulted in a loss of biodiversity (Soussana and Duru, 2007). At the same time, overexploitation of natural grasslands in rangeland systems in semi-arid regions may have also diminished the species richness of these ecosystems. Moreover, the increase in atmospheric CO_2 concentration and the associated climate changes should have an impact on the species composition of most grassland areas through the equilibrium between C3 and C4 species, grasses and legumes, grasses and forbs or shrubs. In turn, these changes in species composition have a powerful effect on the main ecosystem functions and then on the main ecosystem services such as primary productivity and herbivore feeding, carbon sequestration and regulation of biogeochemical cycles. Soussana and Duru (2007) stressed that, within the CAB Abstract database, there is abundant literature combining 'grasslands' with one of three keywords ('biodiversity', 'climate change' and 'carbon sequestration and greenhouse gas') but only five references dealt with the combination of all keywords. So it is important to stress that even if the conservation aspect of biodiversity of grassland has to be analysed per se, as an important ecosystem service, its functional role through the interconnections with other ecosystem services has to be questioned. In this view, the loop of interactions between grassland biodiversity and soil organic matter dynamics, biogeochemical cycles and regulation of environmental fluxes to hydrosphere and atmosphere has to be studied with long-term perspectives and should be considered in relation to designing management practices. Moreover, grassland management not only affects the biodiversity locally hosted by the grassland ecosystem itself, but also the landscape biodiversity at larger scale, by means of the trophic network that can be sustained and the habitat provided for vertebrate populations such as birds or mammals. Thus the contribution of grassland areas to landscape ecology has to be studied in conjunction with other land-use systems, such as forests and arable cropping, as a source of diversity for regulating biotic and abiotic fluxes. In this way, the functional and conservation roles of biodiversity have to be analysed together, in an interactive manner, at the different levels of organization, from populations and communities at field plot level to the mosaic assemblage of crop fields, grasslands, forests and hedges created by human activities at farm and landscape level. This must necessarily take into account the socio-economic forces that determine the land-use system and its evolution.

Interactions of Grassland Areas with Other Land-use Systems: from Local to Global Scale

As stated above, some of the impacts of grasslands on biodiversity can be analysed only at the landscape level and are determined by the spatial and temporal interactions of grassland areas with other land-use systems. For some other environmental impacts such as soil erosion, quality of surface and ground water, and air quality, where spatially explicit processes are important, it is impossible to develop only an aggregative approach from field plots to catchments and landscapes. In these situations the regional impact is not the sum or the average of the local impacts because of the topographical interactions between the different spatial components. Moreover, a great part of the human activity at landscape or catchment level is devoted to the structuring and the dynamic of the spatial and temporal interactions between the different land-use components. Most of the grassland areas in the world, except some regions such as the Central Asian steppe, are associated locally with other land-use systems, mainly forest and arable cropping. So the contribution of grasslands to ecosystem services is the result of both local impacts of grassland areas and regional effects of grasslands through spatial and temporal interactions with other land-use systems. These interactions are mainly driven by human activities through livestock production systems. In this way the beneficial role of grasslands in the environment can be emphasized. This is the case for mixed farming systems, where the introduction of grassland areas within arable cropping systems can result in a more sustainable agricultural production system than arable cropping and livestock production separately. So the re-evaluation of ley farming systems in the light of the ecosystem service concept is probably a way to reconcile efficiency of food production with sustainable rural development within many regions of the world. For that we have to be innovative, and not only to re-adapt old agriculture systems. Even if the specialization of farms is strongly determined by socio-economic constraints, it should be possible to avoid excessive specialization at landscape or regional level by favouring local interactions between specialized neighbouring farms, re-creating at landscape level a meta-agroecosystem. In the same way, interactions between forest and grassland areas should be mediated through relevant livestock production systems, using these two ecosystems in synergistic ways for optimizing both their own productivity and their overall capacity to provide ecosystem services. In more fragile systems, such as rangelands or semi-arid pastoral systems, the sustainability of resources facing excessive human pressure and climate changes could also require more substantial integration at regional level with some local cropping or forest areas. All these integrated systems have to be conceived on the basis of well-documented knowledge on the different processes driving the interactions between landscape components and how they can lead to environmental and production enhancement. Then the place and role of grasslands as a component of these systems could be well defined according to local and regional conditions. These kinds of integrated systems require strong socio-economic, ethnic and political organization, which are beyond the scope of this book, but they have to be studied and investigated in connection with ecological, environmental and agronomic aspects to give any chance of successful initiatives.

The Place of Grassland as a Resource for Livestock Production for Human Diet

As stated above, grassland ecosystems and livestock are usually inseparable. Nowadays, livestock production is strongly increasing worldwide under two driving forces: the increase in human population and the increase in the demand for animal products within human diet in emerging countries. Since the conversion from plant to animal products is a relatively inefficient process in terms of demands for energy and water, projections of the development of livestock systems to meet these demands are considered to be untenable. Moreover, the domestic herbivores that are associated with grasslands have two further handicaps: they are less efficient than monogastric species for transforming plant protein into animal protein, and they produce a large quantity of CH_4 through enteric fermentation that has a four times greater climate warming effect than CO_2. Dealing with these issues requires an integrated view of the contribution of grassland ecosystems to efficient human food production and to other ecosystem services. Domestic herbivores are not necessarily in competition with humans since they utilize plant material unsuitable for the human diet. Moreover, most of the grassland areas in the world are not directly in competition with cropping areas for land usage because they are located on soils and in climatic conditions unsuitable for efficient cropping. Also, when they are mixed with arable cropping systems, grasslands contribute to increase the efficiency of these systems, to mitigate their negative impacts on environment and biodiversity, and help to sustain a greater overall productivity. A major challenge to grassland farming in recent years has been the spread of maize and soybean production round the world, with adverse consequences on soil stability and biodiversity. Recovery of the potential for grassland systems and livestock production in these areas will depend upon information on the energetic efficiency of the alternative livestock production systems and their contribution to greenhouse gas emissions. Analysis will be needed at both global and local scales in order to anticipate regional transfers of resources and also to promote local approaches, so that interactions and synergies between different land-use systems can be used to allow grassland ecosystems to play their full role in global development.

References

Lauenroth, W.K. (1979) Grassland primary production: North American grasslands in perspective. In: French, N.R. (ed.) *Perspectives in Grassland Ecology. Ecological Studies.* Springer-Verlag, New York, pp. 3–24.

Lemaire, G., Wilkins, R. and Hodgson, J. (2005) Challenges for grassland science: managing research priorities. *Agriculture, Ecosystems and Environment,* 108, 99–108.

Millenium Ecosystem Assessment (2005) *Ecosystems and Human Well-being: Synthesis.* Island Press, Washington, DC.

Soussana, J.F. and Duru, M. (2007) Grassland science in Europe facing new challenges: biodiversity and global environmental change. *CAB Reviews: Perspectives in Agriculture, Veterinary Science, Nutrition and Natural Resources,* 1, 72 (http://www.cababstractsplus.org/cabreviews).

Part I:

Productivity and Use of Grassland Resources

Editor: John Hodgson

This book emphasizes the interactive nature of the relationships between the soil, plant, animal and environmental components of grassland systems, both natural and managed. The objective in Part I is to focus on interactions between the plant and animal components, with emphasis on primary production and tissue flows in plant communities and the influence of foraging strategy on harvesting efficiency and nutrient intake in grazing animals.

Chapter 1 focuses on the influence of defoliation strategy on tissue flows and primary production in grasslands, and the scope for manipulating the balance between tissue growth and loss, soil C and N resources, and the potential impacts of climate change on the sustainability of grassland systems. Chapter 2 is concerned with the constraints in the nutritive value of forages and the extent to which these can be modified by selective grazing related to the physical and biochemical characteristics of plants, while Chapter 3 focuses on the influence of vegetation structure on the mechanics of the grazing process. These themes are elaborated in Chapter 4, dealing with the implications of variations in plant and animal diversity to productivity and stability in grazing systems, and in Chapter 5, where emphasis is placed on the influence of scale on the management of grassland resources. Chapter 6 considers the contrasts between the requirements for productivity and sustainability in grassland, and the extent to which the two can be made compatible.

1 Primary Production of Grasslands, Herbage Accumulation and Use, and Impacts of Climate Change

Anthony Parsons, Jacqueline Rowarth, John Thornley and Paul Newton

1.1 Principles of Growth and Utilization of Grass

The key to understanding the principles of the growth and utilization of grass is that the harvestable components are the photosynthetic organs themselves – predominantly the leaves. Unlike in annual cereals, maximizing sustainable yield means considering how to achieve the optimal trade-off between the amount of leaf removed and the amount left to photosynthesize, to replace tissues and so sustain that rate of harvest. The way grazing (defoliation) intensity and nutrient supply affect the balance of the major carbon fluxes, and the dynamic optimum solution, are therefore best characterized in terms of the mean leaf area sustained over a growing season.

Mean leaf area index (LAI) is derived from the changes in C and matter fluxes during each and every defoliation and regrowth. Immediately following defoliation, low canopy photosynthesis rates, due to low residual leaf area (Fig. 1.1a), are compounded by low photosynthetic potentials in the remnants of older leaves that had expanded in relative shade in the high leaf area prior to defoliation (Parsons *et al.*, 1988a). The expansion of new leaf area may depend critically (although not substantially in total mass) on the re-mobilization of reserve carbohydrates and proteins from the stubble, leading to an initial net loss in weight. Canopy photosynthesis soon increases (Fig. 1.1a) as new leaves of high photosynthetic capacity expand, but reaches a plateau as the canopy comes to intercept all available light, and also as the photosynthetic potential of leaves declines, both with leaf age, and as new leaves again expand in relative shade (Woledge, 1972, 1978). 'Losses' of matter in respiration (approximately 40–50% of 'gross' photosynthesis) increase in keeping with the gross C uptake, of which 25% is associated with the energy and mass inefficiency of synthesis of new tissues and the remaining 25% with the 'maintenance' of existing biomass (see Thornley and Johnson, 1990). A ceiling yield of biomass is achieved in many grassland species due more to the rapid turnover (short lifespan) of leaves (rate of leaf death (*ds*) in Fig. 1.1a) than to the rise in respiratory 'costs' (Robson, 1973a,b). In *Lolium perenne*, the longevity of leaves is approximately 30 days at ~15°C. The mass flux from the death rate of leaves (*ds*) is clearly low, initially, but, after a lag related to the longevity of leaves, it rises rapidly as the size of leaves involved increases, and soon comes to equal the no longer increasing rate of leaf biosynthesis (gross tissue production, GTP). The net

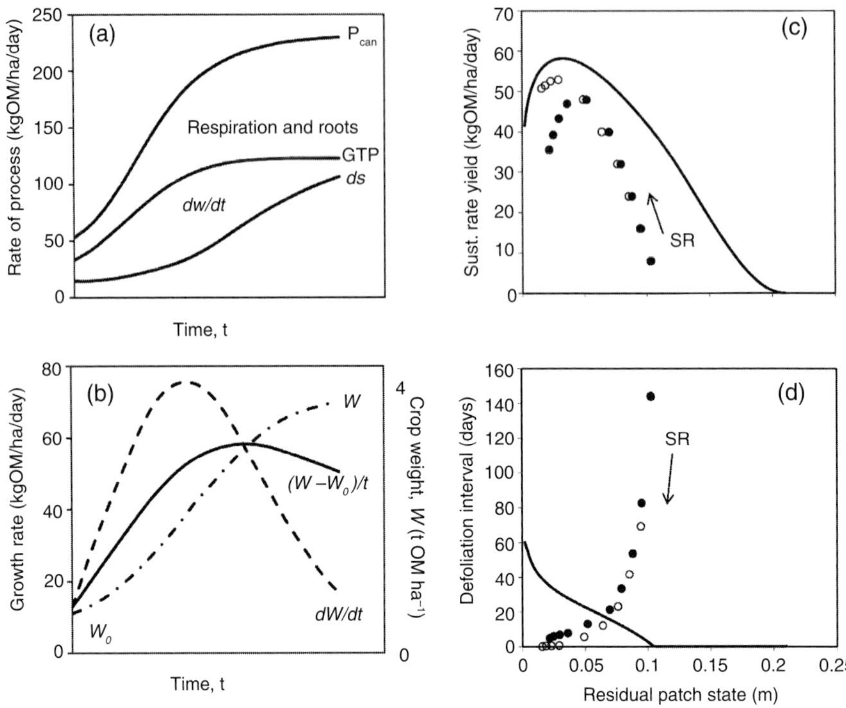

Fig. 1.1. The use of a model (a) to show the changes in the flows of C (here in units of organic matter, OM) during regrowth from a low initial (cf. 'residual') state: P_{can}, canopy gross photosynthesis; GTP, gross shoot tissue production; *ds*, rate of leaf death; and (b) to illustrate the basis for the optimal timing of harvest (when the average growth rate, $(W - W_0) / t$, is maximum). (The model is shown for different growth conditions to when first published). Using a patch-scale model (see Schwinning and Parsons, 1999), the theoretical maximum sustainable rate of yield is calculated for all possible initial ('residual') states, solid line in (c), and the duration of regrowth required to achieve this, solid line in (d). A spatial, stochastic, bite-patch scale foraging analysis (Parsons *et al.*, 2000, 2001) is used to propose the combinations of defoliation intervals and yields that arise (emergent properties) for a sequence of increasing stocking rates (SRs) under continuous (●) and rotational (○) grazing (30 days regrowth, 1 day defoliation). Values shown are the arithmetic means. Under this rotation, most defoliations at high SRs are less than one day, but yield is improved relative to continuous grazing because the dominant interval, 30 days, takes the system closer to the required optimum.

accumulation rate (*dW/dt*) falls to zero (see reviews by Robson *et al.*, 1988; Parsons, 1994; Parsons and Chapman, 2000).

Regrowth from a low leaf area is therefore often depicted as a form of 'logistic' (Thornley and Johnson, 1990; Cacho, 1993), as shown by the plot of '*W*' versus time since defoliation, *t*, (Fig. 1.1b) with the corresponding changes in the instantaneous growth rate, *dW/dt*. Long-standing controversy over how to maximize the sustainable growth rate of grass (see Parsons and Penning, 1988; Parsons *et al.*, 1988b), focusing on changes in *W* and *dW/dt*, overlooked the fact that the goal was actually to maximize the net amount harvested $(W - W_0)$, averaged over the time elapsed, *t*. In keeping with the marginal value theorem (MVT), which is a central tenet in ecology (Charnov, 1976) regarding the optimal time for a forager to move on to the next 'patch', the optimum timing of harvest following any defoliation is the time when $(W - W_0)/t$ is maximum. The pattern of changes in this averaged growth rate, $(W - W_0)/t$, is also plotted in Fig. 1.1b.

The maximum average growth rate may be envisaged graphically as the steepest tangent to the growth curve W versus t, and the optimum timing of harvest can be defined as when $dW/dt = (W - W_0)/t$ (the value of t where these lines cross).

Regrowth from higher residual leaf area (after defoliation) is aided by the greater residual photosynthesis, and hence a lesser dependence on consumption of 'reserves', but is burdened by a higher residual rate of leaf death, leading in many cases to a lower net growth rate, that declines sooner during regrowth (Parsons *et al.*, 1988b). The average growth rate may be constant or decline with time, implying lenient defoliation should encourage more frequent defoliation.

Using a simple model of plant growth (Schwinning and Parsons, 1999), we can calculate and plot (see solid line, Fig. 1.1c) the maximum sustainable rate of yield $(W - W_0)/t$ that could be achieved, for all possible initial patch states (depicted here in terms of a residual patch height). The solid line represents what could be achieved, albeit only likely by cutting, and only with perfect knowledge of when to harvest. The optimal timing of harvest (cf defoliation interval) required to achieve this is plotted in Fig. 1.1d (solid line). Figure 1.1c and d match agronomic observation, that the best way to harvest 'grass' is by relatively infrequent, severe defoliation (see the position of the global maximum of the solid line in Fig. 1.1c). In practice, the primacy of infrequent severe defoliation is greatly re-enforced by the time taken during regrowth to replenish any 'reserves' consumed at the outset (see Rasmussen *et al.*, 2009), and by substantial deleterious changes in the morphology of the grass canopy when sustained at high leaf area (Hunt, 1965; Tainton, 1974).

Although there has been legendary controversy over the merits of continuous versus intermittent (rotational) grazing and cutting, defoliation can always be conceived as intermittent at the 'patch' or 'bite' scale. In well-managed grassland, the mean 'frequency' and 'severity' of defoliation of individual plants following a bite might differ little between continuous and rotational grazing (Hodgson and Ollerenshaw, 1969; Clark *et al.*, 1984). Indeed, differences between continuous and rotational grazing might be regarded more as a matter of scale, and of how defoliation events are distributed and co-located in space and time.

When defoliation is by animals, however, there is no simple relationship between the number of animals (stocking rate (SR)) applied, nor the way these are allocated at a paddock scale, and the actual residual pasture states and durations of regrowth at the patch-bite scale. These fundamental drivers to plant growth remain emergent properties, despite best management efforts, of animal bite numbers and foraging behaviour (Parsons *et al.*, 2000, 2001); see predicted values for a sequence of increasing SR on Fig. 1.1c,d. Hence it is widely recognized that harvesting grass with animals rarely matches the potential yield achievable under cutting, and, as indicated (Fig. 1.1c), it is only at high SR that there is a fundamental advantage (to yield) of using rotational grazing rather than controlled continuous grazing (e.g. continuous variable stocking). None of this detracts from the practical advantage of rotational grazing as a means to allocate feed strategically.

1.2 How Pasture Utilization Affects the Uptake, Cycling and Fate of C and N

Although the fundamental determinants of plant growth and utilization are the initial (residual) leaf area and the regrowth duration of 'patches', at a pragmatic level the fundamental drivers of C and nitrogen (N) fluxes that a farmer can manipulate are SR, fertilizer (N) input, and choice of defoliation/grazing system. Manipulating these drivers affects production, utilization and environmental impacts.

First it is critical to recognize the extent to which grazing animals 'uncouple' the C and N cycles. For every 100 units of C eaten by an animal in organic matter (OM) each day, a substantial proportion (approximately 70%) is released back to the atmosphere as carbon dioxide (CO_2) in respiration. In contrast, for every 100 units of N eaten, a similar proportion (approximately 70%, decreasing with higher feed C:N ratio) is excreted to the soil surface as urine. Only approximately 25% of the OM ingested from the vegetation remains in the ecosystem with C and N coupled in OM, and this is returned to the soil surface as dung.

The environmental consequences of just a further 3–5% of the C eaten, emitted as methane, and the substantial elevation of labile nitrogen (reduced C:N ratio) in the soil, on leaching and gaseous N loss, are described by Ledgard *et al.*, Chapter 9 and Buchmann, Chapter 10, in this volume. Less well recognized, in terms of their role in the C and N cycles, are the implications of grazing animals 'fast cycling' C. Such animals return C rapidly to the atmosphere that would otherwise have remained for a longer time in the biosphere. The speed of C cycling has relevance for the concentrations of CO_2 in the atmosphere. For simplicity, we can imagine the planet as being engaged in a conserved cycle of C between two state variables, 'atmosphere' and 'biosphere'. In a conserved cycle, where inputs from one equal outputs from the other, and vice versa at equilibrium, the relative distribution of C between atmosphere and biosphere (the amounts in each) can be shown to be in direct proportion to the relative longevity of C in each. The more time C is out of the atmosphere in biosphere organic material, the smaller the amount of C there is in the atmosphere.

In the absence of animals, the uptake of C from the atmosphere is generally greater than when animals are present (but see McNaughton, 1983), as herbivores remove leaf area and reduce the capacity for photosynthesis (see left side, for zero animals, in Fig. 1.2a–d). Moreover, any C that escapes grazing can have appreciable longevity: leaves in cool temperate pastures have longevities of weeks; roots have longevities (live) of months (Garwood, 1967, in Parsons, 1988); the OM entering the soil as leaf and root litter has a range of longevities extending from days to months, and a small proportion has longevities of decades to centuries (see Rumpel, Chapter 7, this volume). In comparison, in the presence of animals, grazing reduces the proportion of any C eaten that remains in the biosphere to approximately 25% (that which is returned to the soil as dung) as well as reducing the fixation of C into the biosphere in photosynthesis. The size of these effects depends, of course, on the number of animals present.

The effects of SR on C and N cycling fluxes are illustrated in Figures 1.2 and 1.3, using output from a grassland ecosystem dynamics model (Thornley, 1998a). Effects are shown for two contrasting levels of N input, representing 'low' and 'higher' fertility.

At any one level of fertility, as SR is increased there is generally a concomitant decrease in the uptake of C in photosynthesis (P_{can}), and the fluxes of C into and through the system, as the pasture is sustained on average at a lower vegetation cover and LAI. Figures 1.2a and b match the general pattern of fluxes observed empirically (Bircham and Hodgson, 1983; Parsons *et al.*, 1983). As canopy photosynthesis declines with SR, there is a concomitant decrease in shoot respiration. Partitioning to roots is conceived, here, using a transport-resistance analogy (see review of approaches in Thornley 1998b). In Fig. 1.2, the changes in the proportion of assimilates (C) partitioned below ground, within the plants, to roots, initially offsets the decline in the total amount of C to be partitioned. The net effect is a decrease in the gross rate of production of new shoot material as SR increases, and a lesser (relative) decline in C allocation below ground. Of the above-ground C flux, the increase in SR leads initially to a linear increase in the rate of removal of C in shoot material, as animal intake per hectare. Intake (per hectare) plateaus, and may decline, once the resulting mean vegetation state has fallen below the point at which each animal can maintain its maximum daily C requirement. Changes in animal behaviour, and difficulties in foraging for a

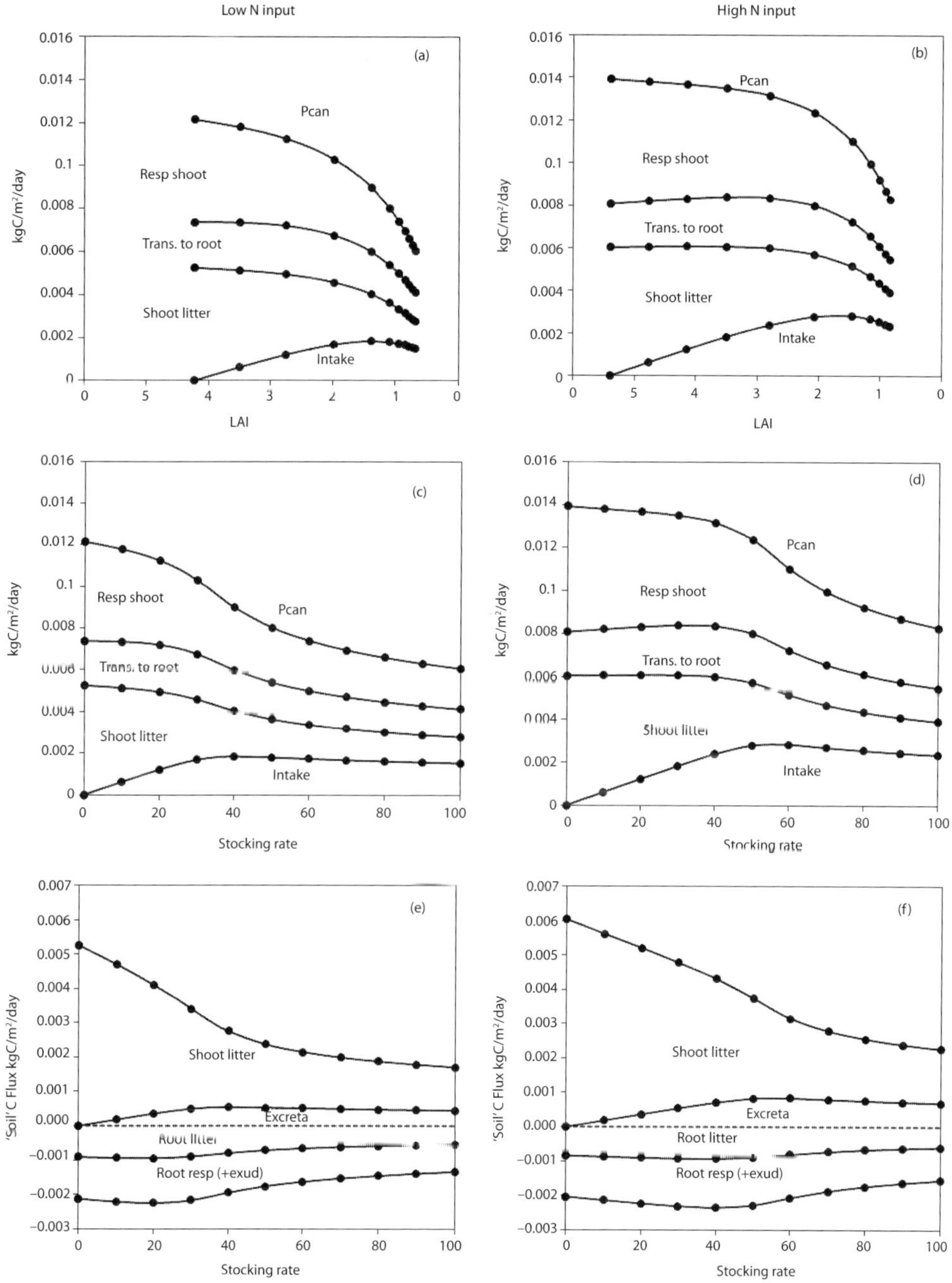

Fig. 1.2. The effects, on the flows of C, of key drivers of growth and utilization: N input (see columns) and stocking rate (SR) (of dry sheep units) (see c–f) as predicted by the Hurley Pasture Model (based on Thornley, 1998a). All values shown are steady-state dynamic equilibria, above ground (a–d) and below ground (e, f). In (a) and (b), the fluxes are shown in relation to the Leaf Area Index, LAI, that arises at each SR. In (e) and (f) flows are depicted above, and beneath, a 'soil surface' (drawn at the origin) for ease of comparison. Total 'soil' respiration potentially would include all fluxes from the upper- to lower-most line in (e) and (f), depending on the method of measurement used in practice.

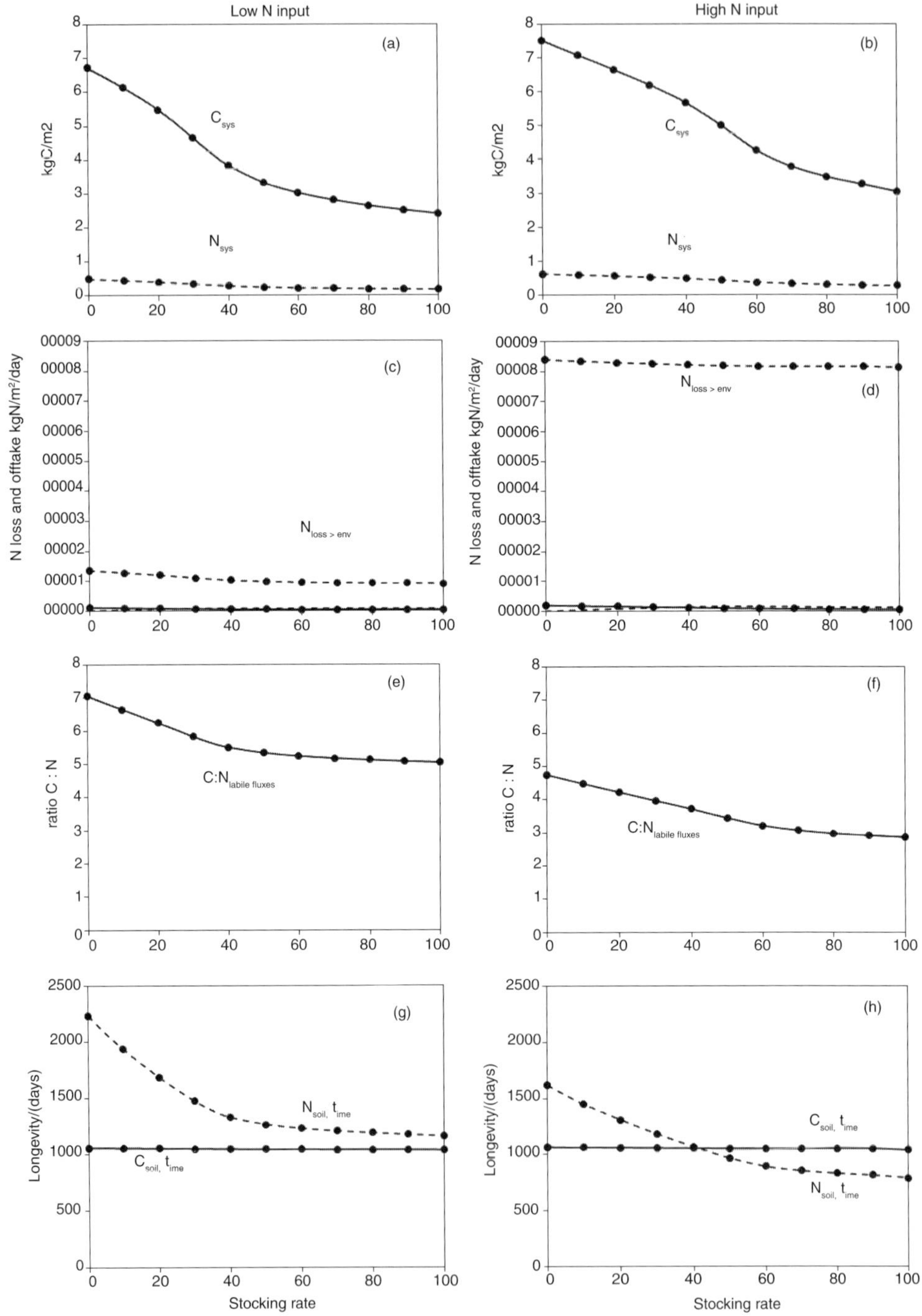

Fig. 1.3. The effects, corresponding to the flows in Fig. 1.2, of N input (see columns) and stocking rate (*x*-axis of all graphs) on total system sequestered C and N (C_{sys}, N_{sys}); the rate of total losses of N to the environment (sum of volatilization, denitrification, nitrification and leaching), and the minor off-take of N in animal products and plant-derived gaseous N emission. In (e) and (f) are shown the ratio of fluxes of labile C and N, to depict resource supply ratios for many soil microorganisms; and in (g) and (h) the emergent rates of turnover of C and N in the soil.

very short prostrate food supply in cool temperate pastures, lead to a marked depression in intake per animal, which avoids the pasture being grazed to extinction at high SR, and so leads to greater stability (Schwinning and Parsons, 1999).

With increases in SR reducing the gross rate of production of new shoot, and increasing the proportion of this declining amount being removed as intake, SR has a major effect in decreasing the flow of C to the soil surface of high C:N ratio shoot litter (see Fig. 1.2e,f), but conversely increases the relative flow of C to the soil surface of dung (C:N ratio close to that of initial OM), both as a quantity and as a proportion of surface 'litter'.

C partitioned below ground (see below 'origin' in Fig. 1.2e,f) contributes directly to the soil C flows into root litter, exudation and root respiration. Note: root and shoot live masses are not represented on these flux diagrams, which represent steady states, at which changes in root and shoot mass are 'zero'. Ultimately all the fluxes shown in Fig. 1.2e,f (see the sum of all components) are decomposed to CO_2 and respired (cf. 'soil' respiration) to the extent that respiration of CO_2 from the 'soil' almost equals the total supply of C to the soil. The balance is lost in C leaching (not depicted here), leaving the entire pastoral system approaching an equilibrium of close to C (but not global warming potential, GWP) neutral balance. It may take decades to centuries to approach this equilibrium, and in practice grassland soils undergo long periods of gradual C sequestration, accumulating and sustaining at equilibrium very substantial C stocks below ground (Goudriaan, 1992). There is much debate over the relative role of shoot litter compared with root litter (Rasse *et al.*, 2005; Denef and Six, 2006) for C in soils, but studies of C balances of pastures, and the widespread acceptance of how surface applied crop residues (and dung/manure) contribute substantially to soil OM suggest in many climates (arguably the wetter ones, and with surface active soil fauna and fungi) surface applied organic matter can contribute substantially to soil OM (see proposed potential fluxes in Fig. 1.2e,f).

An increased supply of nutrients, e.g. nitrogen, undeniably stimulates the photosynthesis and growth, and so total fluxes of C, into the whole system (Fig. 1.2, right-side column). This is predominantly due to a boost to specific activity (i.e. of each gram) of shoot material, due to greater carboxylase content, and greater per gram activity in meristems (with consequent increases in leaf expansion and C capture). This enables a greater SR to be supported (increased carrying capacity) even for any one given LAI (and greater LAI for a given SR) (Fig. 1.3a,d). Using the same concepts for shoot/root partitioning, added nutrients reduces the proportion of C partitioned to roots, but the increase in total C supply (at greater leaf N) means the amount (flux) of C to roots may be little altered compared with that seen for lower N inputs.

The overall impact of SR and N on total system C (C_{sys}) is conceived to be that shown in Fig. 1.3a,b. Clearly there can be no one simple relationship between SR and the pools and fluxes of C, and C sequestration, as different fluxes are obtained when compared at the same SR, depending on N input rate, and vice versa. The total C sequestered at the optimum SR for yield at low N (~35) may be very similar, as here, to that sequestered at the optimal SR at higher N (~55). It is a major cause for confusion and difficulty in interpretation of field observations that these two components of pastoral 'intensification' are so often confounded. Few field trials exploring impacts of intensification have added N without simultaneously increasing SR. This apparent requirement to emulate what is practical on a farm has led to the situation where it is still unclear what impact N inputs per se have on, for example, soil C (OM) levels and sustainability.

None of the fluxes (of C) above can be conceived without attention to the corresponding fluxes of N, and the multitude of changing C:N ratios, through the multiple transformations and states. The predicted effects of SR and N on the total ecosystem N (Fig. 1.3a,b), and on the total rates of losses/emissions of N (Fig. 1.3c,d), are shown. Counter-intuitively, greater SR does not lead to greater N losses (despite the greater number of animals) because, at equilibrium, N losses at all times equal the rate of N input, which, for any one fertilizer input rate, is fixed (the slight 'drop' is because the diagrams include an element of non-legume bacterial N fixation). A shift from a low SR to a high SR will lead to a transient increase in N losses, 'vented' partly by animals, but this increased loss originates from the rundown in N (and C) from soil (Fig. 1.3a,b), not from the animals.

There is a very pressing need to confirm if the synthesis depicted by classic models, such as shown here, totally captures overall responses of pastures to the key drivers, SR and fertilizer (notably N) input rate.

1.3 Growing Need for New Paradigms of Impacts of Plant Growth and Utilization on Soil Organic Matter Dynamics

It is difficult to overstate the significance of the supply of C and energy by plants (photo-autotrophs obtaining C and energy directly from the environment) in the function of the grassland ecosystem, given the size of the plant-derived fluxes above and to below ground, and the way grassland management affects these via light interception, photosynthetic efficiency and leaf area. Many conceptual models of the drivers of growth, utilization and impacts are therefore appropriately plant-C-supply focused, often predominantly bio-physical, but with a cascade of functional dependencies by heterotrophs (organisms that must get C and energy from others), using both structural (body composition) and dynamic (process requirement) stoichiometry to drive the association of C and N cycles, for example. The models readily reflect the substantial extra demand for C, over and above structural requirements, in heterotrophs (such as ruminants and many soil organisms, including bacteria and fungi) for 'consumption' to provide energy via oxidative respiration (Hessen *et al.*, 2004). However, as such models have become coupled to models of soil ecosystem function, there is growing concern how well such plant-C-driven concepts accommodate the function of some classes of soil organisms involved critically in nitrogen transformations.

In this section, we briefly consider some major categories of soil microbes, distinguished by their differing trophic roles, and so resource demands, as a way of re-connecting knowledge of their biology with the principles of grass growth and utilization that control the supply of C and N to the soil.

To recap, in soil, working along a scale of decreasing gaseous oxygen availability (cf. increasing water content), first nitrifiers and then (overlapping) denitrifiers process mineral nitrogen compounds in a way that reflects the decreasing availability of oxygen. Hence, nitrifiers can 'add' sufficient oxygen to convert NH_4 to NO_3 in highly aerobic environments, and act jointly with denitrifiers at 'medium' oxygen availability, using a range of mineral N compounds, to produce lower oxides of nitrogen (e.g. N_2O), whereas, in anaerobic conditions, denitrifiers predominate to convert NO_3 to N_2 (note the products have decreasing oxygen content). Denitrifiers are predominantly heterotrophs, and so plant-derived C dependence is not inappropriate for driving this process conceptually, the only stochiometric issue being that denitrifiers use oxygen atoms from NO_3 to 'combust' C in oxidative respiration, when gaseous oxygen is unavailable in wet soils. This represents a high demand for N relative to C (low functional C:N stoichiometric ratio). Although denitrification has long been regarded as a function of bacteria, many fungi have now been shown to contribute to denitrification, and can be the predominant source of nitrous oxide emissions in grassland (Laughlin and Stevens, 2002; Ma *et al.*, 2008).

Where plant-C-driven conceptual models might face more challenge is with chemoautotrophs, such as those that play a significant role in nitrification. Chemoautotrophs are able to obtain C direct from CO_2 in the soil, and obtain energy from their nitrogen transformations (Poughon *et al.*, 2001). These organisms (processes) may be expected to have a very low functional C:N stoichiometric ratio, and are not (in any direct sense) plant-derived C driven. Nitrification per se is not limited to chemoautotrophic bacteria, but exhibited by many heterotrophic bacteria and fungi (Pederson *et al.*, 1999; Hayatsu *et al.*, 2008; Laughlin *et al.*, 2008) and some modifications to soil microbial function models may be needed to accommodate heterotrophic nitrification (Cookson *et al.*, 2006). Conceptually, however, the balance between autotrophs and heterotrophs, consequent on perturbations of the grassland ecosystem (e.g. SR, N use, climate change), may be an important category distinction for mathematical models that currently presume dependence heavily on plant-derived C and energy. Greater attention is needed whatever the scale of the populations (or mass transfer of C and N involved), because the function of these organisms is of substantial significance as a potent source of greenhouse gases (N_2O), mediates issues of the leaching versus availability of NO_3, and is critical to the integrity of the N cycle.

To consider how the principles of the growth and utilization of grassland might affect the relative supply of C and N to the soil, and so drive the prospects for different classes of organism with their contrasting dynamic stoichiometry, and trophic niche, we plot (Fig. 1.3e,f) the relative rate of replacement of the labile pools of C and nitrogen for the same two management drivers, SR and fertilizer input. Clearly as SR is increased, grazing decreases plant C uptake and supply to soil, and in contrast, the large herbivores and soil heterotrophs 'fast cycle' nitrogen that is present in plant and soil OM in excess of their stoichiometric requirements. Hence, the longevity of N in the system (rate of output per unit state) declines (Fig. 1.3g,h), N availability rises (in absolute terms) and this causes the decline in system labile C:N flux ratios. The addition of N fertilizer inputs intuitively compounds this effect. The most notable change in C:N ratio occurring in the grassland biosphere is the process of leaf senescence (from C:N of approximately 12:1 to 50+:1), as proteins are remobilized as leaves age. These changes are included in all the fluxes depicted here, and are detailed by Thornley (1998a).

There is widespread recognition that such changes in C versus N availability, and the form (organically bound versus labile) of C and N supply, might shift the balance between different 'classes' of soil micro-organism, and/or different soil functions. One widely stated premise (consistent with many observations) is that high N input, and so low C:N ratio (cf. intensive grassland management), leads to high bacterial:fungal decomposition ratios and low total microbial biomass. The converse is proposed under less intensive, 'organic' or low N input systems (Bardgett *et al.*, 1993, 1999; Bardgett and McAlister 1999; Frey *et al.*, 1999).

What is clearly of critical importance is to resolve under what circumstances certain agricultural impacts on plant growth and utilization can sometimes add to, but sometimes detract from, soil C sequestration, greenhouse gas emissions and nitrogen flow to the environment. The outcome of stimulating soil micro-organisms would logically be to decrease soil OM if decomposers were stimulated, offset only if the process of decomposition itself led to an increase in soil OM (C) of compounds (e.g. fungal chitin) that had a greater longevity in the biosphere than the cellulose/lignin being decomposed. Positive feedback to plants from stimulating soil organisms, as a consequence of making soil nutrients re-available (positive microbial loop, e.g. Raynaud *et al.*, 2006), as often claimed, is likely to be evident only in low nutrient environments where nutrients had otherwise been limiting. In high N input systems, it has been proposed that the observed reduction in microbial activity (but with associated changes in microbial community composition and efficiency, and a change to producing more recalcitrant compounds) is beneficial for C sequestration (Ågren *et al.*, 2001). Hence there is much uncertainty over whether stimulating soil microorganisms (or the converse) per se is always beneficial to soil C sequestration. One major challenge is that differences in pastoral 'intensification' (SR and N input) are often confounded with such factors as plant species composition. It has been shown that different individual grassland species, alone, can greatly influence the size and activity of the microbial biomass, through having specific eco-physiological traits (Wardle *et al.*, 1998; Bardgett and Shine, 1999; Bardgett *et al.*, 1999). Other notably challenging phenomena are 'priming' (Fontaine *et al.*, 2007; Kuzyakov, 2010), where a stimulus (prompted by an increased plant C supply) to soil microorganisms may accelerate decomposition and lead to a major loss of soil C, rather than an increase, and progressive nutrient limitation (Luo *et al.*, 2004), discussed below. There is still, however, substantial evidence that the increase in plant C supply to soil consequent on the addition of N fertilizer can lead to greater soil C (see discussion by Johnston *et al.*, 2009; Glendining and Powlson, 1995; Sousanna *et al.*, 2004), as proposed by the current model (Fig. 1.3a,b).

1.4 Proposed Effects of Climate Change on C and N Fluxes and Fate

There have been extensive reviews (e.g. Newton *et al.*, 2007; Soussana and Lüscher, 2007) and model predictions (e.g. Thornley and Cannell, 1997) of the impact of multiple components of climate change, singly and together, on the fluxes of C and N described. We briefly recap on two aspects here: elevated CO_2 and temperature.

As the major resource for photosynthesis, the expectation is that elevated CO_2 per se will increase all plant derived C fluxes into the system. This is borne out by field experimentation that shows C uptake is increased at elevated CO_2 despite down-regulation of carboxylase content (see Leakey *et al.*, 2009 for review and further references). An increased uptake and availability of C due to elevated CO_2 has been shown to lead to an increase in C:N ratio in tissues within any one pasture species (Poorter *et al.*, 1997) and to increase the exudation from roots of labile (soluble) C (e.g. Allard *et al.*, 2006). These factors arguably lead to a stimulation of soil biota populations, leading to the demand and incorporation into organic compounds of nutrients (N) that would otherwise have been available for higher plants (Drigo *et al.*, 2010). This feedback, on plant N availability, has been called progressive nitrogen limitation (PNL) (Luo *et al.*, 2004) and is a potentially negative feedback on grassland responses to elevated CO_2. During 11 years of CO_2 enrichment of a grazed pasture, Newton *et al.* (2010) found a gradual transfer of N from plant to soil pools, indicative of PNL, and undoubtedly a constraint on plant growth responses to elevated CO_2.

Models in general nonetheless predict higher total soil C sequestration in grasslands at elevated CO_2 (Thornley and Cannell, 1997; Thornley, 1998a; Pepper *et al.*, 2005) and observations made using natural gradients near 50+-year-old natural CO_2 springs (Ross *et al.*, 2000; Kool *et al.*, 2007) support this view. As mentioned earlier, an increase in C relative to N may shift the balance in the soil towards fungal rather than bacterial abundance and it is the case that fungal abundance frequently increases at elevated CO_2 (Carney *et al.*, 2007).

Continued (let alone 'progressive') nutrient limitation is not universally accepted, though it is more widely accepted that sustained elevated inputs of C at elevated CO_2 would lead, after a transition period of increased N incorporation in soil biota, to a higher sustained steady-state C and N content in soils, as shown under very long-term exposure to elevated CO_2 around a natural CO_2 spring (Ross *et al.*, 2000).

An additional complexity of grasslands response to climate change, compared with monoculture crops and forests, is again the impact of botanical composition on C and N fluxes (Kardol *et al.*, 2010). Botanical composition is sensitive to elevated CO_2. In particular, there is frequently a stimulation of legume abundance, which can provide additional inputs of N (Zanetti *et al.*, 1996) that may partially offset PNL (Newton *et al.*, 2010). Botanical composition will also strongly influence decomposition rates and this may be a greater influence on decomposition processes at elevated CO_2 than changes in the chemical composition within individual species (Allard *et al.*, 2004).

The effects of a sustained increase in temperature on grassland ecosystem fluxes and states are also not totally resolved. It is widely anticipated that an increase in temperature will increase all enzymatic processes, notably respiration, more than it will increase net photosynthesis. At the organ level, the effect of raised temperatures in increasing cell division, and the appearance and expansion of new leaves (and so growth in the colloquial sense), reduces the time taken for leaves to regrow following defoliation. At the same time, raised temperature reduces more rapidly the level of reserve carbohydrates available for regrowth, and reduces the time in which these must be replenished (see Rasmussen *et al.*, 2009), and the time before the onset of a major rise in the rate of leaf death, hence shortening the optimal duration of regrowth. These aspects are readily accommodated in process-based models. Greater concern has been expressed over the impacts of elevated temperatures in soil: the possibility of a substantial loss through respiration of sequestered soil C (Cramer *et al.*, 2001), and the apparent differences evident between model predictions and experimental data (Kirschbaum, 2006; Thornley and Cannell, 2001). Although experimentally raising soil temperature does greatly increase soil respiration, the effect is not sustained long term. Kirschbaum (2006) hypothesized this may be due to a decline in substrates for respiration, in effect a decline by the system to a new lower C supply steady state. Thornley and Cannell (2001) have proposed an alternative hypothesis, that a rise in temperature may stimulate the physico-chemical reactions that lead to C stabilization as well as the microbial processes leading to enhanced respiration. Any change in soil C sequestered would then depend upon the relative sensitivity of each process, with equilibrium soil C increasing if the physico-chemical reactions are over 50% more sensitive than microbial respiration (Thornley and Cannell, 2001).

As regards PNL, the little experimental evidence available suggests that temperature may play an important regulatory role. In an experiment on native grassland in Tasmania, Hovenden *et al.* (2008) found a PNL effect at elevated CO_2 but the addition of a 2°C warming of the canopy completely removed this effect. It is of interest that warming on its own had no impact on soil N availability.

In the absence, yet, of long-term data from the future climate world, refining process-based models is the best approach to foresee potential ecosystem impacts. Some authors see value in regarding ecosystems as a single complex organism, and its eco-enzymatic stoichiometry as a basis for understanding its wider responses to the environment (Sinsabaugh *et al.*, 2009; Taylor and Townsend, 2010; Yvon-Durocher *et al.*, 2010). There is much credit to this philosophy, given the dependence of heterotrophs ultimately on photoautotrophic C uptake, but with a major exception – the role of critical chemautotrophs (mentioned previously), with their plant-derived C and energy independence. Progress in coupling more advanced soil models to the well-developed plant and animal process models is increasingly important (Soussana *et al.*, 2010).

1.5 Does Intensification Counter Environment?

What is of major concern here is not what is the C and N balance of grazed pastures, which is a variant of a natural ecosystem that has functioned benignly for millions of years, but how grazing pastures compare with other systems for human food production (agriculture) and what is the least malevolent pattern for change in land use (with respect to pastures) to satisfy the burgeoning demand. For those who feel indignant about the mere notion of acceptably burgeoning agriculture, there is one simple alternative: restricting population growth.

Improvements in grassland defoliation management (e.g. a shift from 'continuous grazing'/'set stocking' to the use of 'rotational grazing') have focused specifically on the principle of intercepting leaves at a stage, during regrowth, to minimize the flow of material to surface litter, where, in this yield context, the flow of OM would have been 'wasted' feeding bugs and earthworms in the soil. In this sense, advances in pasture defoliation management might be expected to decrease soil OM content, with associated concerns for loss of water and nutrient retention, and reduced sustainability (see Lemaire and Agnusdei, 2000; Lemaire *et al.*, 2009). However, it is not a simple matter to resolve whether this has been the case. From the principles of fluxes captured in the previous figures, to sustain greater flows of C to soil, and so support C sequestration, pastures need to be maintained at a greater mean vegetation cover/state, e.g. a greater mean LAI. This principle applies regardless of what method is used to distribute defoliations, in space and time. However, as intake (and so performance) per hectare are relatively insensitive to SR and grazing intensity, close to the sustainable yield optimum (Fig. 1.2a,b), and yet fluxes of at least shoot litter increase very substantially with even small increases in vegetation cover, then relatively small reductions in pasture-grazing pressure could lead to valuable increases in the supply of C to soils, with relatively small production losses. More practicably, any grazing defoliation scheme that involves monitoring pasture vegetation states, and so avoids periods of overgrazing and sustained reduction of vegetation cover, would greatly increase both yield/performance per hectare and improve prospects for soil C sequestration.

Understanding these flows and the processes that influence them in both depth and detail is essential if we are to have the capacity to change our world and its capacity to support us. The alternative is to observe and endure the consequences.

References

Ågren, G.I., Bosatto, E. and Magill, A.H. (2001) Combining theory and experiment to understand effects of inorganic nitrogen on litter decomposition. *Oecologia* 128, 94–98.

Allard, V., Newton, P.C.D., Lieffering, M, Soussana, J.-F., Grieu, P. and Matthew, C. (2004) Elevated CO_2 effects on decomposition processes in a grazed grassland. *Global Change Biology* 10, 1553–1564.

Allard, V., Robin, C., Newton, P.C.D., Lieffering, M. and Soussana, J.F. (2006) Short-term and long-term effects of elevated CO_2 on *Lolium perenne* root exudation and its consequences on soil organic matter turnover and plant N yield. *Soil Biology and Biochemistry* 38, 1178–1187.

Bardgett, R.D. and McAlister, E. (1999) The measurement of soil fungal: bacterial biomass ratios as an indicator of ecosystem self-regulation in temperate meadow grasslands. *Biology and Fertility of Soils* 29, 282–290.

Bardgett, R.D. and Shine, A. (1999) Linkages between litter diversity, soil microbial biomass and ecosystem function in temperate grasslands. *Soil Biology and Biochemistry* 31, 317–321.

Bardgett, R.D., Frankland, J.C. and Whittaker, J.B. (1993) The effects of agricultural practices on the soil biota of some upland grasslands. *Agriculture, Ecosystems and Environment* 45, 25–45.

Bardgett, R.D., Mawdsley, J.L., Edwards, S., Hobbs, P.J., Rodwell, J.S. and Davies, W.J. (1999) Plant species and nitrogen effects on soil biological properties of temperate upland grasslands. *Functional Ecology* 13, 650–660.

Bircham, J.S. and Hodgson, J. (1983) The influence of sward condition on rates of herbage growth and senescence in mixed swards under continuous stocking management. *Grass and Forage Science* 38, 323–331.

Cacho, O.J. (1993) A practical equation for pasture growth under grazing. *Grass and Forage Science* 48, 387–394.

Carney, K.M., Hungate, B.A., Drake, B.G. and Megonigal, J.P. (2007) Altered soil microbial community at elevated CO_2 leads to loss of soil carbon. *Proceedings of the National Academy of Sciences USA* 104, 4990–4995.

Charnov, E.L. (1976) Optimal foraging, the marginal value theorem. *Theoretical Population Biology* 9, 129–136.

Clark, D.A., Chapman, D.F., Land, C.A. and Dymock, N. (1984) Defoliation of *Lolium perenne* and *Agrostis* spp. tillers and *Trifolium repens* stolons in set-stocked and rotationally grazed hill pastures. *New Zealand Journal of Agricultural Research* 27, 289–301.

Cookson, W.R., Muller, C., O'Brien, P.A., Murphy, D.V. and Grierson, P.F. (2006) Nitrogen dynamics in an Australian semiarid grassland soil. *Ecology* 87, 2047–2057.

Cramer, W., Bondeau, A. and Woodward, F.I. (2001) Global response of terrestrial ecosystem structure and function to CO_2 and climate change: results from six dynamic global vegetation models. *Global Change Biology* 7, 357–373.

Denef, K. and Six, J. (2006) Contributions of incorporated residue and living roots to aggregate-associated and microbial carbon in two soils with different clay mineralogy. *European Journal of Soil Science* 57, 774–786.

Drigo, B., Pijl, A.S., Duyts, H., Kielak., A.M., Gamper, H.A., Houtekamer, M.J., Boschker H.T.S., Bodelier, P.L.E., Whiteley, A.S., van Veen, J.A. and Kowalchuk, G.A. (2010) Shifting carbon flow from roots into associated microbial communities in response to elevated atmospheric CO_2. *Proceedings of the National Academy of Sciences USA* 107, 10,938–10,942.

Fontaine, S., Barot, S., Barre, P., Bdioui, N., Mary, B. and Rumpel, C. (2007) Stability of organic carbon in deep soil layers controlled by fresh carbon supply. *Nature* 450, 277–280.

Frey S.D., Elliott, E.T. and Paustian, K. (1999) Bacterial and fungal abundance and biomass in conventional and no-tillage agroecosystems along two climatic gradients. *Soil Biology and Biochemistry* 31, 573–585.

Garwood, E.A. (1967) Studies on the roots of grasses. In: *Annual Report of Grasslands Research Institute, 1966*. Grasslands Research Institute, Hurley, UK, pp. 72–79.

Glendining, M.J. and Powlson, D.S. (1995) The effects of long continued application of inorganic nitrogen fertilizer on soil organic nitrogen – a review. In: Lal, R., Stewart, B.A. (eds) *Soil Management: Experimental Basis for Sustainability and Environmental Quality*. CRC Press, Boca Raton, FL, pp. 385–446.

Goudriaan, J. (1992) Biosphere structure, carbon sequestering potential and the 14C carbon atmospheric record. *Journal of Experimental Botany* 43, 1110–1120.

Hayatsu, M., Tago, K. and Saito, M. (2008) Various players in the nitrogen cycle: diversity and functions of the microorganisms involved in nitrification and denitrification. *Soil Science and Plant Nutrition* 54, 33–45.

Hessen, D.O., Ågren, G.I., Anderson, T.R., Elser, J.J. and de Ruiter, P.C. (2004) Carbon sequestration in ecosystems: the role of stoichiometry. *Ecology* 85, 1179–1192.

Hodgson J. and Ollerenshaw, J.H. (1969) The frequency and severity of defoliation of individual tillers in set-stocked swards. *Journal of the British Grassland Society* 24, 226–234.

Hovenden, M.J., Newton, P.C.D., Carran, R.A., Theobald, P., Wills, K.E., Vander Schoor, J.K., Williams, A.L. and Osanai, Y. (2008) Warming prevents the elevated CO_2-induced reduction in available soil nitrogen in a temperate, perennial grassland. *Global Change Biology* 14, 1018–1024.

Hunt, L.A. (1965) Some implications of death and decay in pasture production. *Journal of the British Grassland Society* 20, 27–31.

Johnston, A.E., Poulton, P.R. and Coleman K. (2009) Soil organic matter: its importance in sustainable agriculture and carbon dioxide fluxes. *Advances in Agronomy* 101, 1–57.

Kardol, P., Cregger, M.A., Campany, C.E. and Classen, A.T. (2010) Soil ecosystem functioning under climate change: plant species and community effects. *Ecology* 91, 767–781.

Kirschbaum, M.U.F. (2006) The temperature dependence of organic-matter decomposition – still a topic of debate. *Soil Biology and Biochemistry* 38, 2510–2518.

Kool, D.M., Chung, H., Tate, K.R., Ross, D.J., Newton, P.C.D. and Six, J. (2007) Hierarchical saturation of soil carbon pools near a natural CO_2 spring. *Global Change Biology* 13, 1282–1293.

Kuzyakov, Y. (2010) Priming effects: interactions between living and dead organic matter. *Soil Biology and Biochemistry* 42, 1363–1371.

Laughlin, R.J. and Stevens, R.J. (2002) Evidence for fungal dominance of denitrification and codenitrification in a grassland soil. *Soil Science Society of America Journal* 66, 1540–1548.

Laughlin R.J., Stevens, R.J., Muller, C. and Watson, C.J. (2008) Evidence that fungi can oxidize NH_4^+ to NO_3^- in a grassland soil. *European Journal of Soil Science* 59, 285–291.

Leakey, A.D.B., Ainsworth, E.A., Bernacchi, C.J., Rogers, A., Long, S.P. and Ort, D.R. (2009) Elevated CO_2 effects on plant carbon, nitrogen and water relations: six important lessons from FACE. *Journal of Experimental Botany* 60, 2859–2876.

Lemaire, G. and Agnusdei, M. (2000) Leaf tissue turnover and efficiency of herbage utilization. In: Lemaire, G., Hodgson, J., de Moraes, A., de Faccio Carvalho, P. and Nabinger, C. (eds) *Grassland Ecophysiology and Grazing Ecology*. CAB International, Wallingford, UK, pp. 265–287.

Lemaire, G., Da Silva, S. C., Agnusdei, M., Wade, M. and Hodgson, J. (2009) Interactions between leaf lifespan and defoliation frequency in temperate and tropical pastures: a review. *Grass and Forage Science* 64, 341–353.

Luo, Y., Su, B., Currie, W.S., Dukes, J.S., Finzi, A., Hartwig, U., Hungate, B.A., McMurtrie, R.E., Oren, R.E., Parton, W.J., Pataki, D.E., Shaw, M.R., Zak, D.R. and Field, C.B. (2004) Progressive nutrient limitation of ecosystem responses to rising atmospheric carbon dioxide. *Bioscience* 54, 731–739.

Ma, W.K., Farrell, R.E. and Siciliano, S.D. (2008) Soil formate regulates the fungal nitrous oxide emission pathway. *Applied and Environmental Microbiology*, November, 6690–6696.

McNaughton, S.J. (1983) Compensatory plant growth as a response to herbivory. *OIKOS* 40, 329–336.

Newton, P.C.D., Carran, R.A., Edwards, G.R. and Niklaus, P.A. (2007) *Agroecosystems in a Changing Environment.* CRC Press, New York.

Newton, P.C.D., Lieffering, M., Bowatte, W.M.S.D., Brock, S.C., Hunt, C.L., Theobald, P.W. and Ross, D.J. (2010) The rate of progression and stability of progressive nitrogen limitation at elevated atmospheric CO_2 in a grazed grassland over 11 years of Free Air CO_2 enrichment. *Plant and Soil,* 336, 433–441.

Parsons, A.J. (1988) The effects of season and management on the growth of grass swards. In: Jones, M.B. and Lazenby A. (eds) *The Grass Crop: the Physiological Basis of Production.* Chapman and Hall, Bury St Edmunds, UK, pp. 129–177.

Parsons, A.J. (1994) Exploiting resource capture - grassland. In: Monteith, J.L., Scott, R.K. and Unsworth, M.H. (eds) *Resource Capture by Crops*, Nottingham University Press, Nottingham, UK, pp. 315–349.

Parsons, A.J. and Penning, P.D. (1988) The effect of the duration of regrowth on photosynthesis, leaf death and the average rate of growth in a rotationally grazed sward. *Grass and Forage Science* 43, 15–27.

Parsons, A.J. and Chapman, D.F. (2000) The principles of pasture growth and utilisation. In: Hopkins A. (ed.) *Grass: its Production and Utilisation.* 3rd edn. British Grassland Society. Blackwell Scientific Publications, Oxford, UK, pp. 31–89.

Parsons, A.J., Leafe, E.L., Collett, B., Penning, P.D. and Lewis, J. (1983) The physiology of grass production under grazing. 2. Photosynthesis, crop growth and animal intake of continuously grazed swards. *Journal of Applied Ecology* 20, 127–139.

Parsons, A.J., Johnson, I.R. and Williams, J.H.H. (1988a) Leaf age structure and canopy photosynthesis in rotationally and continuously grazed swards. *Grass and Forage Science* 43, 1–14.

Parsons, A.J., Johnson, I.R. and Harvey, A. (1988b) Use of a model to optimize the interaction between frequency and severity of intermittent defoliation and to provide a fundamental comparison of the continuous and intermittent defoliation of grass. *Grass and Forage Science* 43, 49–59.

Parsons, A.J., Carrere, P. and Schwinning, S. (2000) Dynamics of heterogeneity in a grazed sward. In: Lemaire, G., Hodgson, J., de Moraes, A., de Faccio Carvalho, P. and Nabinger, C. (eds) *Grassland Ecophysiology and Grazing Ecology*. CAB International, Wallingford, UK, pp. 289–315.

Parsons, A.J., Schwinning, S. and Carrere, P. (2001) Plant growth functions and possible spatial and temporal scaling errors in models of herbivory. *Grass and Forage Science* 56, 21–34.

Pedersen, H., Dunkin, K.A. and Firestone, M.K. (1999) The relative importance of autotrophic and heterotrophic nitrification in a conifer forest soil as measured by ^{15}N tracer and pool dilution techniques. *Biogeochemistry* 44, 135–150.

Pepper, D.A., Del Grosso, S.J., McMurtrie, R.E. and Parton, W.J. (2005) Simulated carbon sink response of shortgrass steppe, tallgrass prairie and forest ecosystems to rising $[CO_2]$, temperature and nitrogen input. *Global Biogeochemical Cycles* 19, 1–20.

Poorter, H., van Berkel, Y., Baxter, B., Bel, M., den Hertog, J., Dijkstra, P., Gifford, R.M., Griffin, K.L., Roumet, C. and Wong, S.C. (1997) The effect of elevated CO_2 on the chemical composition and construction costs of 27 C3 species. *Plant, Cell and Environment* 20, 472–482.

Poughon, L., Dussap, C.-G. and Gros, J.-B. (2001) Energy model and metabolic flux analysis for autotrophic nitrifiers. *Biotechnology and Bioengineering* 72, 416–433.

Rasmussen, S., Parsons, A.J., Xue, H. and Newman, J.A. (2009) High sugar grasses – harnessing the benefits of new cultivars through growth management. *Proceedings of the New Zealand Grassland Association* 71, 161–166.

Rasse, D.P., Rumpel, C. and Dignac, M.-F. (2005) Is soil carbon mostly root carbon? Mechanisms for a specific stabilisation. *Plant and Soil* 269, 341–356.

Raynaud, X., Lata, J.-C. and Leadley, P.W. (2006) Soil microbial loop and nutrient uptake by plants: a test using a coupled C:N model of plant–microbial interactions. *Plant and Soil* 287, 95–116.

Robson, M.J. (1973a) The growth and development of simulated swards of perennial ryegrass. I. Leaf growth and dry weight changes as related to the ceiling yield of a seedling sward. *Annals of Botany* 37, 487–500.

Robson, M.J. (1973b) The growth and development of simulated swards of perennial ryegrass. II. Carbon assimilation and respiration in a seedling sward. *Annals of Botany* 37, 501–518.

Robson, M.J., Ryle, G.J.A. and Woledge, J. (1988) The grass plant – its form and function. In: Jones, M.B. and Lazenby A. (eds) *The Grass Crop: the Physiological Basis of Production.* Chapman and Hall, Bury St Edmunds, UK, pp. 25–83.

Ross, D.J., Tate, K.R., Newton, P.C.D., Wilde, R.H. and Clark, H. (2000) Carbon and nitrogen pools and mineralization in a grassland gley soil under elevated carbon dioxide at a natural CO_2 spring. *Global Change Biology* 6, 779–790.

Schwinning, S. and Parsons, A.J. (1999) The stability of grazing systems revisited: spatial models and the role of heterogeneity. *Functional Ecology* 13, 737–747.

Sinsabaugh, R.L., Hill, B.H. and Follstad Shah, J.J. (2009) Ecoenzymatic stoichiometry of microbial organic nutrient acquisition in soil and sediment. *Nature* 462, 795–798.

Soussana, J.-F. and Lüscher, A. (2007) Temperate grasslands and global atmospheric change: a review. *Grass and Forage Science* 62, 127–134.

Soussana, J.-F., Loiseau, P., Vuichard, N., Ceschia, E., Balesdent, J., Chevallier, T. and Arrouays, D. (2004) Carbon cycling and sequestration opportunities in temperate grasslands. *Soil Use and Management* 20, 219–230.

Soussana, J.-F., Graux, A.-I. and Tubiello, N. (2010) Improving the use of modelling for projections of climate change impacts on crops and pastures. *Journal of Experimental Botany* 61, 2217–2228.

Tainton, N.M. (1974) Effect of different grazing rotations on pasture production. *Journal of the British Grassland Association* 29, 191–202.

Taylor, P.G. and Townsend, A.R. (2010) Stoichiometric control of organic carbon–nitrate relationships from soils to the sea. *Nature* 464, 1178–1181.

Thornley, J.H.M. (1998a) *Grassland Dynamics: an Ecosystem Simulation Model.* CAB International, Wallingford, UK.

Thornley, J.H.M. (1998b) Modelling shoot:root relations: the only way forward? *Annals of Botany* 81, 165–171.

Thornley, J.H.M. and Cannell, M.G.R. (1997) Temperate grassland responses to climate change: an analysis using the Hurley Pasture Model. *Annals of Botany* 80, 205–221.

Thornley, J.H.M. and Cannell, M.G.R. (2001) Soil carbon storage response to temperature: an hypothesis. *Annals of Botany* 87, 591–598.

Thornley, J.H.M. and Johnson, I.R. (1990) *Plant and Crop Modelling.* Clarendon Press, Oxford, UK.

Wardle, D.A., Barker, G.M., Bonner, K.I. and Nicholson, K.S. (1998) Can comparative approaches based on plant ecophysiological traits predict the nature of biotic interactions and individual plant species effects in ecosystems? *Journal of Ecology* 86, 405–420.

Woledge, J. (1972) The effect of shading on the photosynthetic rate and longevity of grass leaves. *Annals of Botany* 36, 551–561.

Woledge, J. (1978) The effect of shading during vegetative and reproductive growth on the photosynthetic capacity of leaves in a grass sward. *Annals of Botany* 42, 1085–1089.

Yvon-Durocher, G., Jones, J.I., Trimmer, M., Woodward, G. and Montoya, J.M. (2010) Warming alters the metabolic balance of ecosystems. *Philosophical Transactions of the Royal Society of London B* 365, 2117–2126.

Zanetti, S., Hartwig, U.A., Lüscher, A., Hebeissen, T., Frehner, M., Fischer, B.U., Hendrey, G.R., Blum, H. and Nösberger, J. (1996) Stimulation of symbiotic N_2 fixation in *Trifolium repens* L. under elevated atmospheric CO_2 in a grassland ecosystem. *Plant Physiology* 112, 575–583.

2 Nutritional Constraints for Grazing Animals and the Importance of Selective Grazing Behaviour

Dennis P. Poppi

2.1 Introduction

The chapters in Part I of this book focus on the productivity of grassland systems, with particular emphasis on the behaviour of plants and animals in such systems. The objective in this chapter is to focus attention on the nutrition of grazing animals, with implications for the effects of forage characteristics on nutrient intake, and the effects of animal–sward interactions on nutrient flows.

2.2 Forage Intake

Intake and digestibility are the two factors that determine total nutrient supply or metabolizable energy intake by an animal, whether housed or grazing. Allden and Whittaker (1970) showed that intake was limited by the herbage mass on offer and the allowance or pasture dry matter (DM) offered/head/day. Both these parameters (herbage mass and allowance) have an underlying outcome, that of utilization, i.e. the amount or proportion of the herbage on offer that is removed during a grazing event, usually expressed on a daily basis.

In practical terms, intake and animal performance (live weight gain or milk production) follow a curvilinear asymptotic function in response to allowance, which is used in managing animals at pasture either to achieve maximum performance or maximum herbage utilization (Allden and Whittaker, 1970; Milligan *et al.*, 1987). Where herbage mass and allowance are not limiting, forage intake may increase up to levels of digestibility in excess of 80% (Hodgson, 1968). Thus rules for the minimum amount of herbage mass on offer to maximize intake (1800 kg DM/ha; Allden and Whittaker, 1970) and the need to offer animals at least four times more feed than they could consume on a daily basis to maximize intake (Milligan *et al.*, 1987) evolved and became part of management plans. Poppi *et al.* (1987) identified nutritional constraints as limits to intake when allowance was high (at plateau for the response curve) and harvesting constraints as limits to intake for the ascending part of the response curve of intake to allowance. This pattern appears to hold for temperate and tropical grasses and legumes (Allden and Whittaker, 1970; Stobbs, 1973; Chacon and Stobbs, 1976; Hendricksen and Minson, 1980) but recently A. White (personal communication) has shown that whilst a similar relationship holds for utilization of rangeland pastures in Northern Australia in the wet season there is no such relationship in the dry season. In his experiments, intake declined after 13–15% utilization of the pasture in the wet season but there was no decline in intake with increasing pasture utilization in the dry season

Rumen fill and the rate of disappearance of digesta from the rumen are important factors that control intake of forage-based diets by ruminants. Cruickshank (1986) demonstrated that the rate of increase in rumen fill during a grazing event was affected by the rate of intake and the retention time of digesta in the rumen. Rate of intake, as outlined above, is largely influenced by diet selection and bite size whilst retention time is largely influenced by the inherent rate of digestion and rate of particle disintegration set by the anatomical features of the leaf (C4 versus C3 anatomical differences) and its degree of maturity. In these studies (Cruickshank, 1986), forages such as white clover, with high rates of intake but also with very low retention time characteristics (high rate of disappearance from the rumen), would not fill the rumen to the same extent during a grazing event as that observed in lambs grazing ryegrass (Cruickshank, 1986). There was a marked diurnal pattern of rumen fill as sheep grazed pastures, with maximum fill only occurring during the afternoon grazing period when rate of intake and duration of grazing appeared to be higher. Gregorini *et al.* (2006) demonstrated similar features in cattle strip grazing ryegrass and have used this diurnal concept of selection and rate of intake to modify intake and diet selection, with positive effects on animal performance.

2.3 Diet Selection

It is commonly accepted that ruminants select a diet that is different to that on offer, often higher in digestibility, crude protein and mineral content. Some key principles have emerged, the importance of which may vary depending on the animal species. Leaf is selected in preference to stem and dead material, the reason for which is not certain but may simply be a function of breaking strain associated with **prehension** (Hendricksen and Minson, 1980), the sugar content of the plant part including selection for stem (Leury *et al.*, 1999) or the position of leaf or plant species in the upper accessible horizon, on leaf or species selection (Milne *et al.*, 1982). Leaf is selected from the upper horizon of the sward where there are dense swards (high bulk density) with high leaf density and little stem (Milne *et al.*, 1982). However, where swards are more horizontally dispersed, animals will select from horizons or patches of highest accessible leaf density and that may be at the base of the sward (L'Huillier *et al.*, 1986). Across the landscape of rangeland pastures, similar principles apply for site selection and plant species, both of which may vary markedly in the horizontal plane. The drive to select leaf results in a decline in bite size as leaf mass or accessibility declines (Stobbs, 1973; Chacon and Stobbs, 1976; Hendricksen and Minson, 1980). Animals try to accommodate for a smaller bite size by increasing grazing time but there is a maximum grazing time available to an animal on a daily basis and the consequence of a reduced bite size is frequently a reduction in intake. The extra effort of harvesting leaf may in fact lead to a reduced grazing time. Stems have a negative influence in that they are not selected and may be regarded as partial aversion, often (but not always) lower in nutrient content, and may pose a barrier to the animals ability to harvest leaf (Stobbs, 1973; Benvenutti *et al.*, 2006, 2008a,b, 2009). Sheep, with their smaller incisor width, are more selective than cattle, in that they select leaf at all levels in the sward, whereas cattle and goats tend to select from the sward surface (Nicol, 1997).

Grassland communities are heterogeneous in terms of sward structure and the composition of the plants. Monospecific or limited-species pastures have dominated intensive animal production systems based on pasture. However, for reasons such as biodiversity, there are extensive grassland areas of the world that maintain a very heterogeneous plant community, and in Europe there is more emphasis on returning grasslands to this state. Diet selection has primarily focused on attributes of sward structure but recently more emphasis has turned to other animal factors influencing diet selection, which become more apparent in these heterogeneous grasslands (Rutter, 2010). These may be related to nutrient requirement of the animal and behavioural and welfare considerations of the need of any animal to continually test novel foods (Villalba *et al.*, 2010). Rutter (2010) has reviewed a series of experiments where diet selection for ryegrass or white clover was affected by variation in the requirements of the animal, by the spatial distribution of individual plants, and by the time of day. These principles may be extended to heterogeneous grassland pastures, with implications for the production of the animal, the maintenance of plant diversity and the nutrient recycling that occurs with movement across the landscape.

Plants differ in nutrient composition between species and stages of maturity. The most variable parameters are digestibility (related to fibre and net energy content) and crude protein content (and so metabolizable protein supply). Traditionally, these have been manipulated by selection of plant species for cultivation and managing these intensively to maximize animal performance. In very intensive production systems, this can have negative environmental consequences (e.g. nitrogen levels in groundwater) and limited biodiversity for a region. In extensive systems, these issues do not arise because the cost of intensifying the system and the ability of the region (largely as rainfall) to support such a system generally preclude this type of development. The application of intensive principles of plant and animal management to these extensive systems often leads to overgrazing and degradation of the grasslands.

Some general principles of nutrient composition are apparent. Immature plants have higher digestibility and crude protein and fibre content than mature plants. C4 grasses have relatively low digestibility and crude protein content and relatively high fibre content. Legumes are higher in crude protein and lower in fibre, and may be higher in digestibility, than grasses. Temperate plants are generally higher in digestibility and crude protein content than tropical ones. Plant species differ in instantaneous intake rate (IIR) as a function of sward structure, and temperate species have a higher IIR than tropical species. All plant species accumulate soluble carbohydrate over the day, and highest intake occurs during the afternoon grazing period when energy content is highest. Faced with this diversity of nutrient composition, animals have developed strategies for diet selection to maximize total nutrient intake and specific nutrient intake linked to continual testing of novel foods and the consequences of such. Rutter (2010) has outlined, in a simple two-pasture system, that selection for white clover and ryegrass is modified by time of day, spatial distribution of the two plant species and nutrient requirement of the cow (stage of lactation and potential milk yield). Selection for protein in ruminants is difficult because of the N transactions that occur in the rumen. Free-choice pen experiments provide dietary extremes in metabolizable protein content that are useful experimentally (Kyriazakis and Oldham, 1993) but are not so apparent with temperate pasture species in terms of the protein/energy ratio of absorbed substrates or the metabolizable protein content of the diet (Cruickshank *et al.*, 1992; Poppi *et al.*, 1997). Intake, rather than the metabolizable protein content of the grasses and legumes, had a much more significant effect on total nutrient and metabolizable protein supply to the animal (Poppi and McLennan, 1995). Despite this, there are differences in selection between grasses and legumes species that cannot be explained simply by a selection process for crude protein or metabolizable protein, as the species are not sufficiently different in metabolizable protein content for the animal to distinguish them on this factor alone. Other factors no doubt play a role: fibre content of the diet may indirectly affect time spent in night grazing and hence the risk of predation (Villalba *et al.*, 2010). Tropical pastures and rangeland pastures are more diverse, especially with respect to protein content, and so selection processes have more immediate nutrient feedback, especially with condensed tannin-containing legumes (Poppi and McLennan, 1995). Petty *et al.* (1998) showed, with a leucaena/pangola grass pasture grown in strips, that there was preferential selection for the high crude protein leucaena on initial entry into a rotationally grazed paddock system but that this shifted over the 7-day grazing period as leucaena declined in green leaf mass, supporting the concept that green legume leaf mass was a major driver for selection. However the steers selected some grass within a day of entry, indicating that crude protein per se was not the simple driver and that a mixture provided a compromise to maximize nutrient intake.

2.4 Diet Selection and Animal Performance

There are consequences of the animals' desire to select leaf when grazing that affect animal performance. Selecting leaf results in a specified diet quality (primarily digestibility and crude protein (CP)) which ultimately determines the metabolizable energy (ME) or net energy (NE) value of the diet. The degree of selection determines the deviation from the average of the total sward (Milne *et al.*, 1982; CSIRO, 2007). Selecting leaf determines the bite size of the animal and hence instantaneous rate of intake. These factors influence the time the animal is prepared to harvest DM in this manner: daily grazing time (Stobbs, 1973, 1974; Gregorini *et al.*, 2009). Both of these features determine total energy supply to the animal and hence its performance.

Tropical swards and rangeland pastures are characterized by grass species that have rapid rates of stem elongation and inflorescence after rainfall, with negative consequences on intake. The principles of diet selection hold that animals select for green leaf but the stem and inflorescence provide barriers to selecting a bite, and hence result in a smaller bite size and rate of intake (Benvenutti *et al.*, 2006, 2008a,b, 2009). These authors demonstrated that there was also a very wide range in breaking strain of stems from different tropical grasses, and were able to group species based on breaking strain of the stem. Some stems did not pose a barrier to prehension. There are some consequences of this for subsequent grazing events. For all pasture species (temperate and tropical), the residual pasture after a grazing event is high in stem and dead material when pasture allowance is high so as to achieve maximum intake. Regrowth leads to a pasture with undesirable sward characteristics to maximize intake in subsequent grazing events. In intensive temperate systems this is managed by further grazing of the pasture, down to low residuals, with other stock with low performance targets. In intensive tropical swards, many of the grass species cannot be grazed to low residuals for subsequent regrowth of ideal sward structure because of the position of growth points of the plant above ground level. Thus there is sensitivity of the plant to the degree of defoliation, which must be taken into account.

2.5 Implications for Grazing Management

The experiments outlined above have provided a basis for managing the animal at pasture to achieve a desired level of performance. At certain times of the year and with certain classes of stock, maximum performance is required (e.g. lamb growth, flushing for reproduction) and a high allowance or low utilization is used. At other times of the year, maintenance is all that is needed or pasture needs to be budgeted during periods of feed deficit, so low allowance is used. This needs to be balanced across the year to achieve high annual utilization of the pasture that is grown. In temperate pasture systems, a reasonable target is to utilize 70% of what is grown on an annual basis, but in tropical rangelands a target of 20% utilization is desirable based on the need to maintain desirable species in the sward and to maintain adequate ground cover for intermittent high-rainfall events, ensuring high water infiltration rather than run off (Cowley *et al.*, 2007). Intensive tropical swards have an intermediate target level of utilization.

Simple two-pasture-species systems are relatively easy to manage intensively and provide high levels of production/ha but there can be negative effects on biodiversity and the environment, which provide constraints on how far the intensification can proceed. Extensive grassland systems in temperate and tropical environments are more difficult to manage and maintain and have lower production/ha. Nevertheless, the diet selection drivers that are more apparent in extensive systems appear to operate under both pen and intensive grazing scenarios and these can be manipulated in the intensive grazing systems to increase intake and animal production, as exemplified by Gregorini *et al.* (2006) and Rutter (2010).

Another feature that emerges during grazing of heterogeneous swards, such as rangeland pasture systems, is preferred grazing of some species, especially the perennial grass species that characterize these plant communities. These species are necessary for maintaining the rangeland and also to sustain high animal performance. Repeated regrazing of preferred species, through diet selection of that species, can reduce root reserves and the ability of that species to survive or maintain a reasonable proportion of the sward (Ash and McIvor, 1998) but can be controlled by intermittent protection from grazing during the wet season (A. White, personal communication). This provides a mechanism whereby diet selection and low utilization can avoid overgrazing of the preferred plant species. White further observed that there was a shift from defoliation of preferred to non-preferred species as the residual height of the preferred species declined during grazing (personal communication). Benvenutti *et al.* (2006, 2008a,b, 2009) showed that this appeared largely to be a function of the stem barrier and breaking strain of the stem, which differed between species. This knowledge can be used to derive management systems that maximize animal performance yet maintain rangeland pastures in a desired state. It follows that animal species differences in selection also impact on the defoliation events of a rangeland pasture ecosystem and its subsequent stability, with examples of grazer and browser systems in Africa and in Australia, the latter involving introduced ruminants and Macropod foregut fermentors (Edwards *et al.*, 1996).

Diet selection also drives nutrient transfer to the soil and across the landscape. The bulk of the daily intake appears to occur during the afternoon grazing event in continuously grazed swards (Cruickshank, 1986; Betteridge *et al.*, 2010). This coincides with maximum soluble carbohydrate accumulation from photosynthesis and so enhances the ability of the rumen microbes to capture the excess degraded CP in temperate pasture types. This is not an issue with C4 grasses where, although the same soluble carbohydrate accumulation occurs, the CP levels are not high enough for there to be significant losses of N from the rumen. In fact, the diurnal sequence may enhance N recycling to the rumen.

In temperate plants there is an excess of degradable protein (as distinct from metabolizable protein) and that extra N has to be excreted primarily in urine. In tropical plants there may be shortages of degradable protein, leading to poor microbial protein production within the rumen and greater N recycling. Animals consuming tropical plants must also process much greater amounts of fibre, with negative physical consequences in terms of rumen fill and retention time of fibre in the rumen. Plants also differ in sward structure, with short term effects on IIR, as outlined earlier, and this must also be accommodated within the framework of trying to maximize nutrient intake and minimize negative consequences or processing costs. Protein and fibre are the two 'nutrient' types that have most effect, whilst IIR and grazing time are the behavioural factors that have most effect.

The transfer of N in the plant and deposited in the animal is reasonably inefficient, with only 11-16% of ingested N deposited as protein N in animals grazing temperate high-quality pastures (Cruickshank *et al.*, 1992). A selection process that drives towards metabolizable protein will always have this as a negative consequence of the system if the plants and parts have an excess of N relative to fermentable metabolizable energy. With tropical pastures and rangelands these are not important issues, as there is usually no excess of degradable protein and there is usually a wide variety of species for selection. Here, the problem is the limitations imposed by other aspects of these species, i.e. shortage of N, high fibre and low IIR.

In all cases, the transfer of N and P from the pasture to specific urine patches and defecation sites across the landscape occurs. Betteridge *et al.* (2010) have mapped this in hill country pastures in New Zealand with sheep and cattle. Sheep urinated more times than cattle and both species had a greater frequency of urination associated with 'easy hill country', indicating a shift towards these sites for urinary N transfer. Burle *et al.* (2003) have shown that approximately 65% of the N in leucaena strips is recycled via urine and faeces to the grass. Vallis *et al.* (1985) showed that 32% of urinary N in a tropical environment was volatilized within 7 days, indicating the large atmospheric losses through this route of N excretion. The main route of excess N excretion in ruminants is via urine, with little difference in faecal excretion (Marini and Van Amburgh, 2003). The maintenance of vigorous and well managed pastures in terms of stocking rate and N fertilizer application can have positive aspects on soil C accumulation from roots and earthworm activity within deeper horizons, especially with deep-rooted grasses (Fisher *et al.*, 1994). The necessity for pasture spelling on tropical rangeland pastures would have similar consequences, although the objective here is to maintain plant species that are preferred by cattle.

2.6 Understanding Intake

The interactions between grazed swards and grazing animals are complex, and are frequently difficult to control. Substantial progress has been made in understanding the quantitative effects of specific plant characteristics on ingestive behaviour (e.g. Prache and Delagarde, Chapter 3, this volume) but it is still difficult to define and explain animal behaviour in practical situations, where a range of qualitative and quantitative factors would be expected to vary simultaneously. Selection for nutrients, especially net energy, protein and some forms of carbohydrate, is now a commonly accepted concept across a wide range of animal species, from caterpillars (Lee *et al.*, 2002) to sheep (Kyriazakis and Oldham, 1993) and horses (Edouard *et al.*, 2010). The Geometric Framework for nutrient selection provides a basis for intake and diet selection, with selection for protein as a key driver (Lee *et al.*, 2002).The minimal total discomfort theory of Forbes (2001, 2007) improves on the concept of optimal foraging strategy as a driver to intake regulation (Cosgrove and Edwards, 2007; Edwards *et al.*, 2008) by accommodating metabolic, physical and behavioural parameters. Poppi (unpublished observations) has shown that the Forbes' model can explain the decline in intake as pasture is grazed down and also predict intake of animals in pens consuming diets varying markedly in retention time. Its use in future studies on selective behaviour and forage intake is recommended.

2.7 Conclusions

Diet selection drives key parameters known to influence intake and provides the nutrients that result in a level of animal performance. Animals select for nutrients and to minimize the physical and behavioural consequences of the mixed diet.

Diversity in diet selection is a preferred option with animals but intensive systems of production minimize this, in seeking to maximize performance. In extensive systems, diet selection of specific plant species can have negative consequences on the maintenance of the system if defoliation events of the individual plant are too frequent. Most intensive systems have inefficiency of N use, which leads to greater recycling of N via the urine with potential negative consequences. This is not an issue with low soil N and/or tropical and rangeland pasture systems, where N is limiting in most circumstances.

Acknowledgements

I wish to acknowledge the technical help received from S.R. McLennan and D.A. Poppi.

References

Allden, W.G. and Whittaker, I.A.McD. (1970) The determinants of herbage intake by grazing sheep: the interrelationship of factors influencing herbage intake and availability. *Australian Journal of Agricultural Research* 21, 755–766.

Ash, A.J. and McIvor, J.G. (1998) How season of grazing and herbivore selectivity influence monsoon tall-grass communities of northern Australia. *Journal of Vegetation Science* 9, 123–132.

Benvenutti, M.A., Gordon, I.J. and Poppi, D.P. (2006) The effect of the density and physical properties of grass stems on the foraging behaviour and instantaneous intake rate by cattle grazing an artificial reproductive tropical sward. *Grass and Forage Science* 61, 272–281.

Benvenutti, M.A., Gordon, I.J. and Poppi, D.P. (2008a) The effects of the stem density of tropical swards and age of grazing cattle on their foraging behaviour. *Grass and Forage Science* 63, 1–8.

Benvenutti, M.A., Gordon, I.J., Poppi, D.P., Crowther, R. and Spinks, W. (2008b) Foraging mechanics and their outcomes for cattle grazing reproductive tropical swards. *Applied Animal Behaviour Science* 113, 15–31.

Benvenutti, M.A., Gordon, I.J., Poppi, D.P., Crowther, R., Spinks, W. and Moreno, F.C. (2009) The horizontal barrier effect of stems on the foraging behaviour of cattle grazing five tropical grasses. *Livestock Science* 126, 229–238.

Betteridge, K., Costall, D., Balldur, S., Upsdell, M. and Umemura, K. (2010) Urine distribution and grazing behaviour of female sheep and cattle grazing a steep New Zealand hill pasture. *Animal Production Science* 50, 624–629.

Burle, S.T.M., Shelton, H.M. and Dalzell, S.A. (2003) Nitrogen recycling in degraded *Leucaena leucocephala-Brachiaria decumbens* pastures on acid infertile swards in south-east Queensland, Australia. *Tropical Grasslands* 37, 119–128.

Chacon, E. and Stobbs, T.H. (1976) Influence of progressive defoliation of a grass sward on the eating behaviour of cattle. *Australian Journal of Agricultural Research* 27, 709–727.

Cosgrove, G.P. and Edwards, G.R. (2007) Control of grazing intake. In: Rattray, P.V., Brookes, I.M., and Nicol, A.M. (eds) *Pasture and Supplements for Grazing Animals.* New Zealand Society of Animal Production Occasional Publication No. 14, Hamilton, New Zealand, pp. 61–80.

Cowley, R.A., McCosker, K.D., MacDonald, R.N. and Hearnden, M.N. (2007) Optimal pasture utilisation rates for sustainable cattle production with a commercial Brahman herd in the Victoria River Downs region of the Northern Territory. In: Pattie, B. and Restall, B. (eds) *Proceedings of the Northern Beef Research Update Conference.* Northern Australia Beef Research Council, pp. 34–44.

Cruickshank, G.J. (1986) Nutritional constraints to lamb growth at pasture. PhD thesis, Lincoln College, New Zealand.

Cruickshank, G.J., Poppi, D.P. and Sykes, A.R. (1992) The intake, digestion and protein degradation of grazed herbage by early-weaned lambs. *British Journal of Nutrition* 68, 340–364.

CSIRO (2007) *Nutrient Requirements of Domesticated Ruminants.* Freer, M., Dove, H. and Nolan, J.V. (eds), CSIRO Publishing, Collingwood, Australia.

Edouard, N., Duncan, P., Dumont, B., Baumont, R. and Fleurance, G. (2010) Foraging in a heterogenous environment – an experimental study of the trade-off between intake rate and diet quality. *Applied Animal Behaviour Science* 126, 27–36.

Edwards, G.P., Croft, D.B. and Dawson, T.J. (1996) Competition between red kangaroo (*Macropus rufus*) and sheep (*Ovis aries*) in the arid rangelands of Australia. *Australian Journal of Ecology* 21, 165–172.

Edwards, G.R., Parsons, A.J. and Bryant, R.H. (2008) Manipulating dietary preference to improve animal performance. *Australian Journal of Experimental Agriculture* 48, 773–779.
Fisher, M.J., Rao, I.M., Ayarza, M.A., Lascano, C.E., Sanz, J.I., Thomas, R.J. and Vera, R.R. (1994) Carbon storage by introduced deep-rooted grasses in the South American savannas. *Nature* 371, 236–238.
Forbes, J.M. (2001) Consequences of feeding for future feeding. *Comparative Biochemistry and Physiology Part A* 128, 463–470.
Forbes, J.M. (2007) A personal view of how ruminant animals control their intake and choice of food: minimal total discomfort. *Nutrition Research Reviews* 20, 132–146.
Gregorini, P., Eirin, M., Refi, R., Ursino, M., Ansin O.E. and Gunter S.A. (2006) Timing of herbage allocation in strip grazing: Effects on grazing pattern and performance of beef heifers. *Journal of Animal Science* 84, 1943–1950.
Gregorini, P., Gunter, S.A., Beck, P.A., Caldwell, J., Bowman, M.T. and Coblentz, W.K. (2009) Short-term foraging dynamics of cattle grazing swards of different canopy structures. *Journal of Animal Science* 87, 3817–3824.
Hendricksen, R. and Minson, D.J. (1980) The feed intake and grazing behaviour of cattle grazing a crop of *Lablab purpureus* cv Rongai. *Journal of Agricultural Science (Cambridge)* 95, 547–554.
Hodgson, J. (1968) The relationship between the digestibility of a sward and the herbage consumption of grazing calves. *Journal of Agricultural Science (Cambridge)* 70, 47–51.
Kyriazakis, I. and Oldham, J.D. (1993) Diet selection in sheep: the ability of growing lambs to select a diet that meets their crude protein (nitrogen x 6.25) requirements. *British Journal of Nutrition* 69, 617–629.
Lee, K.P., Behmer, S.T., Simpson, S.J. and Raubenheimer, D. (2002) A geometric analysis of nutrient regulation in the generalist caterpillar *Spodoptera littoralis* (Boisduval). *Journal of Insect Physiology* 48, 655–665.
Leury, B.J., Siever-Kelly, C., Gatford, K.L., Simpson, R.J. and Dove, H. (1999) Spray-topping annual grass pasture with glyphosate to delay loss of feeding value during summer. IV. Diet composition, herbage intake, and performance in grazing sheep. *Australian Journal of Agricultural Research* 50, 487–495.
L'Huillier, P.J., Poppi, D.P. and Fraser, T. (1986) Influence of structure and composition of ryegrass and prairie grass–white clover swards on the grazed horizon and diet harvested by sheep. *Grass and Forage Science* 41, 259–267.
Marini, J.C. and Van Amburgh, M.E. (2003) Nitrogen metabolism and recycling in Holstein heifers. *Journal of Animal Science* 81, 545–552.
Milligan, K.E., Brookes, I.M. and Thompson, K.F. (1987) Feed planning on pasture. In: Nicol, A.M. (ed.) *Livestock Feeding on Pasture*. New Zealand Society of Animal Production Occasional Publication No. 10, Hamilton, New Zealand, pp. 75–88.
Milne, J.A., Hodgson, J., Thompson, R., Souter, W.G. and Barthram, G.T. (1982) The diet ingested by sheep grazing swards differing in white clover and perennial ryegrass content. *Grass and Forage Science* 37, 209–218.
Nicol, A.M. (1997) The application of mixed grazing. In: *Proceedings of the XVIII International Grassland Congress*. Winnipeg, Canada, pp. 525–534.
Petty, S.R., Poppi, D.P. and Triglone, T.R. (1998) Effect of maize supplementation, seasonal temperature and humidity on the liveweight gain of steers grazing irrigated *Leucaena leucocephala/Digitaria eriantha* pastures in north-west Australia. *Journal of Agricultural Science, Cambridge* 130, 95–105.
Poppi, D.P. and McLennan, S.R. (1995) Protein and energy utilisation by animals at pasture. *Journal of Animal Science* 73, 278–290.
Poppi, D.P., Hughes, T.P. and L'Huillier, P.J. (1987) Intake of pasture by grazing ruminants. In: Nicol, A.M. (ed.), *Livestock Feeding on Pasture*. New Zealand Society of Animal Production Occasional Publication No. 10, Hamilton New Zealand, pp 55–63.
Poppi, D.P., McLennan, S.R., Bediye, S., de Vega, A. and Zorrilla-Rios, J. (1997) Forage quality: strategies for increasing nutritive value of forages. In: *Proceedings of the XVIII International Grassland Congress*. Winnipeg, Canada, pp. 307–322.
Rutter, S.M. (2010) Review: grazing preferences in sheep and cattle: implications for production, the environment and animal welfare. *Canadian Journal of Animal Science* 90, 285–293.
Stobbs, T.H. (1973) The effect of plant structure on intake in tropical pastures. I. Variation in the bite size of grazing cattle. *Australian Journal of Agricultural Research* 24, 809–819.
Stobbs, T.H. (1974) Components of grazing behaviour of dairy cows on some tropical and temperate pastures. *Proceedings of the Australian Society of Animal Production* 10, 299–302.
Vallis, I., Peake, D.C.I., Jones, R.K. and McCown, R.L. (1985) Fate of urea-N from cattle urine in a pasture-crop sequence in a seasonally dry tropical environment. *Australian Journal of Agricultural Research* 36. 809–817.
Villalba, J.J., Provenza, F.D. and Manteca, X. (2010) Links between ruminants' feed preference and their welfare. *Animal* 4, 1240–1247.

3 The Influence of Vegetation Characteristics on Foraging Strategy and Ingestive Behaviour

Sophie Prache and Rémy Delagarde

3.1 Introduction

In addition to their primary function in food production, pasture-based animal production systems contribute to the management of climate change and landscape diversity, implications for biodiversity and the socio-economic sustainability of agriculture, and the ethics, quality and safety of natural animal products. In the more productive pasture-based systems, the challenge is to combine a high animal productivity with high pasture utilization and a moderate use of fertilizers, all these factors contributing to reducing the carbon footprint of the milk and meat produced, water pollution, reliance on bought-in concentrate and sensitivity to input costs volatility (Benoit and Laignel, 2010; Dillon *et al.*, 2005). These economic and environmental concerns have renewed interest in increasing the use of grass/legume swards. In the more extensive grassland systems, farmers have to manage larger flocks on larger and more diversified areas and pastures, and to reconcile production with environmental objectives such as maintaining open landscapes and contributing to landscape biodiversity. The lower grazing pressure is favourable to pasture structural heterogeneity and therefore to pasture biodiversity. Facing this diversity in grassland systems, management objectives and forage resources, the scientific challenge is to provide an understanding of the biological principles that govern system behaviour in a broad range of conditions. The objective of this chapter is to give an overview of the state of knowledge on the influence of vegetation characteristics on foraging strategy and ingestive behaviour in animals grazing both temperate and tropical swards.

3.2 Conceptual Representation of Foraging Behaviour

Herbivores selectively graze the vegetation and choose a diet of better quality than the average vegetation on offer. Foraging behaviour is therefore of major importance because it determines the nutrient intake of the animals as well as the location and intensity of their impact on the vegetation. Information on these determinants is central to understanding the dynamics of the grazed ecosystem and to providing the basis for the management of both animals and vegetation. This selective behaviour prompted a description of the foraging process, based on the hierarchy of scales, where animals make decisions (from selecting a bite, to the choice of feeding patches and the length of time to spend feeding) that may be further integrated to daily intake. This representation allows reconciliation of the ways of looking at different grassland types, from homogeneous swards to grasslands that may be heterogeneous in terms of height, phenological stage and botanical composition. The theoretical framework of Optimal Foraging Theory (OFT), proposed in the landmark paper of Stephens and Krebs (1986) for predicting foraging decisions, is based on cost/reward principles. It has received considerable attention and refinements in the last decade, its application being expanded to trade-offs faced by herbivores grazing heterogeneous pastures and to include diverse nutrients and the balance between them; it has proved a useful approach (Illius *et al.*, 1999; Prache *et al.*, 2006; Edouard *et al.*, 2010).

3.3 The Influence of Vegetation Characteristics on Foraging Decisions

3.3.1 Short-term foraging decisions

Patch level

The lowest level of decision making for a herbivore is the selection of bites within patches. Great progress has been made in understanding the interactions between the morphology and behaviour of the grazing herbivore and the structure of the sward, determining diet selection, bite characteristics and the instantaneous intake rate (IIR) within a patch. On the patch, IIR is represented as the ratio of bite mass to the time required to harvest and masticate it. These variables mainly depend on the ease with which herbage is gathered into the mouth, the resistance of herbage to fracture and mastication, and the presence of barrier components. Bite mass varies with the volume of the bite, which is limited by the bite area, the depth to which the animal inserts its mouth into the sward, and the bulk density of the vegetation in that volume. Time per bite depends on the searching constraints for the animal to select the preferred forage items, and the ease with which these items can be harvested and then masticated. Breakthroughs in the development of 'constructed microswards' allowed precise control of sward structure in order to overcome difficulties encountered in 'real swards', such as their inherent variability and the strong correlations between sward height, bulk density and barrier effects.

On vegetative patches, IIR is mostly determined by bite mass and by the functional relationship relating time per bite to bite mass. Here, sward height and bulk density are the major determinants of bite mass. In sheep, where prehension and mastication jaw movements are exclusive, time per bite is represented as the sum of prehension time, which is considered to be constant, and mastication time, which increases linearly with bite mass, the slope of the regression increasing with herbage fibre content (Baumont *et al.*, 2004). In cattle, where prehension and mastication jaw movements may overlap to form compound jaw movements, the relationship between time per bite and bite mass is rather curvilinear, the total number of jaw movements required per bite being determined by the greater of the prehension and the mastication requirements. The functional response (i.e. the relationship of IIR with herbage mass) is then most frequently a saturation curve characterized by its baseline slope and asymptote (maximum potential IIR) (Prache *et al.*, 1998; Hirata *et al.*, 2010).

Sward height affects bite mass mainly through its effect on bite depth. A number of studies, in both temperate and tropical swards, have shown that bite depth is almost a constant proportion, about 50%, of pasture height. Bite area results from the interactions between mouth size, herbage-grasping movements, sward height and density. It is related to mouth size, but cattle are able to collect herbage from an area larger than that of the open mouth, by using their tongue. In tall swards, bite area scales with the product of incisor arcade breadth and mouth gape, but on short swards, it scales with incisor arcade breadth, as only a narrow band of tillers can be prehended. In cattle, the relation of bite area to sward height depends on sward density, because animals reduce the amplitude and the number of tongue movements with increasing sward density (Laca *et al.*, 1992). This interaction has not been quantified for small ruminants, although it is probably different because of the differences in herbage-gathering movements between animal species.

Bite mass may be represented as the product of bite depth, bite area and bulk density of the grazed horizon. It increases with sward height and density, each variable acting independently and additively (Laca *et al.*, 1992), the effects being generally linear for cattle and curvilinear for sheep. Studies using 'constructed microswards' demonstrated that bite mass cannot be predicted solely on the basis of herbage mass, but that both height and density have to be taken into consideration. Actually, herbage mass is a two-dimensional description of a sward whereas bites are taken in three dimensions: a given herbage mass is more accessible to the grazing animal when it is tall and sparse than when it is short and dense.

On reproductive patches, stems may be structural foraging deterrents, acting as vertical and/or horizontal barriers that may affect bite dimensions and mass. IIR may be further impaired by an increased time searching for bites, cropping and masticating them, and the compensation in diet quality through selective grazing may be insufficient to maintain digestible IIR (Prache *et al.*, 1998, Benvenutti *et al.*, 2008, 2009). In mature cocksfoot swards, for example, the close intermixing of vegetative and thick reproductive tillers decreases the accessibility of interspersed leaves, leading to a decrease in IIR and a change in the functional response in sheep (Prache *et al.*, 1998). This effect is mediated via a decrease in bite area and an increase in time per bite, bite depth remaining unaffected (Prache *et al.*, 1998; Benvenutti *et al.*, 2009). The degree of avoidance of stems, however, depends on their tensile resistance to grazing. In a study comparing five tropical grasses, cattle avoided tough stems of high tensile resistance and reduced bite area, bite mass, IIR and digestible IIR as their density increased. However they did not avoid better-quality stems of lower tensile resistance, showing no change in foraging behaviour or in diet quality when the density of such stems increased (Benvenutti *et al.*, 2009). A detailed study using stems and leaves of *Panicum maximum* of variable tensile resistance in various arrangements investigated the mechanisms by which cattle regulate bite dimensions when faced with a stem barrier either in a vertical or horizontal plane (Benvenutti *et al.*, 2008). In swards with a stem-dominated lower stratum and a leaf-dominated upper stratum (leaves longer than stems), the presence of stems in the grazeable stratum can form a vertical barrier altering bite dimensions and mass. Animals restrict bite depth to the leafy horizon or graze only a shallow section of the top of the stems, and bite area is further impaired by the increased stiffness of the plant material, these effects increasing with the resistance of stems. The responses of both bite area and bite force to the grazing resistance of the sward in this case supports the summit bite force hypothesis as a regulator of bite area. When leaves and stems are intermixed within the grazed stratum (stems taller than the leaves), the stems act as a horizontal barrier reducing bite area, the bite depth remaining unaffected. The decrease of both bite area and bite force in that case suggests that the tongue force required to bend the stems may regulate bite area. In both arrays of stems and leaves, bite mass, IIR of dry matter (DM) and digestible DM all decreased with the increase in stem tensile resistance, this negative effect being stronger in larger animals.

Although pseudostems are generally of lower tensile resistance than stems, it has also been suggested that they form a barrier to bite depth. However, further quantification of this effect is required. As in the case for stems, the effects of pseudostems on grazing behaviour probably depend on their structural and physical properties and upon the force required to harvest different sward strata. Finally, the presence of dead material within the sward may also reduce bite dimensions and IIR, but, to our knowledge, its effect has never been quantified, probably because of experimental difficulties.

Because of these various foraging deterrents, it is not surprising that green leaf mass proves to be a better predictor of bite mass and IIR than total herbage mass or sward height in sheep and cattle across various phenological stages of the sward (Flores *et al.*, 1993; Prache *et al.*, 1998) or sward management (Penning *et al.*, 1994). The increased time spent searching for and cropping selected bites changes the shape of the functional response curve relating IIR to green leaf mass, particularly by depressing its asymptotic maximum. In cocksfoot swards grazed by sheep, the maximum potential IIR decreased by 33% on a mature sward containing 15% reproductive tillers compared with a vegetative sward of similar green leaf availability (Prache *et al.*, 1998). However, it is most likely that there should be an interaction between stemmy sward characteristics and animal species because of different mouth dimensions and gathering motions, which require further experimental evaluation.

Selection of feeding patches

A higher level of decision is the choice of a new patch on which to graze. In heterogeneous swards, herbivores frequently face a choice between patches that may differ in botanical composition, as well as in vegetation structure and quality. The last decade has witnessed important developments for experimentally testing the relevance of the theoretical framework of optimal foraging for predicting patch choice by grazing herbivores. This holistic approach has been proposed to predict which food components would be included in the diet on the basis of their relative profitability.

The first experiments on vegetative temperate swards used a combination of short-term paired feeding tests of the same plant species, by varying the height and density but maintaining the quality of the sward; in this case, the animals preferred the patches where they maximised their dry matter IIR (Distel *et al.*, 1995; Edouard *et al.*, 2010). The short-term preference for the higher dry matter IIR alternative was confirmed in a subsequent study using pairs of vegetative turves of different temperate grasses (Illius *et al.*, 1999). However, methodological refinements allowing testing of optimality principles at a daily timescale on free-ranging animals showed the sensitivity of the animals to diet quality. At a daily timescale, preferences better fitted the relative values of digestible organic matter (OM) intake rate (IR) across the dietary components than the relative values of OM intake rate (Prache *et al.*, 2006). This study investigated the relationship between preference and relative IR in sheep grazing down monocultures of ryegrass and tall fescue offered at the same initial height and 0.50:0.50 by area. The initial marked preference for ryegrass was associated with a relative value of digestible OM intake rate between ryegrass and tall fescue of 1.5; the preference for ryegrass then declined with ryegrass depletion, towards no preference for either species, at which time the relative value of digestible OM intake rate between the species was close to unity, which is in line with the OFT.

Further studies explored more complex scenarios, where the animals faced trade-offs between nutrients. On grasslands composed of a mosaic of tall reproductive patches allowing a high IIR of low quality within a short vegetative background of higher quality but lower IIR, animals frequently favoured the shorter alternative of higher quality (Ginane *et al.*, 2003; Edouard *et al.*, 2010). In this case, the relative IIR of DM, of digestible DM or of net energy no longer correctly predicted patch selection, and the best predictor of patch preference then became the relative digestible crude protein IIR, although some over-selection for the taller sward was observed (Edouard *et al.*, 2010). This shift in behaviour reflected changes in the constraints faced by the animals during the sward maturation process. When facing a trade-off between nitrogen and energy, animals primarily select for nitrogen, probably because they cannot synthesize essential amino acids, while they are able to mobilize their body reserves to compensate to some extent for a lower energy supply (Edouard *et al.*, 2010). Little comparable information is available for tropical pastures, although animals grazing them are prone to face trade-offs between forage quantity and quality. Hirata *et al.* (2006) investigated patch preference by cattle offered two alternative patches of tropical grasses differing in availability and quality. Although the relative IR was not measured, the outcome of preference-currency models indicated that preference for a patch increased as the relative crude protein (CP) content and the relative sward bulk density of that patch increased over the alternative.

One of the most consistent findings from these studies, however, is that animals offered a choice repeatedly switch between patches and show partial preferences. In the short term, the necessity for the animal to sample the different patches to update their profitability and the difficulty in discriminating have been proposed as explanations. The pattern of switches observed by our group between ryegrass and tall fescue patches when the relative value of digestible OM intake rate decreased from 1.5 to 1.0 actually reflects the level of difficulty for the animal in choosing between species, the greatest difficulty occurring when alternatives offered a similar IR (Prache *et al.*, 2006). On a daily scale basis, the search for a nutritional and toxicological balance (Provenza *et al.*, 2007), the preference for novelty, and the preference for more fibrous forage in the evening for maintaining rumen fill at night have all been proposed as explanations for a diurnal pattern of preferences and mixed diets. Moreover, if OFT stresses the importance of the efficiency of foraging in the reproductive success of an animal species, the ability to diversify its diet under food scarcity may also contribute to the strategy of a species to increase its reproductive output.

3.3.2 From short-term behaviour to daily intake

The integration from short-term behaviour to daily intake necessitates linking instantaneous behaviour with satiation processes, together with the animal's motivation to eat. Post-ingestive signals coming from feed contribute to the satiation process. Ingestion and digestion are thus linked by the nutritive value and the fill effect of the ingested forages. Progress in integrating ingestion and digestion has been achieved by modelling, but our ability to predict feeding time (FT) remains a critical area because it greatly depends on the animal's volition.

Bulk in the rumen constrains both FT and intake. There is also clear evidence that animals can increase FT to compensate for a low rate of nutrient intake, but this occurs only up to a certain limit. In rotational grazing, it is typically observed that animals initially increase FT to compensate for a decrease in IIR, but that FT may drop sharply during the later stages of sward depletion, despite continued reduction in IIR (Prache *et al.*, 1998). Monitoring FT remotely has thus been proposed to be used for management purposes to provide graziers with information as when to move animals to a new paddock (Gordon and Benvenutti, 2006). In contrast, a 20% increase in DM intake rate, consecutive to a partial drying of perennial ryegrass, has been shown to be associated with a decrease of 10–15% in FT, thus leading only to a 5–10% increase in daily herbage DM intake (Cabrera Estrada *et al.*, 2004).

Food diversity has been shown to increase an animal's motivation to eat and its level of food intake by 10–15%, this effect being mediated via an increase in FT (Champion *et al.*, 2004; Côrtes *et al.*, 2006). There is convincing evidence that grazing herbivores prefer diversity and that offering them food diversity may allow them to enhance their grazing behaviour. From a practical point of view, this behavioural response may be used for management purposes to improve animal performance. However, further research is needed to explore the extent to which these results could be generalized to situations where time available to forage is an important constraint, such as in lactating animals or when herbage availability is low.

Beyond the physical and searching constraints imposed by the vegetation that have been reviewed above, the animal is thus able to use its behaviour as a flexible means to face both vegetation and physiological constraints and to overcome these constraints to some extent. Intake rate not only depends on vegetation characteristics but is also partly determined by energy requirements, fasting and level of supplementation. After a period of food deprivation or when the time of access to food is decreased, for example, the animal is able to increase its rate of food intake (Prache *et al.*, 1998; Pérez-Ramírez *et al.*, 2008) by increasing bite mass and/or decreasing the time taken to masticate each unit of DM. Harvesting and mastication constraints may thus be overcome to some extent by hungry animals. Consequently, though the idea of vegetation constraints has helped to build a general framework from which to represent the interactions between the grazing animal, the vegetation and the control of intake, there is convincing evidence that these constraints do not always have fixed values. In the same way, there is no absolute threshold for the different post-ingestive signals involved in the satiation process, i.e. rumen fill and nutrients (Sauvant *et al.*, 1996; Baumont *et al.*, 2004).

This adaptability in behaviour questions the validity of the theories of absolute thresholds and fixed constraints. Beyond the conventional view of feeding behaviour being governed by fixed ingestive and digestive constraints, an alternative view has emerged that represents feeding behaviour as being the result of all the positive and negative stimuli to feeding at a specific time. Feeding continues as long as positive stimuli have a higher intensity than negative stimuli, and when the intensity of the negative stimuli exceeds that of the positive stimuli, feeding is inhibited. The model by Sauvant *et al.* (1996) is a first attempt to manage two antagonist forces, the motivation to eat and the satiation process. It consists of two interconnected sub-models: a ruminal digestion sub-model together with a feeding decision sub-model; the choice of feeding depending on the relative functions of intake motivation and satiety. This model was further adapted to grazing situations by adding both a sward-horizons sub-model, partitioning the sward into grazed horizons, each characterized by its structure, quality and potential IIR, and a foraging decision sub-model describing preferences between horizons, based upon their relative availability and potential IIR. Actual intake rate then integrates the animal's choice between horizons and its satiety status. In this way, interactions between vegetation constraints and the animal's internal state are dynamically integrated from the scale of the bite over several days. Satisfactory validations were obtained on data sets recorded under either rotational or continuous grazing, and further developments will be to adapt this model to other ruminant species and to plurispecific swards.

3.4 The Influence of Vegetation Characteristics on Daily Intake

The prediction of grazing-herbivore performance, of pasture-use efficiency and of the dynamics of pasture regrowth primarily depends on the ability to predict daily pasture intake in a large range of grazing conditions. The regulation of daily pasture DM intake is complex and multi-factorial. It is primarily determined by nutritional factors, including animal characteristics (energy requirements and intake capacity), herbage chemical composition and nutritive value (digestibility and ingestibility) and the nature and level of supplemental feeds. These factors are not specific to grazing conditions and they have been quantified indoors using animals fed ad libitum. At pasture, however, herbage intake also depends on the foraging processes. The vegetation characteristics determine, at least partly, herbage mass and accessibility for the grazing animal, through their effects on searching and harvesting constraints. These latter characteristics can be viewed as non-nutritional factors that may summate to a relative factor affecting daily pasture DM intake (Poppi *et al.*, 1987; Delagarde and O'Donovan, 2005).

On a daily scale, our ability to understand the relationships between vegetation characteristics, foraging behaviour and DM intake is limited by methodological difficulties. Estimating daily pasture DM intake and feeding time needs several days of measurements to achieve sufficient precision, due to high between-day variability. For that reason, the effects of vegetation characteristics on daily intake have been investigated more via nutritional studies using markers than via pure behavioural studies. Here, mean daily IR and bite mass are calculated from integrated measurements, by dividing DM intake by feeding time or bite number. The advantage of such an approach is probably more robust predictive relationships, the disadvantage being a limited description of the short-term foraging processes, making it difficult to link short-term behaviour and daily intake.

At a daily scale, the quantitative relationship between vegetation characteristics and pasture DM intake should not be regarded independently of the grazing system. The primary difference between grazing systems is the instantaneous stocking rate, i.e. the area available each day for each animal. Under continuous stocking, a low herbage mass, leading to a low IR, can be compensated, at least partly, by an increase in both the area grazed and in feeding time. It is also the case during the grazing down process under rotational grazing, where the area available per animal remains high. For example, in rotational grazing systems that use a 4–5 day residence time within paddocks, the cows' daily walking movements are high and the frequency of tiller defoliation per day is below two (Wade, 1991). On the contrary, under strip-grazing, instantaneous stocking rate is so high that sward state changes rapidly within a day, with two to four daily defoliations per tiller, and no possibility for the grazing animal to explore additional areas for harvesting food. In that sense, strip-grazing represents an integrative view of all the grazing down processes under rotational grazing, i.e. the average behaviour and intake from the first to the last day within a paddock.

Considering 'no daily limiting area' systems, sward height, herbage mass or sward bulk density may be insufficient to predict daily intake variations. For continuous stocking maintained at a given sward height, a characteristic sward structure develops and herbage availability may be simply described in terms of sward height or herbage mass. In this case, herbage intake increases asymptotically with sward mass or height, a maximum intake being reached for a sward height of 8–9 cm in dairy cows and from 6 cm onwards for lactating ewes. However, when pooling experiments under continuous and rotational stocking systems, Wade (1991) and Penning *et al.* (1994) demonstrated the importance of leaf proportion on the regulation of daily intake across grazing managements and types of animal. In sheep, Penning *et al.* (1994) concluded that the best predictor of daily intake was green leaf mass, independently of total mass or sward height. In dairy cows, Wade (1991) demonstrated that variation in daily intake and milk production were related to free leaf lamina length, independent of total pasture height. Moreover, the motivation of ruminants to increase feeding time at very low herbage mass largely depends on the remaining leaf proportion within the sward (Penning *et al.*, 1994; Delagarde *et al.*, 2010). During the last days under severe rotational grazing, feeding time may decrease even though IR and daily intake are very low (no leaf remaining); on the contrary, under continuous stocking, feeding time may increase at a very low sward height, because of a still high leaf proportion within the sward.

Considering ‘daily limiting area’ systems (i.e. strip-grazing), all grazing processes from the first to the last defoliated stratum occur within a day. Sward-state changes are much more rapid than the theoretical scale of intake regulation (day), and the vegetation characteristics can no longer be regarded as homogeneous within a day. Short-term grazing behaviour variables, such as bite mass and IIR, strongly vary during the course of a day according to the characteristics of the stratum being defoliated (Barrett *et al.*, 2001; Baumont *et al.*, 2004). Under strip-grazing, sward structure and availability are therefore characterized by both pre-grazing and post-grazing descriptors, the best parameters, i.e. those that best fit daily intake, being the pre-grazing green leaf allowance and the post-grazing leaf lamina length (Delagarde *et al.*, 2001a,b). This is in line with the barrier role of stem, pseudostem and dead material described previously. It is worth noting that post-grazing leaf lamina length and daily intake are not correlated to pre-grazing sward height nor to post-grazing sward height, but to the proportion of sward height or herbage mass defoliated. Studies have suggested that intake is limited when more than 50% of the pre-grazing pasture mass (measured at ground level) has been defoliated, irrespective of the pre-grazing sward conditions (Delagarde *et al.*, 2001b). This degree of sward depletion is also indicative of daily herbage allowance, the product of the pre-grazing herbage mass and the area offered daily, which has a primary effect on herbage intake (Delagarde and O’Donovan, 2005).

Finally, no pre-grazing vegetation characteristic considered alone can be a good predictor of daily intake under strip-grazing, because pasture availability is closely related to the proportion of the sward depleted. A recent review of the literature indicates that, in temperate pastures, herbage allowance measured at 2–3 cm above ground level may be a good predictor of daily herbage intake irrespective of pre-grazing sward conditions (Delagarde *et al.*, 2011).

3.5 Conclusions

Intake, diet composition and the impact of grazing on the vegetation are the results of complex interactions between animals and vegetation. Through selective grazing, the animals select a diet of higher nutrient quality than that on offer, and distribute their impact on the environment. Facing this complexity, a conceptual representation of the grazing process supported by theoretical bases has been proposed, which organizes the processes into a hierarchy of scales where animals make decisions, and also provides for their integration to intake. A great deal is now known about the physical interactions between the grazing animal and the sward structure at the patch level in temperate swards that determine bite characteristics and instantaneous intake rate. Great progress has been made recently in understanding these interactions in tropical pastures. These studies demonstrate the importance of both the availability of green leaves, as a key determinant of grazing behaviour and intake, and of the presence and density of high tensile resistance stems, as foraging deterrents. Beyond the nutritive value of the forage, these results demonstrate the huge effect of the pasture phenological stage on grazing behaviour and intake, and can inform plant breeders about criteria of importance in the selection of plant species and varieties. The choice by herbivores of the patch on which to feed in pastures composed of a mosaic of patches involves rate maximizing subject to nutrient constraints, including selecting complementary nutrients. In vegetative temperate swards where protein requirements can be easily met, animals tend to maximize their rate of digestible organic matter intake. In lower quality pastures, such as tropical swards or temperate swards containing reproductive patches, protein requirements become the driving force of behaviour, and animals then tend to maximize their rate of crude protein intake. However, one of the most consistent findings from these studies is that animals always show partial preferences and prefer mixed diets. Moreover, allowing animals to express this behaviour by offering them diversity at pasture may increase their motivation to eat and stimulate their intake. Further research is needed to test the generality of these results, to investigate their practical utilization and to establish the respective roles of behavioural and digestive factors.

References

Barrett, P.D., Laidlaw, A.S., Mayne, C.S. and Christie, H. (2001) Pattern of herbage intake rate and bite dimensions of rotationally grazed dairy cows as sward height declines. *Grass and Forage Science* 56, 362–373.

Baumont, R., Cohen-Salmon, D., Prache, S., Sauvant, D. (2004) A mechanistic model of intake and grazing behaviour in sheep integrating sward architecture and animal decisions. *Animal Feed Science and Technology* 112, 5–28.

Benoit, M. and Laignel, G. (2010) Energy consumption in mixed crop–sheep farming systems: what factors of variation and how to decrease? *Animal* 4, 1597–1605.

Benvenutti, M.A., Gordon, I.J., Poppi, D.P., Crowther, R. and Spinks, W. (2008) Foraging mechanics and their outcomes for cattle grazing reproductive tropical grasses. *Applied Animal Behaviour Science* 113, 15–31.

Benvenutti, M.A., Gordon, I.J., Poppi, D.P., Crowther, R., Spinks, W. and Moreno, F.C. (2009) The horizontal barrier effect of stems on the foraging behaviour of cattle grazing five tropical grasses. *Livestock Science* 126, 229–238.

Cabrera Estrada, J.I., Delagarde, R., Faverdin, P. and Peyraud, J.L. (2004) Dry matter intake and eating rate of grass by dairy cows is restricted by internal, but not external water. *Animal Feed Science and Technology* 114, 59–74.

Champion, R.A., Orr, R.J., Penning, P.D. and Rutter, S.M. (2004) The effect of the spatial scale of heterogeneity of two herbage species on the grazing behaviour of lactating sheep. *Applied Animal Behaviour Science* 88, 61–76.

Côrtes, C., Damasceno, J.C., Jamot, J. and Prache, S. (2006) Ewes increase their intake when offered a choice of herbage species at pasture. *Animal Science* 82, 183–191.

Delagarde, R., and O'Donovan, M. (2005) Modelling of herbage intake and milk production by grazing dairy cows. In: Murphy, J.J. (ed.) *Utilisation of Grazed Grass in Temperate Animal Systems. Proceedings of a Satellite Workshop of the XXth International Grassland Congress*, July 2005, Cork, Ireland. Wageningen Academic Publishers, Wageningen, pp. 89–104.

Delagarde, R., Prache, S., D'Hour, P. and Petit, M. (2001a) Ingestion de l'herbe par les ruminants au pâturage. *Fourrages* 166, 189–212.

Delagarde, R., Peyraud, J.L., Parga, J., and Ribeiro Filho, H.M.N. (2001b) Caractéristiques de la prairie avant et après un pâturage: quels indicateurs de l'ingestion chez la vache laitière? *Rencontres autour des Recherches sur les Ruminants* 8, 209–212.

Delagarde, R., Peyraud, J.L. and Wade, M.H. (2010) Daily pattern of feeding activities of dairy cows in an 8-d rotational grazing system. *Grassland Science in Europe* 15, 931–933.

Delagarde, R., Faverdin, P., Baratte, C. and Peyraud, J.L. (2011) GrazeIn: a model of herbage intake and milk production for grazing dairy cows. 2. Prediction of intake under rotational and continuously-stocked grazing management. *Grass and Forage Science* 66, 45–60.

Dillon, P., Roche, J.R., Shalloo, L. and Horan, B. (2005) Optimising financial return from grazing in temperate pastures. In: Murphy, J.J. (ed.) *Utilisation of Grazed Grass in Temperate Animal Systems. Proceedings of a Satellite Workshop of the XXth International Grassland Congress*, July 2005, Cork, Ireland, Wageningen Academic Publishers, Wageningen. pp. 131–147.

Distel, R.A., Laca, E.A., Griggs, T.C. and Demment, M.W. (1995) Patch selection by cattle: maximization of intake rate in horizontally heterogeneous pastures. *Applied Animal Behaviour Science* 45, 11–21.

Edouard, N., Duncan, P., Dumont, B., Baumont, R. and Fleurance G. (2010) Foraging in a heterogeneous environment – an experimental study of the trade-off between intake rate and diet quality. *Applied Animal Behaviour Science* 126, 27–36.

Flores, E.R., Laca, E.A., Griggs, T.C. and Demment, M.W. (1993) Sward height and vertical morphological differentiation determine cattle bite dimensions. *Agronomy Journal* 85, 527–532.

Ginane, C., Petit, M. and D'Hour, P. (2003) How do grazing heifers choose between maturing reproductive and tall or short vegetative swards ? *Applied Animal Behaviour Science* 83, 15–27.

Gordon, I.J. and Benvenutti, M. (2006) Food in 3D: How ruminant livestock interact with sown sward architecture at the bite scale. In: Bels, V. (ed.) *Feeding in Domestic Vertebrates: from Structure to Behaviour.* CAB International, Wallingford, UK, pp 263–277.

Hirata, M., Kanemaru, E. and Tobisa, M. (2006) Patch choice by cattle grazing tropical grass swards: a preliminary study. *Applied Animal Behaviour Science* 97, 134–144.

Hirata, M., Kunieda, E. and Tobisa, M. (2010) Short-term ingestive behaviour of cattle grazing tropical stoloniferous grasses with contrasting growth forms. *Journal of Agricultural Science (Cambridge)* 148, 615–624.

Illius, A.W., Gordon, I.J., Elston, D.A. and Milne, J.D. (1999) Diet selection in goats: a test of intake-rate maximization. *Ecology* 80, 1008–1018.

Laca, E.A., Ungar, E.D., Seligman, N. and Demment, M.W. (1992) Effects of sward height and bulk density on bite dimensions of cattle grazing homogeneous swards, *Grass and Forage Science* 47, 91–102.

Penning, P.D., Parsons, A.J., Orr, R.J. and Hooper, G.E. (1994) Intake and behaviour responses by sheep to changes in sward characteristics under rotational grazing. *Grass and Forage Science* 49, 476–486.

Pérez-Ramírez, E., Delagarde, R. and Delaby, L. (2008) Herbage intake and behavioural adaptation of grazing dairy cows by restricting time at pasture under two feeding regimes. *Animal* 2, 1384–1392.

Poppi, D.P., Hughes, T.P. and L'Huillier, P.J. (1987) Intake of pasture by grazing ruminants. In: Nicol, A.M. (ed.) *Feeding Livestock on Pasture.* New Zealand Society of Animal Production Occasional Publication No. 10, Hamiton, New Zealand, pp. 55–63.

Prache, S., Roguet, C. and Petit, M. (1998) How degree of selectivity modifies foraging behaviour of dry ewes on reprodutive compared to vegetative sward structure. *Applied Animal Behaviour Science* 57, 91–108.

Prache, S., Bechet, G. and Damasceno, J.C. (2006) Diet choice in grazing sheep: a new approach to investigate the relationships between preferences and intake-rate on a daily time scale. *Applied Animal Behaviour Science* 99, 253–270.

Provenza, F.D., Villalba, J.J., Haskell, J., MacAdam, J.W., Griggs, T.C. and Wiedmeier R.D. (2007) The value to herbivores of plant physical and chemical diversity in time and space. *Crop Science* 47, 382–398.

Sauvant, D., Baumont, R. and Faverdin, P. (1996) Development of a mechanistic model of intake and chewing activities of sheep. *Journal of Animal Science* 74, 2785–2802.

Stephens, D.W. and Krebs, J.R. (1986) *Foraging Theory.* Princeton University Press, Princeton, NJ.

Wade, M.H. (1991) Factors affecting the availability of vegetative *Lolium perenne* to grazing dairy cows with special reference to swards characteristics, stocking rate and grazing method. PhD thesis, Université de Rennes I, France.

4 Variability, Heterogeneity, Diversity and Productivity in Grazing Systems

Andrew W. Illius, Hervé Fritz and John M. Fryxell

4.1 Introduction

This chapter poses a perennial question in ecology: what determines the abundance and diversity of species in grazing ecosystems? The context of this question is recognition that ecosystems provide a wide range of services, many of which derive from or are related to the system's diversity. For example, grassland and savannah biomes contribute significant amounts of all five of the ecosystem services (surface water supply, water flow regulation, carbon storage, soil accumulation and soil retention) examined in South Africa by Egoh *et al.* (2009). They found significant spatial overlap of ecosystem services with species richness (plants and animals) and vegetation diversity hotspots, consistent with the generally positive relationship between species richness and ecosystem functions at large scale.

Our focus is primarily on large vertebrate herbivores, and how the diversity and coexistence of these consumers may be related to the biodiversity and heterogeneity of the vegetation. We focus on the principles underlying diversity in diet optimization and resource use by herbivores, to examine how they exploit environmental heterogeneity and what effect this may have on species coexistence and abundance. We go to some lengths to consider the consequences of climatic variability, which has profound effects on primary and secondary production, effects which may increase under some climate-change scenarios.

Although biodiversity has important connections with the provision of ecosystems services, the assessment of biodiversity is complex, and the most readily available measures rarely reflect the real attribute of interest for any particular role. For example, to measure biodiversity in terms of visible organisms, such as animals or plants, misses the much greater taxonomic diversity of microorganisms. These dominate the branches of the tree of life, and include most forms that are the main providers of most regulating and supporting services and are keys to many provisioning services. We acknowledge that there is far more to the biodiversity of grazing systems than meets the eye and than is represented by the macrofauna and flora.

McNaughton (1985) draws attention to several important properties of the Serengeti ecosystem, one of the most pristine, extensive and diverse of grazing systems: (i) variability in rainfall, and in the spatial distribution of nutrient availability, grazing and fire, is an inherent feature of many grassland ecosystems; (ii) the resulting pulses in plant growth produce a

(eds G. Lemaire, J. Hodgson and A. Chabbi)

shifting landscape of locally abundant but sometimes ephemeral forage resources; (iii) the mobility of grazers is essential to their ability to exploit such resources, consume a high fraction of primary production and become the dominant fauna; and (iv) grazers alter plant community composition and diversity, which may modify a host of ecosystem functions, and they can also modify forage quality, which may feed back on grazer abundance. McNaughton conjectured that grazers' abilities to buffer temporal stochasticity in primary production, by exploiting spatial heterogeneity and their modification of forage resources, 'produce a much more deterministic and stable grazing food web than would be expected in such a variable environment if vegetation productivity was less responsive to rainfall and grazing, the animals were sedentary, and community composition was uncoupled from its functional properties' (McNaughton, 1985, p. 282).

4.2 Temporal and Spatial Variability in Vegetation

As a food source for animals, vegetation has a number of shortcomings. It has a composition quite different from that of the consumer, much of it is indigestible even to commensal microbes, and the more nutritious components are sparsely distributed across the landscape. Allowing that ruminants are well adapted to coping with the first two limitations, the finding of sufficient food to meet daily and seasonal requirements remains a challenge.

4.2.1 Forage quality

The sources of variation in the quality of forage for herbivores are well known. Immature tissue consists mainly of metabolically active cell contents, which are almost completely digestible, but tissue maturation brings an increasing fraction of cell wall, itself becoming more lignified with age until less than half the forage is digestible. Diversification of herbivore feeding niches arises from the differences in digestibility between plant parts and growth stage (such as newly emerged leaf, maturing stem or the twigs of browse species). These pose herbivores with significant opportunities and constraints on diet selection, and such differences are possibly greater than those offered by choice between plant species compared at similar growth stages. Immature plants take longer to crop than those that are more mature, but mature plants require longer digestive processing, so daily intake by herbivores is frequently a dome-shaped function of plant abundance (Fryxell, 1991).

Spatial heterogeneity in growing conditions affects plant species distribution and phenology. For example, savannah grasses growing in moister locations mature for longer and end up with more lignified and less digestible tissue, explaining why variation in digestibility is greater within species at the top and bottom of a catena than between species (Murray and Illius, 1996). At a landscape scale, there may be sufficient asynchrony in plant phenology to allow herbivores to exploit plants at peak nutritional benefit by ranging widely (Albon and Langvatn, 1992; Fryxell *et al.*, 2004; Mueller *et al.*, 2008; Holdo *et al.*, 2009).

4.2.2 Forage abundance

Most grasslands are subject to marked seasonality of production. Annual cycles of temperature or rainfall impose a cycle of plant growth and phenology that results in a cycle of food abundance and quality. Intra- and inter-annual climatic variability adds further to uncertainty in resource abundance, especially in arid and semi-arid tropical environments, where such variability is extreme. Production is closely related to rainfall, as demonstrated by Le Houérou (1984), and refined by Huxman *et al.* (2004), who derive similar estimates of rain use efficiency (RUE; annual above-ground net primary production (ANPP) per mm of annual precipitation).

Grasslands showed greater inter-annual variability in ANPP than the other North American biomes examined by Knapp and Smith (2001), with coefficient of variation (CV) about 0.3 at the more arid sites. Dryer grasslands are more variable than mesic or humid ones, for two reasons. First, the decreasing mean annual rainfall across a transect from humid grasslands to arid rangelands is associated with greater stochasticity in timing and size of rain events, with the result that the inter-annual CV of rainfall is inversely related to the mean (Davidowitz, 2002). The same has been shown for the resulting primary production, estimated in Argentine range-lands by remote sensing, there being an inverse relationship between the mean and inter-annual CV of normalized-difference vegetation index integrated over the year (NDVI-I) (Oesterheld *et al.*, 1998), with the most arid sites having a CV between 0.4 and 0.5. Second, the inter-annual variability of primary production has generally been found to be greater than the variability of rainfall in arid and semi-arid grasslands. For example, the analysis of 77 series of annual data on production and rainfall by Le Houérou *et al.* (1988) showed the variability in production to be, on average, 1.5 times greater than the variability in rainfall. Comparison of six North American grassland sites showed ANPP CV/rainfall CV ranging from 0.8 (i.e. ANPP is more stable than rainfall) to 1.8 (ANPP is nearly twice as variable as rainfall; Knapp and Smith, 2001). Over a 19-year time series at a semi-arid site in South Africa, with grass plots manipulated to have a species composition and basal cover representative of rangeland under grazing in 'good', 'medium' or 'poor' condition, the inter- annual CV of production was 0.57, 0.68 or 0.71, respectively, compared with a rainfall CV of 0.29 (Wiegand *et al.*, 2004). For ANPP to vary more widely than rainfall does requires some non-linearity or lag effects in the relationship between soil moisture and plant growth, and Weigand *et al.* (2004) were able to demonstrate 'memory' effects of rainfall that 'amplified the variability in precipitation: production was higher during a sequence of several wet years or months and lower during a sequence of several dry years or months' (p. 256). The authors suggest that some grasslands may show shorter memory, which, combined with only short-term fluctuations in rainfall, would not tend to amplify annual variability in ANPP.

There is some evidence (Holm *et al.*, 2003, 2005) that grazing impacts on vegetation structure may reduce the response of primary production to rainfall and hence reduce secondary production, as proposed by Illius and O'Connor (1999). However, Palmer and Ainslie (2007) could find no change in the RUE implied by livestock numbers over a 75-year time series in a semi-arid pastoral region of South Africa, despite apparently radical modification of the vegetation by the conversion of range-land to cultivated farmland. They attributed the resilience of livestock numbers to hotspots of grass production around homesteads and water courses. The absence of an external reference value for RUE under pristine conditions has often hampered interpretation of temporal trends, as in the ongoing controversy over whether the Sahelian grasslands are 'degraded' on the evidence of temporal changes in RUE (Prince *et al.*, 2007).

4.3 Diversity in Resource Use and Diet in Mammalian Herbivory

There is good evidence of diet optimization by herbivores, which show flexibility and adaptive changes in the selection of forage at the level of the plant part, phenological stage and vegetation community, and by their broader movement patterns and range use. Several decades of feeding studies well demonstrate that body size is the primary determinant of dietary options for mammalian herbivores. Large species, such as the plains bison or Cape buffalo, have broad muzzles and flattened dental arcades, are incapable of fine selection of specific plant tissues, but are able to meet their nutrient requirements more readily from mature plant tissues than smaller species (Fryxell *et al.*, 2008). On the other hand, smaller herbivores are more efficient than large herbivores at picking out the more nutritious bits from highly heterogeneous plant communities. Coupled with the dome-shaped intake function that characterizes mammalian herbivores, allometric variation in foraging efficiency dictates that herbivores of different size tend to have different mixes of graminaceous versus herbaceous or woody browse in their diet, and different levels of vegetation abundance offer optimal rates of intake (Wilmshurst *et al.*, 2000; Fryxell *et al.*, 2008).

There is also evidence that mammalian herbivores tend to selectively graze patches that yield the highest energetic gain (Wilmshurst *et al.*, 1995; Fryxell *et al.*, 2004; Mueller *et al.*, 2008). Such variation in feeding constraints and consequently forage preferences creates ample opportunity for niche partitioning (Olff *et al.*, 2002; Fryxell *et al.*, 2008).

4.4 Herbivore Abundance: Landscape Structure and the Seasonal Constraints in Forage Resources

Herbivore biomass in natural grazing systems is clearly related to primary production. Fritz and Duncan (1994) showed that the 100-fold range in biomass density of the ungulate communities across 31 natural ecosystems in East and southern Africa can largely be explained by mean annual rainfall (taken as a proxy for primary production) and a classification of soil nutrient status into high, medium or low. Beneath this regional correlation of primary and secondary production lie sharp contrasts between the relative importance of the growing and dormant seasons in supporting and limiting herbivore abundance. Clearly, the growing season is when primary production occurs, but arguably the dormant season is the most critical time for the support of herbivore populations because that is when forage is most likely to become depleted (Illius and O'Connor, 1999, 2000).

Inter-annual and spatial variability in primary production further complicate the relationship between plant and animal production. Populations of herbivores are likely to do less well in more variable environments. Davis *et al.* (2002) showed that variability in rainfall reduces the long-term growth rate of food-limited herbivore populations, and thus may determine whether or not populations persist. The simple reason for this is that the numerical response of herbivore growth rate to vegetation abundance is convex, and hence the effect of below-average vegetative growth exceeds the effect of above-average growth. There may be a similar asymmetry in the response of ANPP to rainfall: manipulation of the size and frequency of rainfall events showed that increasing the intra-annual variability of rainfall (but not its mean) increased temporal variability in soil moisture in grasslands, and ANPP lost during the longer dry periods was not fully compensated by the extra growth resulting from the wetter but less frequent rainfall events (Knapp *et al.*, 2002). Wang *et al.* (2006) showed that lower winter temperatures and increasing variability in winter severity strengthened the effect of density on elk and bison population growth rates. In contrast, the strength of density dependence was reduced by increasing spatial heterogeneity in vegetation, suggesting that access to heterogeneous vegetation allowed some buffering of temporal variability.

Mammalian herbivores can often persist, in spite of inter-annual variation in rainfall, by taking advantage of spatial heterogeneity in vegetation. Models of adaptive movement by Serengeti herbivores suggest that animal population viability may be dependent on access to resources that vary asynchronously over space (Fryxell *et al.*, 2005) and that free access sustains higher herbivore numbers than would otherwise be supportable (Holdo *et al.*, 2009). The same models suggest that herbivore mobility has cascading effects on other important properties, such as the amount of combustible litter that can build up, woody vegetation cover and plant species diversity, all of which are important contributors to ecosystem services. Northern ungulates living in habitats with greater heterogeneity in elevation had shallower slopes in the relationship between per capita population growth rate and population density (Wang *et al.*, 2009). These studies provide evidence of a positive relationship between habitat heterogeneity and ecosystem carrying capacity for a given ANPP.

Highly variable primary production in rangelands has led some to take the extreme view that herbivore populations are not dynamically coupled to their resources (e.g. Behnke *et al.*, 1993). The concept of carrying capacity is discarded by this 'non-equilibrium' paradigm, because of the fluctuating resource base. Livestock populations are thought to have negligible feedback on the vegetation, as their numbers are reduced by episodes of drought-induced mortality. Degradation due to 'overgrazing' is thus held to be unlikely, as vegetation composition, cover and productivity respond primarily to rainfall. In a critique of this paradigm, Illius and O'Connor (1999, 2000) proposed that herbivore populations in climatically variable and spatially heterogeneous environments are regulated in a density-dependent manner by a particular subset of ('key') resources that are accessible in the dry season, while being largely uncoupled from the remaining, wet season resources. Generalizing from the perception of Scoones (1995) that livestock survival depends on access to 'key resource' areas during the dry season, Illius and O'Connor (2000) defined such resources in relation to the key factor. Given that the key factor determining animal population size is survival over the season of plant dormancy, key resources are those eaten then and whose abundance controls the strength of the key factor. Thus, reduced access to such resources would cause the population to decline. They further demonstrated that: (i) long-term mean herbivore abundance is very largely determined by the quantity of key resources and scarcely at all by resources available in the growing season; (ii) the potential for grazing-induced transformation of the vegetation comprising the wet-season resource increases as the ratio of key resource to wet-season resource (and hence herbivore numbers) increases. The implications for managed or conserved ecosystems are that animal abundance and grazing impact are likely to be influenced markedly by the landscape's endowment of such key resources.

The implication of some subset of resources having a key role in herbivore population dynamics is, of course, that the remaining resources have a lesser or unimportant role. Thus, we can now begin to distinguish resources in terms of whether consumer population dynamics are coupled to them or not. Wet-season rangeland can be classed as a non-equilibrium resource, because the animal population's dynamics are not coupled to it. Superabundance of non-key resources is likely to be observed during the growing season, because the animal population is typically limited by scarcer, high-quality resources during the dormant season. Note also that the key and non-equilibrium resources need not be widely separated in space and may even occur on the same plant. Given a distribution of food quality, as described by Hobbs and Swift (1985), a consumer population will tend towards equilibrium with the quantity of resources above the minimum quality threshold, and will not be in equilibrium with the remaining (= non-equilibrium) resources below the threshold. Diet selection from heterogeneous resources will naturally cause the animal population's dynamics to depend differentially on different resources.

4.5 Herbivore Diversity and Productivity

What determines herbivore species richness in grazing systems, and does an increase in species richness increase secondary production? The effect of the structure of the community of herbivore consumers may well play a role in explaining why the efficiency of energy transfer between primary production and secondary production increases with ecosystem productivity (slope > 1, as shown in Fritz and Duncan, 1994).

African savannahs provide good evidence for the relationship between habitat and herbivore species diversity (reviewed by du Toit and Cumming, 1999). Herbivore species diversity follows the classic hump-shaped relationship with annual rainfall, a proxy of primary production (Olff *et al.*, 2002; Prins and Fritz, 2008). Olff *et al.* (2002) argued that the two components of food supply – plant productivity and plant nutrient concentration – have contrasting responses across gradients of precipitation, temperature and soil fertility. Increasing precipitation and temperature increase plant growth and lignification, leading to high biomass but low quality; richer soils allow increased nutrient concentration without inducing such rapid decrease in digestibility. The implication is that the highest potential herbivore diversity should occur in locations with intermediate moisture and high soil nutrients, because larger herbivore species tolerate lower plant nutrient concentration but require greater plant abundance (Olff *et al.*, 2002). It is the accumulation of medium-sized species that mostly contributes to the increase in diversity at the peak, raising questions about the mechanisms of coexistence. Murray and Baird (2008) also predicted, based on resource competition theory, the coexistence of a maximum number of species in environments with intermediate productivity. They distinguish species that compete by being the most efficient at acquiring the food that most limits their competitors (extraction maximizers) from those that specialize on the resource that most limits themselves (demand minimizers). Wildebeest are typical extraction maximizers, dominating plant communities by creating grazing lawns and excluding competitors, particularly in highly productive environments. Typical demand minimizers are morphologically diverse sympatric species exploiting different niches due to a trade-off in size-related foraging efficiency (Murray and Illius, 1996; Cromsigt and Olff, 2006). Resource heterogeneity combined with different resource utilization thus creates the conditions for coexistence of species in grasslands, with species of smaller body size being dominant at low productivity and larger body size dominating at higher productivity. The differences in nutritive constraints with body size are also associated with differences in the scale at which species perceive and respond to resource heterogeneity in space and time (Ritchie and Olff, 1999). Spatial heterogeneity in resource distribution, generated by rainfall variability interacting with geomorphology and distance to water sources, promotes niche diversification, and hence diversity (du Toit and Cumming, 1999).

4.5.1 Different limiting processes for different components of the community?

Population limitation mechanisms may differ between species in a community. In particular, body size is likely to play a major role in setting the relative roles of predation and resource availability at the population level, hence the distribution of body sizes within communities may determine the relative importance of top-down and bottom-up controls in entire food webs. Accordingly, Hopcraft *et al.* (2010) proposed a general framework to predict the relative contribution of bottom-up and top-down controls on herbivores across environmental gradients, arguing that the functioning of savannahs is primarily determined by: (i) how vegetation quality and productivity directly influence the differential performance of herbivores of different body sizes; and (ii) how carnivores of different body sizes partition their food niches along a herbivore body-size gradient. Fritz *et al.* (2011) confirm the interplay between top-down and bottom-up forces in savannah systems by showing that: (i) predation could be limiting in most sites for small- and medium-sized ungulate species through the prevalence of size-nested predation (i.e. the prey base of small predators is nested within that of larger predators); and (ii) in systems with non-migratory herbivores, the greatest consumption of primary production occurred in the system dominated by mega-herbivores, which are virtually free of predators. These results emphasize the importance of diversity in body size and morphology on ecosystem functioning, rather than species richness per se.

4.5.2 Diversity and the relationship between primary and secondary production

It may be expected from the **species complementary** framework that an increase in species richness makes more complete use of the primary production (Tilman *et al.*, 1997; Loreau, 1998). From an alternative viewpoint, **species functional redundancy** (derived from Neutral Theory; Hubbell, 2001) may result in there being no effect of increasing herbivore species diversity on secondary production. The former view is supported by Fritz and Duncan (1994), who found that species richness may marginally contribute to large herbivore biomass for a given rainfall, especially for low-nutrient sites. These are likely to be those with highest variability in space and time, and thus greater diversity of niches. Further analysis distinguishing between grazers and browsers showed that the metabolic biomass of grazers was not related to the number of species, while for browsers, species richness significantly increased the total metabolic biomass of the mesobrowser guild (Prins and Fritz, 2008). In fact, the number of species was the only explanatory variable for the metabolic biomass of mesobrowsers. For these sites, the overall browser biomass was only influenced by rainfall, as expected from the fact that the biomass of browsers is largely dominated by elephants and that elephant biomass is exclusively explained by rainfall (Fritz *et al.*, 2002). In showing that specialist browsers seem to be complementary in their use of their primary resources, we not only confirm field observations about possible niche complementarity, but also suggest that systems richer in browsers may use primary production more thoroughly, at least in savannah-type ecosystems. These results further underline the necessity for accounting for the relative proportion of various body-size classes in ecosystem functioning, and, in particular, in relating primary production and secondary production.

4.6 Conclusions

Grazing systems involve herbivore populations exploiting plant resources whose composition is heterogeneous in time and space. Seasonal patterns of plant productivity and phenology cause wide variation in the abundance and nutritive value of forage over the year, and climatic variability generates wide inter-annual variation in primary production. Adaptive responses buffer the environmental heterogeneity in resource abundance, but seasonal bottlenecks in nutrition nevertheless regulate herbivore populations. By setting limits on herbivore numbers, seasonal bottlenecks define the aggregate herbivore impact on vegetation as well as determining herbivore community structure. Explaining how that impact is distributed across vegetation components in the landscape, and hence the way plant community composition is modified by herbivory, remains a significant challenge.

There is still poor empirical support for hypotheses about what constrains herbivore diversity, and how far herbivore and plant diversity influence each other. Herbivore species diversity is clearly related to plant species diversity in some studies of insects, which tend to have more specialized requirements than mammalian herbivores. But there is weaker evidence that mammalian herbivore community composition is a function of plant species diversity, and still less that secondary production is increased. In general, does herbivore diversity have any bearing on the services provided or is it enough to address the maintenance of plant diversity alone? How will plant spatial heterogeneity and temporal variability be affected by climate change and are current conservation/ management practices adequate to meet that challenge? Finally, what are the long-term costs associated with sustaining herbivore and plant diversity if, as increasingly seems likely, climate change significantly shifts species ranges?

References

Albon, S.D. and Langvatn, R. (1992). Plant phenology and the benefits of migration in a temperate ungulate. *Oikos* 65, 502–513.

Behnke, R.H., Scoones, I. and Kerven, C. (1993) *Range Ecology at Disequilibrium.* Overseas Development Institute, London.

Cromsigt, J.P.G.M. and Olff, H. (2006) Resource partitioning among savanna grazers mediated by local heterogeneity: an experimental approach. *Ecology* 87, 1532–1541.

Davidowitz, G. (2002) Does precipitation variability increase from mesic to xeric biomes? *Global Ecology and Biogeography* 11, 143–154.

Davis, S.A., Pech, R.P. and Catchpole, E.A. (2002) Populations in variable environments: the effect of variability in a species' primary resource. *Philosophical Transactions of the Royal Society of London B* 357, 1249–1257.

du Toit, J.T. and Cumming, D.H.M. (1999) Functional significance of ungulate diversity in African savannas and the ecological implications of the spread of pastoralism. *Biodiversity and Conservation* 8, 1643–1661.

Egoh, B., Reyers, B., Rouget, M., Bode M. and Richardson, D.M. (2009) Spatial congruence between biodiversity and ecosystem services in South Africa. *Biological Conservation* 142, 553–562.

Fritz, H. and Duncan, P. (1994) On the carrying capacity for large ungulates of African savanna ecosystems. *Proceedings of the Royal Society of London B* 256, 77–82.

Fritz, H., Duncan, P., Gordon, J. and Illius, A.W. (2002) Megaherbivores influence trophic guilds structure in African ungulate communities. *Oecologia* 131, 620–625.

Fritz, H., Loreau, M., Chamaille-Jammes, S., Valeix, M. and Clobert, J. (2011) A food web perspective on large herbivore community regulation. *Ecography* 34,196–202.

Fryxell, J.M. (1991) Forage quality and aggregation by large herbivores. *American Naturalist* 138, 478–498.

Fryxell, J.M., Wilmshurst, J. and Sinclair, A.R.E. (2004) Predictive models of movement by Serengeti grazers. *Ecology* 85, 2429–2435.

Fryxell, J.M., Wilmshurst, J.F., Sinclair, A.R.E., Haydon , Holt, R.D. and Abrams, P.A. (2005) Landscape scale, heterogeneity, and the viability of Serengeti grazers. *Ecology Letters* 8, 328–335.

Fryxell, J.M., Abrams, P.A., Holt, R.D., Wilmshurst, J.F., Sinclair, A.R.E. and Hilborn, R. (2008) Spatial dynamics and coexistence of the Serengti grazer community. In: Sinlair, A.R.E, Mduma, S., Packer, C. and Fryxell, J.M. (eds) *Serengeti III: Human Impacts on Ecosystem Dynamics.* University of Chicago Press, pp. 277–300.

Hobbs, N.T. and Swift, D. (1985) Estimates of habitat carrying capacity incorporating explicit nutritional constraints. *Journal of Wildlife Management* 49, 814–822.

Holdo, R.M., Holt, R.D. and Fryxell, J.M. (2009) Opposing rainfall and fertility gradients explain the wildebeest migration in the Serengeti. *American Naturalist* 173,431–445.

Holm, A.McR., Watson, I.W., Loneragan, W.A. and Adams, M.A. (2003) Loss of patch-scale heterogeneity on primary productivity and rainfall-use efficiency in Western Australia. *Basic and Applied Ecology* 4, 569–578.

Holm, A.McR., Watson, I.W., Speijers, E.J., Allen, R.J., Eliot, G.J., Shackleton, K.R. and Stretch, J.K. (2005) Loss of patch-scale heterogeneity on secondary productivity in the arid shrubland of Western Australia. *Journal of Arid Environments* 61, 631–649.

Hopcraft, J.G.C., Olff, H. and Sinclair, A.R.E. (2010) Herbivores, resources and risks: alternating regulation along primary environmental gradients in savannas. *Trends in Ecology and Evolution* 25, 119–128.

Hubbell, S.P. (2001) *The Unified Neutral Theory of Biodiversity and Biogeography.* Princeton University Press, Princeton.

Huxman, T.E., Smith, M.D., Fay, P.A., Knapp, A.K., Shaw, M.R., Loik, M.E., Smith, S.D., Tissue, D.T., Zak, J.C., Weltzin, J.F., Pockman, W.T., Sala, O.E., Haddad, B.M., Harte, J., Koch, G.W., Schwinning, S., Small, E.E.

and Williams, D.G. (2004) Convergence across biomes to a common rain use efficiency. *Nature* 429, 651–654.
Illius, A.W. and O'Connor, T.G. (1999) On the relevance of non-equilibrium concepts to semi-arid grazing systems. *Ecological Applications* 9, 798–813.
Illius, A.W. and O'Connor, T.G. (2000) Resource heterogeneity and ungulate population dynamics. *Oikos* 89, 283–294.
Knapp, A.K. and Smith, M.D. (2001) Variation among biomes in temporal dynamics of aboveground primary production. *Science* 291, 481–484.
Knapp, A.K., Fay, P.A., Blair, J.M., Collins, S.L., Smith, M.D., Carlisle, J.D., Harper, C.S., Danner, B.T., Lett, M.S. and McCarron, J.K. (2002) Rainfall variability, carbon cycling, and plant species diversity in a mesic grassland. *Science* 298, 2202–2205.
Le Houérou, H.N. (1984) Rain use efficiency: a unifying concept in arid-land ecology. *Journal of Arid Environments* 7, 41–46.
Le Houérou, H.N., Bingham, R.L. and Skerbek, W. (1988) Relationship between the variability of primary production and the variability of annual precipitation in world arid lands. *Journal of Arid Environments* 15, 1–18.
Loreau, M. (1998) Biodiversity and ecosystem functioning: a mechanistic model. *Proceedings of the National Academy of Sciences USA* 95, 5632–5636.
McNaughton, S.J. (1985) Ecology of a grazing ecosystem: The Serengeti. *Ecological Monographs* 55, 259–294.
Mueller, T., Olson, K.A., Fuller, T.K., Schaller, G.B., Murray, M.G. and Leimgruber, P. (2008) In search of forage: predicting dynamic habitats of Mongolian gazelles using satellite-based estimates of vegetation productivity. *Journal of Applied Ecology* 45, 649–658.
Murray, M.G. and Baird, D.R. (2008) Resource-ratio theory applied to large herbivores. *Ecology* 89, 1445–1456.
Murray, M.G. and Illius, A.W. (1996) Multispecies grazing in the Serengeti. In: Hodgson, J. and Illius, A.W. (eds) *The Ecology and Management of Grazing Systems*. CAB International, Wallingford, UK, pp. 429–457.
Oesterheld, M., DiBella, C.M. and Kerdiles, H. (1998) Relation between NOAA-AVHRR data and stocking rate of rangelands. *Ecological Applications* 8, 207–212.
Olff, H., Ritchie, M.E. and Prins, H.H.T. (2002) Global environmental controls of diversity in large herbivores. *Nature* 415, 901–904.
Palmer, A.R. and Ainslie, A. (2007) Using rain-use efficiency to explore livestock production trends in rangelands in the Transkei, South Africa. *African Journal of Range Forage Science* 24, 43–49.
Prince, S.D., Wessels, K.J., Tucker, C.J. and Nicholson, S.E. (2007) Desertification in the Sahel: a reinterpretation of a reinterpretation. *Global Change Biology* 13, 1308–1313.
Prins, H.H.T. and Fritz, H. (2008) Species diversity of browsing and grazing ungulates: consequences for the structure and abundance of secondary production. In: Gordon, I.J. and Prins, H.H.T. (eds) *The Ecology of Large Mammalian Herbivore Browsing and Grazing*. Springer-Verlag, Berlin, pp. 179–200.
Ritchie, M.E. and Olff, H. (1999) Spatial scaling laws yield a synthetic theory of biodiversity. *Nature* 400, 557–560.
Scoones, I. (1995) Exploiting heterogeneity: habitat use by cattle in dryland Zimbabwe. *Journal of Arid Environments* 29, 221–237.
Tilman, D., Lehman, C.L. and Thomson, K.T. (1997) Plant diversity and ecosystem productivity: theoretical considerations. *Proceedings of the National Academy of Sciences USA* 94, 1857–1861.
Wang, G.M., Hobbs, N.T., Boone, R.B., Illius, A.W., Gordon, I.J., Gross, J.E. and Hamlin, K.L. (2006) Spatial and temporal variability modify density dependence in populations of large herbivores *Ecology* 87, 95–102.
Wang, G.M., Hobbs, N.T., Twombly, S., Boone, R.B., Illius, A.W., Gordon, I.J. and Gross, J.E. (2009) Density dependence in northern ungulates: interactions with predation and resources. *Population Ecology* 51, 123–132.
Wiegand, T., Snyman, H.A., Kellner, K. and Paruelo J.M. (2004) Do grasslands have a memory: modeling phytomass production of a semiarid South African grassland. *Ecosystems* 7, 243–258.
Wilmshurst, J.F., Fryxell, J.M. and Hudson, R.J. (1995) Forage quality and patch choice by wapiti (*Cervus elaphus*). *Behavioral Ecology* 6, 209–217.
Wilmshurst, J.F., Fryxell, J.M. and Bergman, C.M. (2000) The allometry of patch selection in ruminants. *Proceedings of the Royal Society of London B* 267, 345–349.

5 Importance of Scale in the Management of Grassland Resources

John G. McIvor

5.1 Introduction

Grasslands vary widely, both spatially and temporally, particularly in more extensive systems. Such variations are caused by both environmental differences (climate and weather, soil type, elevation, etc.), and responses to management (grazing, fertilizer application, fire, etc.). Animals are the major source of saleable products and income from grassland but they can also alter resource condition, both positively and negatively. In contrast to the variable feed supply from grasslands, animals have relatively constant needs for an adequate and balanced diet and develop foraging strategies to increase the intake of energy and nutrients above that from uniform or random grazing, or to reduce energy expenditure. Knowledge of foraging behaviour and diet selection can aid management to ensure efficient forage utilization (via intake and diet composition) and to control the impact of animals on the vegetation (via the location, frequency and severity of their grazing). Using foraging behaviour as an example, this chapter examines the causes of variation at a range of scales (leaves to regions; seconds to years), the foraging responses of animals, and the consequent impacts on resources. This information is then used to discuss how scale affects grazing management and implications for future research.

5.2 Scale and Foraging Behaviour

The quantity and quality of forage available to animals varies in space and time, and hence foraging behaviour also varies. A number of authors (Senft *et al.*, 1987; Stuth, 1991; Bailey *et al.*, 1996; Laca and Ortega, 1996; O'Reagain, 2001) have described hierarchical foraging models based on level of organization or feeding behaviour. All models have multiple levels, with each level having its own behaviour and timescale. Grazing decisions at broader scales can constrain choice at finer scales (Senft *et al.*, 1987; Bailey *et al.*, 1996). At large scale, relatively infrequent (months to years) decisions are made on where grazing will occur in regions; at intermediate scale, animals make decisions over minutes to weeks on feeding areas (patches and landscape units); while at the smallest scale, bites are selected and animals make decisions on selection between plants and plant parts and to move between feeding stations at the scale of seconds to minutes.

Both biotic (primarily relating to the quantity and quality of forage) and abiotic factors (affecting the effort required to find food) influence foraging behaviour (Bailey, 2005).

5.2.1 Biotic factors

At the scale of individual plant parts, leaf has higher quality than stem, and young forage has higher quality than old forage. These differences arise from anatomical and chemical differences between leaf and stem (Wilson, 1994) and changes in anatomy and chemistry as forage matures and dies. Animals intensively select for young green leaf (Arnold, 1964; Stobbs, 1973; Arnold and Dudzinski, 1978) and reject older stems. This selection results in a decline in the quality of the remaining pasture as the highest quality portion is removed, and disadvantages leafier plants.

At the plant scale, species and varieties differ in anatomy, morphology and chemical composition, which leads to differences in yield, quality, structure, accessibility and attractiveness. Grazing animals exploit these differences, selecting for some species while avoiding others, thus altering species composition.

Variation at the patch scale can be one of the most obvious features of grazed areas. At this scale, variation is due to differences in the species composition, morphology, structure and availability of forage. This may be due to differences in soil properties, and hence forage growth, between patches. However, effects of grazing are also important – the patches may reflect past differential defoliation (varying from frequent regrazing to avoidance), and growth responses to dung and urine. As forage in a patch is consumed, an animal must make a trade-off between continuing to graze the patch where intake is diminishing, or moving to another patch, which incurs a time cost (Baumont *et al.*, 2000). Animals avoid low-quality forage and select high-quality patches and may regraze these preferred patches. Over time this continued grazing favours grazing-resistant species, and may reduce yield and ground cover.

At the landscape-unit scale, variation in forage quantity and quality is a function of growing conditions (water and nutrient supply, aspect and temperature) and past and current grazing. Past grazing may have modified the plants present and their characteristics, while current grazing will determine the available forage. Animals graze more at locations with abundant quantities of preferred forages (Senft *et al.*, 1987; Bailey *et al.*, 1996; Launchbaugh and Howery, 2005). While this can benefit animal production, when grazing is concentrated on particular landscape units (e.g. riparian areas) forage yields and ground cover are reduced, with subsequent increases in runoff and soil movement (Bartley *et al.*, 2010). Conversely, animals may ignore some units, where forage is not utilized and potential animal production is foregone.

Climate, elevation, soil and vegetation type lead to differences in the quantity, quality and maturity of forage at the regional scale. Free-ranging animals may exploit this variation to graze in different areas in different seasons (e.g. the well-known migrations of Serengeti wildlife herds). As a consequence of this migration, individual areas are rested from grazing for parts of the year, which can benefit pasture and land condition. Grazing may disperse or concentrate depending on the availability of key resources: when resources are widespread animals disperse but when resources are restricted to limited areas the animals concentrate on favoured locations.

5.2.2 Abiotic factors

Abiotic factors affect forage resources and foraging behaviour but, in contrast to biotic factors, are not affected by forage use. Abiotic factors are the primary determinants of distribution patterns of animals at landscape and regional scales and act as constraints within which biotic factors operate (Bailey *et al.*, 1996). Distance to water is an important determinant of grazing distribution, forage utilization declining with increasing distance from water (Valentine, 1947). Animals prefer gentle slopes and avoid travelling large vertical distances to water (Bailey, 2005). Animals seek shelter from extremes of heat and cold, and differences in temperature within pastures and availability of shelter may constrain where animals graze. Similarly, risk of predation may deter animals from grazing in otherwise suitable and attractive areas.

5.3 Scale and Grazing Management

Just as there are numerous causes of variation in forages and foraging behaviour across scales, there are many influences across all scales that affect how grassland is managed (Table 5.1).

5.3.1 Influences on grazing management

The biophysical causes of variation in forages and animal responses influence management in many ways. At regional scale, the capacity of a grassland to meet the nutrient requirements of different types of livestock determines the location of animal industries, e.g. dairy cows require higher diet quality than wool-producing sheep. At intermediate scales, forage quantity and quality contribute to the attractiveness of a patch or landscape unit to animals and whether they may be under- or overgrazed. At the scale of plants and plant parts, the availability of green leaf strongly influences animal production.

In addition, there are a number of socio-economic influences. Markets determine the type and amount of product. Land tenure and ownership influence planning and attitudes to the use of resources. Ruminants emit methane, and any future charges on greenhouse gas emissions will impact on animal industries. National and international financial policies (exchange rates, interest rates, taxes) influence profitability of production. Socio-economic influences are minor or non-existent at the plant scale.

5.3.2 Grazing management goals and practices

Management objectives vary between managers and over time for a manager. However, it is likely that most grassland managers will have a combination of objectives relating to effective utilization of forages for animal production, and the maintenance or improvement of resources (soil, vegetation, water). Management needs to work across all scales – this could include responding to international markets, moving animals between regions in response to differences in feed supply, altering the distribution of animals between landscape units, and controlling weeds at the patch and plant level.

At large scale, many factors (particularly the socio-economic influences) are beyond the control of the individual manager but they are very important and set the context where management is applied. Managers can respond at this level by moving animals in response to spatially variable forage supply by transhumance (where seasonal changes of forage availability and/or quality are predictable) or nomadism (where forage production is spatially variable but unpredictable), and selling animals before seasonal drought is expected.

Grazing management has most impact at intermediate scale (large patches and landscape units) and a number of land management problems (e.g. overgrazing, degradation of riparian areas, habitat deterioration) also occur at this scale (Bailey *et al.*, 1996). Individual managers have some influence over factors such as land prices, the costs they pay for inputs, e.g. producer co-operatives, and the prices they receive for products, e.g. niche marketing schemes. Management can directly influence how and where animals graze. For instance, evenness of use of available forage is a common goal for grassland managers. Approaches to altering grazing distribution to achieve more uniform use include modifying animal behaviour (strategic placement of feed supplements within paddocks; selecting

Table 5.1. Grazing management in relation to scale: causes of variation, and management goals and practices.

	Small scale (plant parts/plants; seconds to minutes)	Intermediate scale (patch/landscape unit; minutes to weeks)	Large scale (region/world; month to years)
Biophysical causes of variation	Leaf/stem differences and changes with age; species differences	Plant communities; water and nutrient supply; landscape features; grazing (defoliation, dung, urine)	Climate; plant communities; water and nutrient supply; landscape features; grazing
Socio-economic influences on variation		Land prices, costs and returns	Markets; finances (interest rates, exchange rates); land tenure; climate change
Outcomes		Animal production	Animal production
	Pasture composition; biodiversity	Pasture composition; soil health; ground cover; runoff and soil loss; biodiversity	Greenhouse gases; atmosphere; water quality
Management goals	Limit time of grazing to maximize selection of young leaf	Utilize available forage without under- or overgrazing for optimum animal production and maintenance of resources	Exploit spatial variability; minimize offsite resource impacts
Management practices	Frequent changes of pastures in rotations; reasons often not well known but if they are can exploit to favour desired species	Control location, time and intensity of grazing (animal numbers, fencing (smaller paddocks and different land types), install more water points; burning, placement of supplements, fertilizer application, sown pastures, provide shelter, rest to allow patches to recover; fire to increases the attractiveness of ungrazed areas; shepherding)	Transhumance; nomadism; agistment; type of enterprise; type of livestock; spatial separation of breeding and fattening areas; sales

animals (breed, class, individual) showing desired preferences for landscape units and grazing distribution; shepherding or herding cattle away from heavily used areas), or changing the attributes of the paddock or landscape (fencing to separate landscape units; reducing paddock size to distribute grazing more widely across the landscape; more or relocated water points to enable animals to access new areas; providing shelter to make an area more attractive; and improving the quantity and/or quality of forage in particular areas to increase their use (e.g. using fire, applying fertilizer, sowing improved pastures)) (Bailey 2005). It is important to match scale of management to scale of impact (both spatial and temporal) and it is unrealistic to expect one management technique to be effective at multiple scales (Palmer *et al.*, 2003; Hunt *et al.*, 2007). In a study of paddock size (9, 34 or 57 km^2) in northern Australia, Hunt *et al.* (2007) found that while use of the whole area was more even in the small paddock, this did not prevent uneven use of particular patches or landscape units.

Variation at the plant part scale is difficult to respond to, although in intensively managed pastures managers may aim to maintain high green/dead and leaf/stem ratios.

Many grazing methods aim to limit variation: e.g. small paddocks to ensure animals can access all areas and that spatial variation within paddocks is minimized; rapid movement of animals through rotational systems so that the decline in pasture quality during the grazing period is reduced. However, variation may also be exploited at various scales. Animals grazing a patchy pasture may have greater intake and higher diet quality than those grazing a more homogeneous pasture (Ash *et al.*, 2004). A combination of multiple landscape units can increase diet choice (Boone and Hobbs, 2004; Stokes *et al.*, 2006) and quality; Stokes *et al.* (2004) showed crude protein in the diet during the dry season was higher in large paddocks than in less variable, smaller paddocks. This can lead to trade-offs where small paddocks give more even use but may limit diet quality, and exploiting patches to increase diet quality in large paddocks can lead to overuse and degradation of preferred, high-quality patches and landscape units. Transhumance and nomadism to exploit differences in forages at regional scales have already been mentioned.

5.4 Scale and Research

Research over the past four decades has given us a good understanding of foraging behaviour at the bite and feeding-station scale in relatively simple pastures. Much of this study has been in intensive temperate pastures and these studies need to be expanded to a wider range of grassland types, to ensure the generality of the results. Studies need to include the effects of different scales of spatial heterogeneity on diet selection and to look at diet selection in multi-species swards (compared with monocultures and binary mixtures) (Rutter, 2006). Incorporating multiple scales of measurement will reduce the chance of missing the scale at which grazing methods affect the performance of grazing systems (Laca, 2009). While the factors that determine preferences for different landscape units are known, there is less understanding of what determines preferences at the soil type or vegetation community level (O'Reagain, 2001).

Linking models with field experiments has provided insights and this should continue, with more explicit models of foraging behaviour that incorporate multiple scales of decision making (Soder *et al.*, 2009). Mechanistic models complemented by empirical research and real-time information could be used to generate grazing prescriptions (Laca, 2009).

There has been much study of forage characteristics and plant–animal interactions but less study of individual animals. Topics for research include how the origin and extent of individual variation in foraging behaviour impact on intake and diet selection, the metabolic effects of diet selection and consequences for livestock production, and the effect of foraging behaviour on plant communities and landscapes.

Studies of foraging behaviour and diet selection need to be linked to effects on animal production and resource conservation, to ensure they remain relevant to managing our grazing lands both now and into the future. Further research is needed, but, in the past, knowledge has accumulated faster than it has been incorporated into management. We need this incorporation if we are to maximize the benefits of our research.

References

Arnold, G.W. (1964) Factors within plant associations affecting the behaviour and performance of gazing animals. In: Crisp, D.J. (ed.) *Grazing in Terrestrial and Marine Environments.* Blackwell Scientific Publications, Oxford, UK, pp. 135–154.

Arnold, G.W. and Dudzinski, M.L. (1978) *Ethology of Free-ranging Domestic Animals.* Elsevier Scientific Publishing Company, Amsterdam, The Netherlands.

Ash, A., Gross, J. and Stafford Smith, M. (2004) Scale, heterogeneity and secondary production in tropical rangelands. *African Journal of Range and Forage Science* 21, 137–145.

Bailey, D.W. (2005) Identification and creation of optimum habitat conditions for livestock. *Rangeland Ecology and Management* 58, 109–118.

Bailey, D.W., Gross, J.E., Laca, E. A., Rittenhouse, L.R., Coughenour, M.B., Swift, D.M. and Sims, P.L. (1996) Mechanisms that result in large herbivore grazing distribution patterns. *Journal of Range Management* 49, 386–400.

Bartley, R., Corfield, J.P., Abbott, B.N., Hawdon, A.A. and Wilkinson, S.N. (2010) Impacts of improved grazing land management on sediment yields, Part 1: Hillslope processes. *Journal of Hydrology* 389, 237–248.

Baumont, R., Prache, S., Meuret, M. and Morand-Fehr, P. (2000) How forage characteristics influence behaviour and intake in small ruminants: a review. *Livestock Production Science* 64, 15–18.

Boone, R.B. and Hobbs, N.T. (2004) Lines around fragments: effects of fencing on large herbivores. *African Journal of Range and Forage Science* 21, 147–158.

Hunt, L.P., Petty, S., Cowley, R., Fisher, A., Ash, A.J. and MacDonald, N. (2007) Factors affecting the management of cattle grazing distribution in northern Australia: preliminary observations on the effect of paddock size and water points. *The Rangeland Journal* 29, 169–179.

Laca, E.A. (2009) New approaches and tools for grazing management. *Rangeland Ecology and Management* 62, 407–417.

Laca, E.A. and Ortega, I.M. (1996) Integrating foraging mechanisms across spatial and temporal scales. In: West, N.E. (ed.) *Rangelands in a Sustainable Biosphere. Proceedings of the Fifth International Rangeland Congress, Salt Lake City, Utah, 1995* Vol. II. Society for Range Management, Denver, Colorado, pp. 129–132.

Launchbaugh, K.L. and Howery, L.D. (2005) Understanding landscape use patterns of livestock as a consequence of foraging behaviour. *Rangeland Ecology and Management* 58, 99–108.

O'Reagain, P. (2001) Foraging strategies on rangeland: effects on intake and animal performance. In: *Proceedings of the XIX International Grassland Conference, Sao Pedro, Brazil.* Brazilian Society of Animal Husbandry, Piracicaba, Brazil, pp. 277–284

Palmer, S.C.F., Hester, A.J., Elston, D.A., Gordon, I.J. and Hartley, S.E. (2003) The perils of having tasty neighbours: grazing impacts of large herbivores at vegetation boundaries. *Ecology* 84, 2877–2890.

Rutter, S.M. (2006) Diet preference for grass and legumes in free-ranging domestic sheep and cattle: Current theory and future application. *Applied Animal Behaviour Science* 97, 17–35.

Senft, R.L., Coughenour, M.B., Bailey, D.W., Rittenhouse, L.R., Sala, O.E. and Swift, D.M. (1987) Large herbivore foraging and ecological hierarchies. *Bioscience* 37, 789–799.

Soder, K.J., Gregorini, P., Scaglia, G. and Rook, A.J. (2009) Dietary selection by domestic grazing ruminants in temperate pastures: current state of knowledge, methodologies, and future direction. *Rangeland Ecology and Management* 62, 389–398.

Stobbs, T.H. (1973) The effect of plant structure on the intake of tropical pastures. II. Differences in sward structure, nutritive value and bite size of animals grazing *Setaria anceps* and *Chloris gayana* at various stages of regrowth. *Australian Journal of Agricultural Research* 24, 821–829.

Stokes, C.J., Ash, A.J. and McAllister, R.R.J. (2004) Fragmentation of Australian rangelands: risks and trade-offs for land management. In: Bastin, G., Walsh, D. and Nicolson, S. (eds) *Living in the Outback. Proceedings of the 13th Biennial Conference of the Australian Rangeland Society, Alice Springs, Northern Territory, Australia.* Australian Rangeland Society, Aldgate, Australia, pp. 39–48.

Stokes, C.J., McAllister, R.R.J. and Ash, A.J. (2006) Fragmentation of Australian rangelands: processes, benefits and risks of changing patterns of land use. *The Rangeland Journal* 28, 83–96.

Stuth, J.W. (1991) Foraging behaviour. In: Heitschmidt, R.K. and Stuth, J.W. (eds) *Grazing Management An Ecological Perspective.* Timber Press, Portland, Oregon, pp. 65–83.

Valentine, K.A. (1947) Distance from water as a factor in grazing capacity of rangeland. *Journal of Forestry* 45, 749–754.

Wilson, J.R. (1994) Cell wall characteristics in relation to forage digestion by ruminants. *Journal of Agricultural Science (Cambridge)* 122, 173–182.

6 Livestock Production Styles and Managing Grassland Ecosystems

David R. Kemp and David L. Michalk

6.1 Introduction

Grasslands occupy about 40% of the world's terrestrial area excluding Antarctica and Greenland (Suttie *et al.*, 2005). These grasslands are used for grazing and fodder production to support livestock industries as well as significant numbers of wildlife. Ecologically, grasslands had been sustained for millennia by their environments and large numbers of herbivores, which restricted the growth and development of trees and shrubs. Many grasses (Harper, 1977) evolved under grazing that encouraged herbivores to maintain grassy ecosystems. These processes are still evident today. For example, in southern Africa the dominance of grassy ecosystems was attributed to megaherbivore grazing (Bond, 2003), whereas the expansion of woody weeds in drier areas of pastoral Australia has been attributed to the absence now of megaherbivores (Flannery, 1994). The area of grassland has decreased in many locations due to the expansion of cropping and urban development, while increasing in other areas where pastures have replaced forests and shrublands cleared primarily for timber and then used for grazing. Where land-use changes are substantial, the environmental impacts are generally greater than where grasslands remain in some form for livestock production.

The evolutionary framework wherein grassy ecosystems evolved has implications for their ecological function and the ecosystem services they now provide. Grasslands exist where soils, climate and their interactions with other plant and animal species, including humans, all act to sustain those ecosystems. Today, however, those actions may not always apply. Grasslands are no longer sustained solely by 'natural' processes. Most ecosystems suffer from moderate to high levels of disturbance, beyond that encountered in previous eras. This means interventions by humans are required to sustain plant and animal species, so that the ecosystem remains a grassland and functions to optimize use of natural resources, e.g. carbon capture and utilization, water use, nutrient cycling and maintenance of biodiversity (Kemp and Michalk, 2007). A major challenge faced by land managers is to optimize levels of utilization so that grasslands remain fully functional to satisfy the needs of today, and also the needs of future generations (Bruntland, 1987).

Humans are biologically omnivores and, throughout human history, animals and their products have been an essential part of the human diet. Societies vary in the types and amount of animal products consumed, reflecting the environments in which they live. There are major parts of the world's land surface where basic food and fibre needs can only be effectively derived from animals. Initially these were wild harvests, but, as human populations increased, animals were domesticated and managed for production. Over the centuries, attempts have often been made to grow crops in grassland areas, but the consequences have been mixed, ranging from marginal at best where some grain and forage production was possible, to disastrous, creating ever-expanding deserts, particularly in the Middle East and western China (Tao, 2001). It seems that many societies need to relearn the consequences of being over-optimistic about natural resource resilience (Flannery, 1994). Statistics indicate an increasing demand for animal products as the world becomes more affluent (FAO, 2009). In most cases, local attitudes to livestock management can determine how successfully that demand is satisfied and, in particular, how the natural resource base is sustained in meeting those demands. How grasslands are managed depends very much on the context and livelihoods of the people involved, who may not be very interested in optimizing grassland ecosystem function, even though maintaining or enhancing their ecological function sustains the livestock production that the communities depend upon.

Critical functions that need attention are minimizing soil erosion from wind or water, delivering clean water into river systems, and maintaining a diversity of plant and associated species such that productivity is optimized and opportunities for weed invasion are limited. At present pastoralists need to generate the income from livestock that covers the cost of maintaining their resource base and ecosystem function. Only the developed world is investigating payments to pastoralists to support wider environmental goals, which means that, elsewhere, incomes generated from livestock need to finance both system maintenance and household needs. The big challenge for many land managers is how to achieve their personal household/survival/production goals while maintaining, and often improving, the resource base upon which they depend and the wider functions of their grassland ecosystem that the world demands.

This chapter provides an overview of some major issues in grassland-based livestock production and management styles, and what that can mean for achieving viable levels of production and for sustaining the natural resource base and ecosystem function of grasslands. Grasslands are broadly defined here to include any lands used by grazing ruminants, where the herbaceous layer is the primary source of forage for those ruminants. Emphasis is primarily on grasslands that are not cultivated for crop production, though they may be re-sown with indigenous or introduced species.

6.2 Satisfying Demand from Today's and Tomorrow's Grasslands

Humans utilize livestock to provide food, fibre and fuel. When human populations were low, the required products could be obtained from low-input extensive systems. But as human populations and pressures on available land have increased, this has placed considerable stress on grasslands to satisfy increasing demand. As societies become more affluent there is a demand for more livestock products, which in turn leads to intensification in various forms that commonly require higher stocking rates and increasing inputs to sustain productivity. As intensification proceeds, the cost per unit of livestock product may initially decrease, e.g. by significantly reducing the age-to-market animals and the proportion of saleable product to liveweight. However, there are limits on that process. While inputs per head may stabilize at lower levels, as the stocking density is increased total inputs per hectare or per farm increase, raising the risk of adverse off-farm effects. Where breeding of livestock is separated from the production phase, this can still mean higher stocking rates to maximize the number of young animals produced per farm and create intense competition for land for feedlots, e.g. for meat or milk production.

The world is facing major problems with rising energy costs and warming climates. These factors will affect the productivity of many grasslands, especially those in middle latitudes where annual precipitation is expected to decline significantly (CSIRO, 2008; FAO, 2009). Rising energy costs mean that many inputs to agriculture will become progressively more expensive. Since agricultural products tend to be traded in open markets, livestock producers are typically price takers, which means reducing input costs is the main mechanism they can use to remain financially viable. Reduced inputs will ultimately lead to reduced outputs as any 'reserves' in the system are depleted. In turn, reduced outputs per farm are likely unless consumers are willing to pay considerably more for livestock products. Efforts by pastoralists to increase prices to cover higher costs could depress demand, leading to some reduction in outputs. Market pressures will see livestock producers responding with increased outputs and demanding higher prices at the expense of increased risk and reduced resilience for grassland ecosystems. In contrast, achieving a sustainable stocking rate and output per hectare often arguably requires a conservative approach, as discussed later.

The likely world demand for food is expected to increase at a faster rate than population growth. A 50% increase in the human population, coupled to anticipated increases in affluence, could mean a 70–100% increase in demand for food (FAO, 2009). Livestock products will constitute a large part of that increase in demand, particularly in developing countries. At present, estimates indicate that, globally, 55–60% of ruminant intake is derived from grasslands (Fischer, 2009). That work also suggests that many grasslands need to be converted to biofuel production, to avoid the need to divert cropland from food to energy production. However, that in turn would require more crops to be grown on grassland areas, to feed the increased demand for livestock products. None of those changes would improve the sustainability of many of the world's already degraded grassland systems.

6.3 Grassland Productivity and Ecosystem Issues

The world's grasslands are under considerable risk of increasing degradation due to pressures imposed by current management practices. In grasslands with high production potential, re-sowing and considerable inputs (e.g. fertilizer, lime, pesticides and herbicides) are practices commonly used to sustain the productivity of plant and animal species. Essentially these grasslands are managed as a crop, a concept traceable back to Stapledon and his successors in the UK. Higher levels of risk and reduced resilience apply to these systems, though off-farm adverse effects can be reduced with good management. As the site potential for production declines, fewer intervention options are available to grassland managers. Many of the world's grasslands, such as the vast grasslands of Eurasia, have low productivity and this limits management input to practices that improve livestock performance and reduce grazing impacts to better shepherding, the development of watering points and the use of fences to protect grasslands during critical growth stages.

Grasslands vary from natural to sown, though all contain some local species, especially at the meso and micro level. In many major grassland ecosystems there are significant concerns about preserving biodiversity, but unfortunately more emphasis is placed on animals than plants. As the focus of this chapter is on those grasslands used for livestock production, the biodiversity issues relate to the range and proportions of plant species and potentially some of the meso- and micro-fauna that reside within them. Grassland biodiversity problems tend to lie within the definition of broad- or weak-sense sustainability (Bruntland, 1987). Here species redundancy applies and trade-offs are possible, such as where alternative grass or forb species can fulfil similar roles. In this case, broad sense is applied only to species and not to others aspects of systems, such as trade-offs between cropping and grasslands. In many natural grasslands the proportions of a range of grass and other species present changes across the landscape and as local resources vary, i.e. they do not disappear. In contrast, a narrow, or strong-sense sustainability problem does not seem to apply very often in existing grasslands used for livestock production, i.e. where there are rare or endangered species from narrow genetic backgrounds and trade-offs are not appropriate.

The level of biodiversity does influence overall levels of grassland productivity (Tilman *et al.*, 1997) but it is important to take into consideration plant functional type because species are not equal in utilizing the same available resources. As a consequence there can be an optimum relationship between species richness and grassland productivity (Kemp *et al.*, 2003). This can arise because, in mixtures where there are few or no plant species with a high-production potential, a suite of minor, low-productive species may become evident. The additional plant species can then be of a similar functional type, implying a reasonable amount of redundancy (e.g. minor forbs or annual grasses) from the same plant family. From the perspective of livestock production, emphasis should arguably be on the proportion of more productive and higher quality forage species rather than on species richness per se.

6.4 Livestock Management Styles

Across the globe livestock are managed in various ways, depending upon a range of factors including local environments, culture and traditions, levels of knowledge, engagement with markets, products desired, draught needs and financial systems. There is a sequence in management evident within many societies that usefully summarizes the level of interaction between humans and the animals and provides some insight into how management of grassland ecosystems varies. In reviewing the ways in which East-African livestock users managed their flocks and herds, Neidhardt *et al.* (1996) developed a sequence that captured what most people were doing, derived from considering the question 'why do people keep animals'? This can be defined as a sequence of styles (Vanclay, 2004):

User ↔ Keeper ↔ Producer ↔ Breeder

'Users' vary from simply exploiting wild animals, to basic practices of simple retention of animals, herding and regarding them as an asset. **'Keepers'** provide more inputs to animals (e.g. storing fodder) and there are stronger cultural influences in owning livestock, often within mixed farming systems. **'Producers'** provide additional inputs to their livestock and have many attributes similar to commercial livestock producers in developed countries. **'Breeders'** engage in intensive livestock management, with controlled breeding and fine control over production and health in both extensive and intensive production systems. The latter two styles are very much focused on improving the biophysical and economic efficiency of livestock production.

This sequence in styles is two-way, as it is clear that pastoralists do vary their practices with circumstances. In Australia, for example, many pastoralists are 'producers ↔ breeders'; but at times some behave more like 'users ↔ keepers.' During adverse seasons, e.g. droughts, pastoralists often switch to a survival mode, reduce inputs and try to optimize the number of animals that can be retained through a drought, based on the assumption that it is more effective to restock after the drought to restore incomes. Grassland quality typically deteriorates under these conditions. There is often the stated goal of retaining 'good genetics', but, in reality, since the better animals can be sold for a higher price and the poorer animals not at all, there seems little evidence that 'better genetics' are actually retained on many commercial farms.

In studies monitoring the performance of individual livestock on the grasslands of Inner Mongolia, Kemp *et al.* (2010a) found that when subsistence herders reduced livestock numbers on the over-stocked grasslands, they sold the best animals and kept the old and least productive ones. Further consideration of herder practices suggested that, on what had been until recently communal grazing lands, livestock were managed almost as one would a wild population – simply harvesting what seemed the better animals at intervals. However, since 'better' animals were generally larger, this meant that herders were selling, for example, ewes that had not been producing any lambs. Renewal of flocks and herds with some better animals probably only occurs in good seasons that produce a surplus of animals, and, while some of the better ones are sold, enough superior animals then remain to restore at least some of the production base.

This sequence in livestock farming styles has important implications for how grasslands are managed. Traditional 'users' rarely own the land, at least as individuals, do not regard the grazing lands as an asset, and from their perspective the only rational management is to exploit any available forage as quickly as possible before someone else does. This is often coupled with livestock numbers being the farmer's main asset, i.e. a 'livestock bank'. In Australia, similar behaviour used to be evident in droughts, where farmers sought to utilize any available forage rapidly, rather than de-stocking and rationing what was available to fewer but more productive animals. Little consideration was given to managing grasslands to ensure they regenerated in a more productive state after the drought. Many programmes in Australia have changed those attitudes so that livestock numbers are now reduced early in a drought and grasslands are retained in much better condition for much longer periods. In recent droughts there has been reduced runoff of water into river systems and fewer dust storms, reflecting the greater amounts of herbage retained on the soil surface. Naturally, when droughts are protracted, herbaceous plants will eventually die even if they are never grazed.

6.5 Managing Production and Ecosystem Services

Managing grassland ecosystems to optimize functionality and to ensure continued output of the 'products' society desires is heavily influenced by the context wherein livestock production occurs. Traditional transhumance systems (Humphrey and Sneath, 1996) have existed for a long time in many parts of the world (e.g. across Africa and the Eurasian grasslands) and are arguably still sustainable. Herders are typically 'users ↔ keepers' and average stocking rates and utilization intensity are low. The primary focus of these subsistence herders is on survival and only small proportions of the flocks and herds are surplus to household needs and traded in occasional markets.

The problems today, with arguments for restoring traditional grazing practices and using traditional knowledge in managing grasslands, are that, in many regions, the human and, as a consequence, livestock populations are considerably greater than used to be the case, and traditional knowledge is based upon finding ways to survive adverse conditions and not to optimize the level of livestock production for market economies. A survival style is essentially an animal focused model wherein grassland condition is a very secondary consideration. The consequences are that per head animal production is often very low and grassland plants are heavily grazed and eventually utilize all their reserves and die. This then leads to poor water and nutrient management (high runoff rates and soil erosion) and high rates of methane production per animal and per unit of production (Dong *et al.*, 2010).

The problems of livestock managers reverting to a 'user ↔ keeper' style under conditions of hardship (such as droughts, cold winters) have meant, in the past, that, because grassland condition was regarded as secondary, de-stocking did not happen until after the ground was bare and damage to ecosystem function was already done. This was a major reason why perennial grasses in south-eastern Australia were rapidly replaced by Mediterranean annuals (Moore, 1970). The recognition of this problem led to a series of national research programmes, starting in the late 1980s (Kemp and Dowling, 2000), which emphasized grassland management as being as important as livestock management. Losing the most productive plant species from grasslands reduces longer-term sustainability of the system and greatly increases the costs of restoring grasslands to productive states. Utilization rates need to optimize survival of the desirable, productive species.

A 'producer ↔ breeder' style of livestock management typically aims to improve the production efficiencies and to sustain the resource base. Livestock production is more efficient when there is more, higher-quality forage available on grasslands. Grasslands in such a state are better able to capture nutrients and other resources for plant growth, and minimize off-farm effects from soil erosion and nutrient loss. A broad summary across ecosystems of how the livestock management style could affect livestock production and ecosystem functions (Table 6.1), drawn from experiences in Australia and north Asia, does suggest that, as livestock farmers move towards 'producer ↔ breeder' styles, there is a higher likelihood of improving grassland management. To achieve optimal outcomes requires specific practices designed to enhance condition, hydrology, biodiversity and resilience. 'Keepers' are arguably a stage where competition for resources becomes more intense and, without inputs, the resources degrade. In each case there are opportunities to improve the efficiencies and resilience of the grassland systems.

6.6 Sustainable Utilization Levels

The pressures to intensify livestock production from grasslands were discussed earlier. However, for permanent grasslands there is the obvious need to balance inputs and outputs in ways that sustain plant and animal species without resulting in any major off-farm effects such as erosion and nutrient flows into rivers. Ideally, farm practices aim to sustain the productive base and minimize losses, but the art and science of doing that are still developing. Response of pastoralists to increasing cost-price pressures is to restrict inputs and opt for less than maximum production. As a consequence, off-farm effects are possibly less than could occur. Problems can arise, however, if livestock managers utilize a higher proportion of the available forage than is sustainable in terms of maintaining nutrient cycling and plant regeneration, leading ultimately to a decline of the grassland condition.

Table 6.1. Consequences of management styles for livestock and grassland ecosystem functions.

Component	User	Keeper	Producer	Breeder
Livestock production per head	Low	Low	High	High
Livestock production per ha	Low	Low	High	High
Livestock fecundity	Low–good	Low	Moderate	Good
Grassland species condition	Poor–good	Low	Moderate	High
System hydrology	Moderate–good	Low	Moderate	Moderate–good
Biodiversity	High	Low	Low–moderate	Low–moderate
Ecosystem resilience	Moderate–good	Low	Low–moderate	Low–moderate

Sustainable utilisation levels have not been estimated for many contexts. In higher productivity areas, livestock producers often seek to maximize consumption of the available forage and, when the desirable species decline, rehabilitate swards by resowing. This typically happens on dairy farms in developed countries, where the life of a pasture may only be 2–3 years, as occurs in south-eastern Australia. Under high utilization rates, pasture survival depends upon local conditions as well as grazing practices. In New Zealand dairy systems, where grazing periods are short and rest periods considerably longer, a high utilization level can be sustained in favourable climates. However, this also requires significant fertilizer and other inputs to control weeds, pests and diseases.

In moderate to low productivity environments, only a low proportion of the forage grown can be utilized sustainably, as low regrowth rates reflect the lower availability of resources. In northern Australia, sustainable utilisation rates only consume 20–30% of the forage grown during the short wet season through the following long dry season (Ash *et al.*). Higher utilization levels result in significant loss of productive species. In central New South Wales, where there is a wide range of pasture types used with a large variation in production and average stocking rates (Table 6.2), pasture productivity varies from 3–12 t dry matter (DM)/ha/year. In 1950, unfertilized native pastures were stocked at 3 DSE/ha (DSE, dry sheep equivalent), but now they only sustain 1 DSE/ha (Kemp and Dowling, 2000), suggesting that initial stocking rates were too high: nutrients were not replenished and plant species could not survive those grazing pressures under the typical continuous stocking practices. In higher productivity environments, newly sown pastures are often stocked at 14 DSE/ha/year, but within a couple of years their average stocking rate is only 7–9 DSE/ha/year (Vere, 1998; Kemp and Dowling, 2000). Sown pastures in this region cost more than AUS$300/ha to establish, especially when there are failures due to droughts and other causes. Consequently, farmers are motivated to utilize as much of the pasture as soon after sowing as possible, and the net effect is a non-sustainable stocking rate. In these low and moderately productive pastures, a utilization rate above 50% does result in a

Table 6.2. Range in potential production, stocking rate and estimated level of pasture utilization for the typical range in natural to sown pastures in central New South Wales.

Grassland type	t DM/ha/year	DSE/ha/year	Percent utilized	Grassland state
Unfertilized	3	1	17	OK
		3	50	Poor
	5	3	30	OK
Fertilized		5	50	Poor
	8	5	31	OK
Sown/fertilized		7	44	Poor
	12	7	29	OK
		9	38	OK
		14	58	Poor

DSE, dry sheep equivalent: (50 kg)/ha/year. DM, dry matter. Utilization based on assuming 1 DSE consumes 0.5 t DM/ha/year + 30% wastage. The fertilizer used is phosphate. Unfertilized and fertilized pastures would comprise native and volunteer species and sown pastures are based on *Phalaris aquatica*. A value judgement is given as to the quality of the grassland state from a livestock production and environmental view.

decline in productivity. The sustainable utilization level on the least productive pastures is only ~20%, rising to ~40% in sown, higher productivity pastures under current grazing practices.

6.7 A Way Forward?

This brief survey, of general styles of livestock management, their interaction with managing grassland ecosystems and the consequences for managing ecosystem services, argues that the context in which livestock farmers and herders view their livestock is as an initial driver of grassland management, with major consequences. Where communities have a primary focus on survival, animal production efficiency is low and, with increasing human and livestock populations, it is likely that grassland conditions will deteriorate, with consequent major impacts on ecosystems and their function. However, as local markets develop, there are increasing opportunities for livestock producers to develop a more commercial focus. Inherent in those trends are the financial benefits of improving the efficiency of production and product quality (Kemp *et al.*, 2010b). Livestock that are slow growing require additional time and fodder to reach market weights, reduce household incomes and importantly result in higher grazing pressures, often beyond what is sustainable. Significant increase in methane output is also incurred by inefficient livestock production systems. The 'livestock bank' needs to be driven less by numbers of animals and more by what they can produce per unit of ingested energy. To achieve higher production efficiencies typically requires increasing the amount of quality forage per animal. That can only be achieved, in many environments, by reducing stocking rates. In other cases the forage available can be increased through rationing forage at appropriate times and using short-graze and long-rest periods, as done in New Zealand. The animals that are retained need to be those with higher potential for producing the products required, including meat, dairy products and fibre.

Livestock productivity is directly related to the amount of quality fodder (i.e. metabolizable energy) available per animal and it is in the interests of livestock producers to maximize the amounts of quality forage available. Greater amounts of quality forage often mean a better suite of plant species that can efficiently utilize all the available resources for growth by capturing nutrients and water, which in turn reduces the risk of losses through erosion and potential pollution problems. From a global perspective, efficient grassland and livestock management leads to more carbon sequestration within the grassland and maintenance of an appropriate level of biodiversity to optimize productivity. This is very much a win–win circumstance, where the goals of improving production efficiency align with goals for improving grassland ecosystem function and minimizing adverse off-farm effects (Michalk *et al.*, 2010). With rising energy costs and a desire to reduce the carbon footprint of livestock production, the incentives are there for livestock producers who operate at the more intense end of livestock management styles, to optimize their inputs and minimize off-farm impacts. This makes economic and environmental sense.

Changing consumer demands in many countries will continue to increase demand for livestock products, especially from environments where livestock are the only realistic option for food, fibre and fuel production. As the economies of many developing countries change to accommodate a market focus, especially those of China and India and, as anticipated, through Africa, there will be a faster growth in demand for livestock products than for other staple foods. As noted, a 50% increase in the world's human population is expected to result in a 70–100% increase in demand for food, a large part of which will be supplied by livestock products (FAO, 2009). At the same time there is concern that methane production from ruminants needs to be reduced (Dong *et al.*, 2006, 2010). The option of slaughtering all ruminants, which should then include the wildlife that also produce methane, and abandoning livestock production across the globe is not feasible. The strategy that is considered more acceptable is to develop and implement practices across the world's grasslands that improve the bio-physical efficiencies of livestock production and thereby reduce the amount of greenhouse gases produced per unit of product per unit of energy consumed.

Livestock production from grasslands can have a sustainable future, but this will depend upon balancing the increasing demand for livestock products, from an increasingly more affluent world, against the increased costs and pressures on landscapes to produce those products while minimizing off-farm effects on greenhouse gas production, nutrient pollution and impacts on biodiversity. The key will be to emphasize improving the efficiencies in production across all the landscapes where livestock are the primary land use for the production of food, fibre and fuel resources.

References

Ash, A., Corfield, J. and Ksiki, T. (n.d.) *The Ecograze Project – Developing Guidelines to Better Manage Grazing Country*. CSIRO, Clayton South, Australia.

Bond, W. (2003) Top-down control of grassy ecosystems. In: Allsopp, N., Palmer, A.R., Milton, S.J., Kirkman, K.P., Kerley, G.I.H., Hurt, C.R. and Brown, C.J. (eds) *Proceedings of the VIIth International Rangelands Congress*. Document Transformation Technologies, Irene, South Africa, p. 4.

Bruntland, G. (1987) *Our Common Future. World Commission on Environment and Development*. Oxford University Press, Oxford, UK.

CSIRO (2008) *An Overview of Climate Change Adaptation in Australian Primary Industries: Impacts, Options and Priorities*. National Climate Change Research Strategy for Primary Industries, CSIRO, Clayton South, Australia.

Dong, H.M., Mangino, J. and McAllister, T.A. (2006) Emissions from livestock and manure management. In: Eggleston, H.S., Buendia, L., Miwa, K., Ngara, T. and Tanabe, K. (eds) *2006 IPCC Guidelines for National Greenhouse Gas Inventories, Prepared by the National Greenhouse Gas Inventories Programme*. Institute for Global Environmental Strategies, Kanagawa, Japan, pp. 10.1–10.87.

Dong, H.M., Li, Y., Kemp, D., Michalk, D. and Takahashi, T. (2010) Reducing methane production from livestock – can more efficient livestock systems help? In: Kemp, D.R. and Michalk, D.L. (eds) *Sustainable Development of Livestock Systems on Grasslands in North-Western China*. ACIAR Proceedings No. 134 Australian Centre for International Agricultural Research, Canberra, pp. 115–127.

FAO (2009) *How to Feed the World in 2050*. Food and Agriculture Organisation, Rome.

Fischer, G. (2009) World food and agriculture to 2030/50: how do climate change and bioenergy alter the long-term outlook for food, agriculture and resource availability? In: *How to Feed the World in 2050*. Food and Agriculture Organisation, Rome.

Flannery, T.F. (1994) *The Future Eaters*. Reed Books, Melbourne, Australia.

Harper, J.L. (1977) *Population Biology of Plants*. Academic Press, London.

Humphrey, C. and Sneath, D. (1996) *Culture and Environment in Inner Asia: 1 the Pastoral Economy and the Environment*. The White Horse Press, Cambridge, UK.

Kemp, D.R. and Dowling, P.M. (2000) Towards sustainable temperate perennial pastures. *Australian Journal of Experimental Agriculture* 40, 125–132.

Kemp, D.R. and Michalk, D.L. (2007) Towards sustainable grassland and livestock management. *Journal of Agricultural Science (Cambridge)* 145, 543–564.

Kemp, D., Han, G.D., Junk, G., Langford, C., Michalk, D., Millar, G., Takahashi, T., Xu, Z. and Zhao, M.G. (2010a) Demonstration farms – what can be done, what are we learning? In: Kemp, D.R. and Michalk, D.L. (eds) *Sustainable Development of Livestock Systems on Grasslands in North-Western China*. ACIAR Proceedings No. 134. Australian Centre for International Agricultural Research, Canberra, pp. 152–161.

Kemp, D., Han, G.D., Michalk, D., Nan, Z.B., Wu, J.P. and Xu, Z. (2010b) China's grassland livestock farming systems: strategies and tactics for improvement. In: Kemp, D.R., and Michalk, D.L. (eds) *Sustainable*

Development of Livestock Systems on Grasslands in North-Western China. ACIAR Proceedings No. 134. Australian Centre for International Agricultural Research, Canberra, pp. 177–189.

Kemp, D.R., King, W.McG., Lodge, G.M., Murphy, S.R., Quigley, P. and Sanford, P. (2003) SGS Biodiversity Theme: the impact of plant biodiversity on the productivity and stability of grazing systems. *Australian Journal of Experimental Agriculture* 43, 962–975.

Michalk, D., Kemp, D., Han, G.D., Hua, L.M., Nan, Z.B., Takahashi, T., Wu J.P. and Xu, Z. (2010) Redesigning livestock strategies to reduce stocking rates and improve incomes on western China's grasslands. In: Kemp, D.R. and Michalk, D.L. (eds) *Sustainable Development of Livestock Systems on Grasslands in North-Western China*. ACIAR Proceedings No. 134. Australian Centre for International Agricultural Research, Canberra, pp. 140–151.

Moore, R.M. (1970) South-eastern temperate woodlands and grasslands. In: Moore, R.M. (ed.) *Australian Grasslands*. Australian National University Press, Canberra, Australia, pp. 169–190.

Neidhardt, R., Grell, H., Schrecke, W. and Jakob, H. (1996) Sustainable livestock farming in East Africa. *Animal Research and Development* 43/44, 45–53.

Suttie, J.M., Reynolds, S.G. and Batello, C. (eds) (2005) *Grasslands of the World*. Plant Production and Protection Series No. 34. Food and Agriculture Organisation, Rome.

Tao, W. (2001) Rangeland desertification in China. In: *Proceedings of the XIXth International Grassland Congress, February 2001, Sao Paulo, Brazil*. Brazilian Society of Animal Husbandry, Piracicaba, Brazil, pp. 953–955.

Tilman, D., Knops, J., Wedin, D., Reich, P., Ritchie, M. and Siemann, E. (1997) The influence of functional diversity and composition on ecosystem processes. *Science* 277, 1300–1302.

Vanclay, F. (2004) Social principles for agricultural extension to assist in the promotion of natural resource management. *Australian Journal of Experimental Agriculture* 44, 213–222.

Vere, D.T. (1998) Investigating improved pasture productivity change on the New South Wales tablelands. *Agricultural Economics* 18, 63–74.

Part II:

Grasslands as Regulating Biogeochemical Cycles and Environmental Fluxes to Hydrosphere and Atmosphere

Editor: Abad Chabbi

The overall objective of this section is to identify the different processes through which biogeochemical cycles are regulated by the multiple interactions between soil, vegetation and living organisms. Such an understanding may allow forecasts of the magnitude of different fluxes to the atmosphere and hydrosphere. The impacts of grasslands on these fluxes have to be analysed in relation to different management practices (grazing versus cutting, stocking density, mineral fertilization, etc.). Interactions between perennial herbaceous vegetation and soil lead to coupling of C and N cycles, whereas grazing animals 'uncouple' these cycles. The uncoupling processes cause severe environmental problems, especially with respect to greenhouse gas (GHG) emissions (Chapter 1).

Soil organic matter (SOM) influences many soil functions and occupies a key position in the global C cycle. SOM stabilization and destabilization is discussed in Chapter 7, in relation to grassland management. Management practices strongly determine the nature, allocation and residence time of SOM. These impacts still need to be quantified. Furthermore, the C and N cycles and their coupling–decoupling processes across grassland management systems are analysed in Chapter 8, with particular focus on the patchiness nature of grassland areas. The authors recommend adaptive cycles and 'panarchy' concepts as promising approaches in grassland studies. One important environmental consequence of grassland intensification, nitrate leaching and its impact on ground and surface water quality, is analysed in Chapter 9. The role of fertilizer N inputs and biological N_2 fixation are discussed in relation to grazing management systems. Mitigation options at plant, animal and farm level are presented and discussed through a whole-system approach. The GHG budget of grasslands in Europe is presented in Chapter 10, with particular emphasis on the different processes leading to GHG emission in relation to management systems: CO_2 emission and sequestration, CH_4 and N_2O fluxes, and possibilities for mitigation options. The use of legume species within grassland vegetation is discussed in Chapter 11 with two very interesting perspectives: (i) the functional role of legume species in biodiversity, and (ii) the adaption of these species to elevated CO_2 conditions by regulating N cycling. The problem of phosphorus losses in surface runoff, and its consequence for quality of surface hydro-systems, is discussed in Chapter 12. This problem must be analysed at the level of the whole livestock production system and cannot be restricted only to the soil–vegetation interface. In addition, the role of grasslands in buffering P fluxes from effluents in intensive livestock production systems is emphasized.

7 Carbon Storage and Organic Matter Dynamics in Grassland Soils

Cornelia Rumpel

7.1 Introduction

Carbon storage in soils has received increasing attention in recent years due to the possibility of counteracting the increasing atmospheric CO_2 concentrations (Houghton *et al.*, 1999). The soil carbon pool amounts to 1500 Pg C and has been estimated at approximately 3.3 times the size of the atmospheric pool and 4.5 times the size of the biotic pool (Lal, 2004). Carbon sequestration in soils may be a cost-effective and environmentally friendly way not only to store C but at the same time ensure food security by improving soil quality (Lal, 2004).

Grasslands, including rangelands, shrublands, pastureland, and cropland sown with pasture and fodder crops, contain about 20% of the world's soil carbon stocks (FAOSTAT, 2009; Ramankutty *et al.*, 2008). Grasslands characteristically have high inherent soil organic matter (SOM) content, averaging 333 Mg/ha (Schlesinger, 1977). The high organic matter content of grasslands is beneficial to soil quality (Shukla *et al.*, 2006) and its presence in soil determines ecosystem health and soil productivity. SOM – an important source of plant nutrients – influences the fate of organic residues and inorganic fertilizers, increases soil aggregation, which can limit soil erosion, and also increases cation exchange and water-holding capacity. It is a key regulator of grassland ecosystem processes. In order to assure good soil functioning, two SOM types are necessary: easily decomposable SOM, to provide plants and the microbial biomass with nutrients and energy, and stable SOM, which is important for carbon storage and physical properties of soils. The proportion to which these SOM types are necessary for good soil functioning is subject to research.

One of the most important characteristics of the soil carbon pool is its reactivity towards human interventions. Soil carbon stocks may be increased as well as decreased by human activity. Land management is therefore most important in determining whether soil is conserving its quality and whether it acts as source or sink of CO_2. Grasslands and forests may be preferable systems in terms of carbon storage, compared with agricultural land, because they are characterized by permanent soil cover (Conant, 2010). Unlike forests, where vegetation is the primary source of carbon storage, most of the grassland carbon stocks are found below ground. It is thus crucial to understand the processes influencing OM stabilization in grassland soil in order to determine management strategies, aiming at increasing carbon stocks. Little is known about the quantitative contribution of the main sources of OM input into grassland soils: (i) senescent plant leaves, which are not consumed by animals; (ii) plant leaves, which escape harvesting; (iii) plant roots and root exudates; (iv) dung and urine of grazing animals; and (v) black carbon in fire affected grasslands. Although grassland is usually characterized by permanent vegetation cover, management differences, such as mowing versus pasture, and grazing intensity may influence the carbon input and dynamics (Parsons *et al.*, Chapter 1, this volume). To evaluate the carbon storage potential of grassland soils, knowledge about the contribution of stable and unstable carbon types as well as the drivers of SOM dynamics is required. The aim of this chapter is to give an overview about possible carbon stabilization and destabilization mechanisms under contrasting grassland management. The specific processes operating in grassland soils will be compared with those under other land uses. Moreover, the different sources of carbon input and their potential to be sequestered in soil will be discussed.

7.2 Organic Matter Input into Grassland Soil

OM input into grassland soil occurs in the form of above- and below-ground plant tissue (Fig. 7.1). In pastures, plant litter is one of the major sources of plant nutrients due to the abscission, lodging and trampling of aerial plant parts, and its decomposition is a fundamental process influencing ecosystem carbon flux and nutrient cycling (Hoorens *et al.*, 2003). In grazed grasslands, leaf tissues are continuously produced, depending on temperature (Lemaire and Agnusdei, 2000), genetics and other factors. They are harvested more or less frequently by herbivores, such as grazing animals. The leaf lifespan is genetically determined in relation to temperature (Lemaire, 1988), and, as it ends, the portions of leaves that have not been harvested by animals senesce. Their colour changes from green to yellow and brown while they remain attached to the plant, before returning to soil as litter. In well-managed grazing situations, the proportion of leaf tissues escaping intake by animals is about 30% of the leaf tissue produced (Mazzanti *et al.*, 1994; Parsons *et al.*, Chapter 1, this volume). This corresponds to about 4000 kg dry matter/ha/year for an average herbage production of 12,000 kg dry matter/ha in most of the intensive grazed grasslands in temperate areas. In grasslands harvested by mowing for forage production, a greater proportion of leaf tissues is harvested before senescence, limiting the addition of leaf litter to soil. Nevertheless, during the harvesting process, green leaves may be lost and returned to soil. Leaf tissue losses during grassland harvest may amount to 20% of the herbage harvested (G. Lemaire, unpublished data). Thus, grassland systems under pasture and those used for forage production receive litter of contrasting composition, which affects decomposition dynamics and carbon cycling. It has been shown that leaf litter of grassland species may be of very different quality depending whether or not it has been affected by the senescence process (Sanaullah *et al.*, 2010). Leaf litter quality also differs between different grassland species as well as between C3 and C4 plant types. The initial litter quality determines its degradation behaviour in soil. Total soluble sugar content is an important controlling factor in the earlier phase of decomposition, whereas the later stages of decomposition may be regulated by the lignin/N ratio of the initial litter. It was shown that litter added to soil at different stages of maturity decomposes at different rates and maintains these differences for several years (Sanaullah *et al.*, 2010).

In temperate grassland systems, more than 80% of the net primary productivity (NPP) is diverted to root production (Swift *et al.*, 1979). However, herbaceous species are less deep rooting than forest vegetation or shrubs, and globally ~ 40% of all soil organic carbon (SOC) in the first meter of grassland soil is found in the uppermost 20 cm (Jobbagy and Jackson, 2000). Root

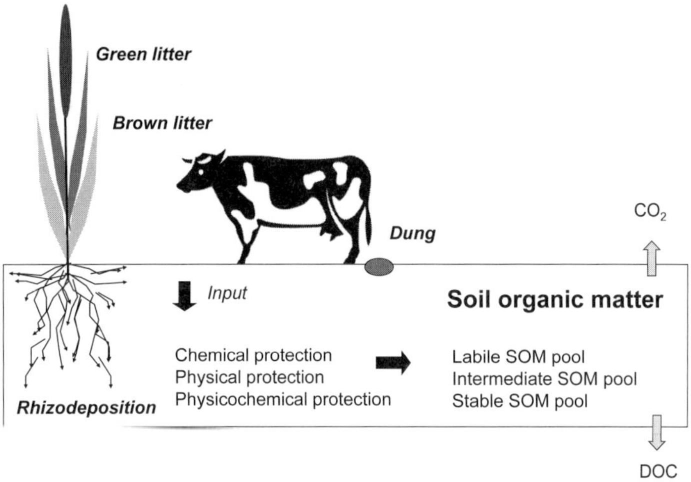

Fig. 7.1. Carbon cycling in grassland soils: SOM, soil organic matter; DOC, dissolved organic carbon.

carbon input into soil may occur as root-cell sloughing, other root debris, mucilage and root exudation. All may contribute greatly to SOM due to preferential stabilization within the mineral matrix (Rasse *et al.*, 2005). Rhizosphere processes, such as root turnover, rhizodeposition, root respiration and microbial response, may be extremely important in grassland soil. These processes may lead to stabilization as well as destabilization of soil carbon (Cheng and Kuzyakov, 2005). Plant roots may affect SOM decomposition through (i) decreasing mineral nutrient availability to soil microorganisms due to plant uptake (Schimel *et al.*, 1989); (ii) changing the physical environment in the rhizosphere (Shields and Paul, 1973); (iii) increasing organic substrate supply; and (iv) enhancing microbial turnover due to fungal grazing. However, interactions at the rhizosphere level between soil microorganisms and roots of different plant communities are complex and remain poorly understood (Cheng and Kuzyakov, 2005; Le Roux *et al.*, Chapter 17, this volume). Rooting *et al.*, behaviour of plants and rhizosphere processes may be strongly influenced by grassland management, which can therefore impact the storage and composition of SOM.

In addition to above- and below-ground litter, dung can constitute an important OM source in grasslands managed by grazing (Jarvis *et al.*, 1996). On a pasture with a grazing intensity of 700 cow days/ha/year, an average of 6.4% of the grazed area is commonly covered by dung patches, which provide a C input equivalent to ~ 22.5 t C/ha (Bol *et al.*, 2000). By grazing and depositing dung and urine, large herbivores remove carbon and nutrients and return them by excretion, leading to cycling and redistribution, which in turn influences the structure and functioning of grassland systems (Haynes and Williams, 1993) as well as OM dynamics through carbon dioxide release and OM incorporation into soil. Dung is a source of labile C, which may increase microbial biomass (Lovell and Jarvis, 1996) and induce priming of native OM. The quantitative retention of dung-derived C in grassland ecosystems was estimated at about 10% (Bol *et al.*, 2000). Molecular studies have shown that carbohydrates constitute part of the dung-derived C in soil (Dungait *et al.*, 2009). Limited information is available about the impact of this labile C on the soil carbon cycle.

Many types of grassland worldwide are managed by regular burning. On these sites, fire-derived black carbon is another SOM source. Quantification of this material in the major grassland ecosystems of the world showed a contribution between 5 and 30% of total SOM (Rodionov *et al.*,

2010). Little is known about the implications of this non-neglectable proportion for carbon storage and SOM dynamics.

7.3 SOM Stabilization and Destabilization Processes in Grassland Soil

SOM is a continuum comprising all stages of degradation from fresh plant litter to stabilized organic matter. It is generally divided into three pools according to mean residence time (Fig. 7.1, Parton *et al.*, 1987). Fresh plant material makes up the labile pool with mean residence times in the range of 1 year. It may be isolated, by density fractionation, as particulate organic matter. The intermediate pool contains OM, which is occluded in soil aggregates and which may persist in soil for decades. The stable SOM pool consists of physicochemically protected OM, which is adsorbed to soil minerals, and highly recalcitrant OM such as charcoal (Krull *et al.*, 2003). The relative contribution of these different pools determines SOM storage and dynamics. Grassland soils may have a high C sequestration potential, which is underlined by the fact that root litters have a high potential to be stabilized in soil (Rasse *et al.*, 2005). Despite their importance as a C source in grasslands, root C fluxes to soil are poorly understood, mainly due to uncertainties associated with the measurement of total root C input, in particular from root exudation and root-cell sloughing. Root litter production can be estimated from root turnover. Root turnover can be measured directly, using observation of roots from birth to disappearance, with micro-rhizotrons. However, micro-rhizotrons are only able to estimate the most dynamic roots (<1 mm) and not roots with larger diameter (>1 mm), for which isotope techniques involving ^{14}C or ^{13}C may be more suitable (Majdi and Anderson, 2005). Depending on the method, the longevity of roots has been estimated to be in the order of 1–18 years (Gaudinski *et al.*, 2001). Root carbon may be affected by all three stabilization processes and therefore remain in soil for much longer.

Stabilized carbon within the soil matrix is not localized homogeneously but occurs within distinct parts of the soil profile (Chabbi *et al.*, 2009). In the topsoil, OM stabilized for more than 100 years has not yet been detected. Older OM compounds occur in general below the A horizon (Rumpel and Kögel-Knabner, 2011). However, even in subsoil, where old SOM is found at high concentrations, preferential flow, root growth and bioturbation may lead to input of fresh OM. Such an input into grassland soils was found to lead to heterogeneous distribution of stabilized SOM, due to dilution with recent OM or destabilization of old SOM (Fontaine *et al.*, 2003). Destabilization of old SOM stored in the subsoil under grassland has also been associated with drying–rewetting processes (Xiang *et al.*, 2008). Grassland management could influence carbon stabilization. This has been demonstrated by measuring C fluxes with the eddy covariance method (Ammann *et al.*, 2007). However, the processes involved, in particular those related to N additions, have not yet been identified.

There are clear differences in molecular composition of SOM under grassland and arable land (Nierop *et al.*, 2001). Plant residues dominate the composition of SOM under grassland, while a higher contribution of microbial residues to SOM of arable land was noticed (Martens *et al.*, 2003). This was explained mainly by increased protection of fresh material through incorporation into soil aggregates, which are destroyed after cultivation, exposing protected SOM to microbial decay (Parton *et al.*, 1987). After grassland conversion the composition of SOM changes and reflects the capability of the system to resist the disturbance or to return to its initial state (Rumpel *et al.*, 2009).

7.4 Influence of Grassland Management on SOM Dynamics and Composition

Soil OM is the most important carbon reservoir on continental surfaces that is largely responsive to human interventions. Management may influence carbon storage and OM dynamics in grassland soils through: (i) management systems (mowing versus grazing, see above); (ii) regulation of grazing intensity; (iii) use of fertilizers; and (iv) land-use change.

Grazing management of grassland ecosystems influences OM input and associated soil properties. Intensive grazing leads to significantly decreased SOC stocks associated with higher bulk density in topsoils (Steffens *et al.*, 2009) and additionally to lower mean-weight diameter of water-stable aggregates, lower monosaccharide contents and lower contribution of microbially derived sugars (Dormaar and Willms, 1998). The grazing regime (high or low intensity) greatly influences the SOM dynamic. Intensively grazed grasslands are dominated by fast-growing species producing litter low in lignin and rich in N, which is rapidly decomposed by bacteria (Bardgett *et al.*, 1998), whereas grasslands adapted to low grazing levels are dominated by slow-growing plant species and slow litter decomposition (Milchunas and Lauenroth, 1993). The two different plant communities have been reported to control SOM stocks through plant roots and their activity (Klumpp *et al.*, 2009). Changes in C dynamics after grazing cessation may be slow and not lead to C sequestration in soils, where the mineral particles are already saturated with OM (Steffens *et al.*, 2009). In contrast, light or moderate grazing does not necessarily decrease SOM contents (Cui *et al.*, 2005). It was even found that long-term grazing with cattle was more favourable for macroaggregation and aggregate stability and thus carbon storage than harvest by mowing (Franzluebbers *et al.*, 2000).

Apart from the presence or absence of grazing animals, management of grassland may include the application of fertilizers. A review of global data sets revealed that fertilized grasslands had, on average, higher SOC stocks than natural or less intensively managed systems (Conant *et al.*, 2001). Accordingly, Nyborg *et al.* (1997) found that increasing levels of fertilization were associated with larger SOC contents and higher productivity in Canadian grasslands. A recent work concluded that conversion of arable land into grassland leads to a positive carbon balance only if the system receives extra nitrogen inputs (Ammann *et al.*, 2007). In contrast, no relationship between the intensity of management and SOC stocks was found for Alpine grasslands (Bitterlich *et al.*, 1999). The mechanisms that lead to these contrasting results still need to be elucidated.

After land-use change, SOM may easily become a carbon source, in particular if land-use change leads to disturbance of long-term equilibria, such as that seen with grassland conversion into arable land. Over the past two centuries, land-use change has led to CO_2 emissions of 136 Pg C (Houghton *et al.*, 1999), about 78 Pg C of them originating from SOM (Lal, 2003). Studies using physical fractionation procedures showed that particulate OM or the light fraction of SOM often respond more rapidly to land-use change than bulk soil (Six *et al.*, 1998). However, several authors also reported rapid response of mineral-associated fractions of SOM upon land-use changes (Leifeld and Kögel-Knabner, 2005). This might be the result of fast diffusion and sorbtive stabilization of small molecules on mineral surfaces during SOM decay. Using molecular characterization of soil fractions, Leifeld and Kögel-Knabner (2005) came to the conclusion that the composition of total SOM may be more sensitive to land-use change than that of free and particulate OM fractions. Biochemical parameters indicated that SOM, even if it is lost quantitatively, may be rather resistant to changes in composition and thus its properties (Rumpel and Chabbi, 2010).

7.5 Conclusions and Research Needs

Despite the wealth of studies on the composition and dynamics of grassland SOM, the mechanisms underlying the response of organic matter pools and single compounds to different management are far from being understood. Future research is needed to determine quantitatively the drivers of SOM stabilization and destabilization for different grassland systems, in order to develop suitable management strategies in response to changing environmental conditions. Differences in grassland management, such as grazing versus mowing, influence soil carbon dynamics through qualitative and quantitative changes in OM input. It is only recently that the impacts of these changes were addressed in SOM studies. The influence of these practices on root carbon input as well as SOM spatial heterogeneity still needs to be evaluated. Moreover, the mechanisms by which grazing management leads to decrease or increase of carbon stocks on different soil types should be determined. Although progress has been made (Bol *et al.*, Chapter 8, this volume), the interaction of carbon and nitrogen, and, in particular, the mechanisms that influence carbon storage and turnover in fertilized grasslands versus unfertilized grasslands and those with legume contribution, need to be elucidated further. More long-term studies are needed to investigate the quantitative impact of land-use change on the contribution of labile and stable SOM types. The optimum ratio of labile versus stable SOM types for different grassland soils has also yet to be determined. In particular, the soil type that supports the grassland has to be taken into account, as SOM stabilization occurs within the mineral matrix of soil.

References

Ammann, C., Flechard, C.R., Leifeld, J., Neftel, A. and Fuhrer, J. (2007) The carbon budget of newly established temperate grassland depends on management intensity. *Agriculture, Ecosystems and Environment* 121, 5–20.

Bardgett, R.D., Wardle, D.A. and Yeates, G.W. (1998). Linking above-ground and below-ground interactions: how plant responses to foliar herbivory influence soil organisms. *Soil Biology and Biochemistry* 30, 1867–1878.

Bitterlich, W., Pöttinger, C., Kaserer, M., Hofer, H., Aicher, M., Tappeiner, U. and Cernusca, A. (1999) Effects of land-use changes on soils along the eastern alpine transect. In: Cernusca, A., Tappeiner, U. and Bayfield, N. (eds) *Land-use changes in European Mountain Ecosystems. ECOMONT – Concepts and Results.* Blackwell, Berlin, pp. 225–235.

Bol, R., Amelung, W. Friedrich, C. and Ostel, N. (2000) Tracing dung-derived carbon in temperate grassland using 13C natural abundance measurements. *Soil Biology and Biochemistry* 32, 1337–1343.

Chabbi, A., Kögel-Knabner, I. and Rumpel, C., (2009). Stabilised carbon in subsoil horizons is located in spatially distinct parts of the soil profile. *Soil Biology and Biochemistry* 41, 256–271.

Cheng, W. and Kuzyakov, Y. (2005) Root effects on soil organic matter decomposition. In: Wright, S. and Zobel, R. (eds) *Roots and Soil Management: Interactions between Roots and the Soil,* Agronomy Monograph No. 48, American Society of Agronomy, Crop Science Society of America, Soil Science Society of America, Madison, Wisconsin, USA, pp. 119–143.

Conant, R.T. (2010) Challenges and opportunities for carbon sequestration in grassland systems. A technical report. *Integrated Cop Management* 9, 1–51.

Conant, R.T., Paustian, K. and Elliott, E.T. (2001) Grassland management and conversion into grassland: effects on soil carbon. *Ecological Application* 11, 343–355.

Cui, X.Y., Wang, W.F., Niu, H.S., Wu, J., Wang, S.P. and Schnug, E. (2005) Effect of long-term grazing on soil organic carbon content in semiarid steppes in Inner Mongolia. *Ecological Research* 20, 519–527.

Dormaar, J.F. and Willms, W.D. (1998) Effect of forty-four years for grazing on fescue grassland soils. *Journal of Range Management* 51, 122–126.

Dungait, A.J., Bol, R., Bull, I.D. and Evershed, R.P. (2009) Tracking the fate of dung-derived carbohydrates in a temperate grassland soil using compound-specific stable isotope analysis. *Organic Geochemistry* 40, 1210–1218.

FAOSTAT (2009) Statistical Database 2007. Food and Agriculture Organisation, Rome.

Fontaine, S., Mariotti, A. and Abbadie, L. (2003) The priming effect of organic matter: a question of microbial competition. *Soil Biology and Biochemistry* 35, 837–843.

Franzluebbers, A.J., Stuedemann, J.A., Schomberg, H.H. and Wilkinson, S.R. (2000) Soil organic C and N pools under long-term pasture management in the Southers Piedmont USA. *Soil Biology and Biochemistry* 32, 469–478.

Gaudinski, J.B., Trumbor, S.E., Devidson, E.A., Cook, A.C., Markewitz, D. and Richter, D.D. (2001). The age of fine-root carbon in three forests of the eastern United States measured by radiocarbon. *Oecologia* 129, 420–429.

Haynes, R.J. and Williams, P.H. (1993) Nutrient cycling and soil fertility in the grazed pasture ecosystem. *Advances in Agronomy* 49, 119–199.

Hoorens, B., Aerts, R. and Stroetenga, M. (2003) Does initial litter chemistry explain litter mixture effects on decomposition? *Oecologia* 137, 578–586.

Houghton, R.A., Hackler, J.L. and Lawrence, K.T. (1999) The US carbon budget: contributions from land-use change. *Science* 285, 574–578.

Jarvis, S.C., Stockdale, E.A., Shephard, M.A. and Powloson, D.S. (1996) Nitrogen mineralisation in temperate agricultural soils: processes and measurement. In: Sparks, D.L. (eds) *Advances in Agronomy*, Vol. LVII, Academic Press, London, pp. 188–237.

Jobbagy, E.G. and Jackson, R.B. (2000). The vertical distribution of soil organic carbon and its relation to climate and vegetation. *Ecological Applications* 10, 423–436.

Klumpp, K., Fontaine, S., Attard, E., Le Roux, X., Gleixner, G. and Soussana, J.-F. (2009) Grazing triggers soil carbon loss by altering plant roots and their contol on soil microbial community. *Journal of Ecology* 97, 876–885.

Krull, E.S., Baldock, J.A. and Skjemstad, J.O. (2003) Importance of mechanisms and processes of the stabilisation of soil organic matter for modelling carbon turnover. *Functional Plant Biology* 30, 207–222.

Lal, R. (2003) Global potential of soil C sequestration to mitigate the greenhouse effect. *Critical Reviews in Plant Sciences* 22, 151–184.

Lal, R. (2004) Soil carbon sequestration impacts on global climate change and food security. *Science* 304, 1623–1627.

Leifeld, J. and Kögel-Knabner, I. (2005) Soil organic matter fractions as early indicators for carbon stock ghanges under different land-use? *Geoderma* 124, 143–155.

Lemaire, G. (1988) Sward dynamics under different management programmes. *Proceedings 12th General Meeting European Grassland Federation, Dublin.* Irish Grassland Association, Belcare, Ireland, pp. 7–23.

Lemaire, G. and Agnusdei, M. (2000) Leaf tissue turnover and efficiency of herbage utilization: In: Lemaire, G., Hodgson, J., de Moraes, A., Nabinger, C. and de F. Carvalho, P.C. (eds) *Grassland Ecophysilogy and Grazing Ecology*, CAB International, Wallingford, UK, pp. 265–287.

Lovell, R.D. and Jarvis, S.C. (1996) Effect of cattle dung on soil microbial biomass C and N in a permanent pasture soil. *Soil Biology and Biochemistry* 28, 291–299.

Majdi, H. and Andersson, P. (2005). Fine root production and turnover in a Norway spruce stand in northern Sweden: effects of nitrogen and water manipulation. *Ecosystems* 8, 191–199.

Martens, D.A., Reedy, T.E. and Lewis, D.T., (2003) Soil organic carbon content and composition of 130-year crop, pasture and forest land-use managements. *Global Change Biology* 10, 65–73.

Mazzanti, A., Lemaire, G. and Gastal, F. (1994) The effect of nitrogen fertilization upon the herbage production of tall fescue swards continuously grazed with sheep. II-Herbage consumption. *Grass and Forage Science* 49, 352–359.

Milchunas, D.G. and Lauenroth, W.K. (1993) Quantitative effects of grazing on vegetation and soils over a global range of environments. *Ecological Monographes* 63, 327–366.

Nierop, K.G.J., Pulleman, M.M. and Marinissen, J.C.Y. (2001) Management induced organic matter differentiation in grassland and arable soil: a study using pyrolysis techniques. *Soil Biology and Biochemistry* 33, 755–764.

Nyborg, M., Molina-Ayala, M., Solberg, E.D., Izaurralde, R.C., Malhi, S.S. and Janzen, H.H. (1997) Carbon storage in grassland soils as related to N and S fertilizers. In: Lal, R., Kimble, J., Follett, R.F. and Stewart, B.A. (eds) *Management of Carbon Sequestration in Soil.* CRC Press, Boca Raton, pp. 421-432.

Parton, W.J., Schimel, C.V., Cole, C.V. and Ojima, D.S. (1987) Analysis of factors controlling soil organic matter levels in Great Plain grasslands. *Soil Science Society of America Journal* 51, 1173–1179.

Ramankutty, N., Evan, A.T., Monfreda, C. and Foley, J.A. (2008). Farming the planet: 1 Geographic dictribution of global agricultural lands in the year 2000. *Global Biogeochemistry Cycles* 22, GB1003.

Rasse, D.P., Rumpel, C. and Dignac, M.-F. (2005) Is soil carbon mostly root carbon? Mechanisms for a specific stabilisation. *Plant and Soil* 269, 341–356.

Rodionov, A., Amelung, W., Peinemann, N., Haumaier, L., Zhang, X., Kleber, M., Glaser, B., Urusevskaya, I. and Zech, W. (2010) Black carbon in grassland ecosystems of the world. *Global Biogeochemistry Cycles* 24, GB3013, doi:10.1029/2009GB003669, 2010

Rumpel, C. and Chabbi, A. (2010) Response of bulk chemical composition, lignin and carbohydrate signature to grassland conversion in a ley-arable cropping system. *Nutrient Cycling in Agroecosystems* 88, 173–182.

Rumpel, C. and Kögel-Knabner, I. (2011) Deep soil organic matter – a key but poorly understood component of terrestrial C cycle, *Plant and Soil* 338, 143–158.

Rumpel, C., Chabbi, A., Nunan, N. and Dignac, M.-F., (2009) Impact of landuse change on the molecular composition of soil organic matter. *Journal of Analytical and Applied Pyrolysis* 85, 431–434.

Sanaullah, M., Chabbi, A., Lemaire, G., Charrier, X. and Rumpel, C. (2010) How does plant leaf senescence of grassland species influence decomposition kinetics and litter compounds dynamics? *Nutrient Cycling in Agroecosystems* 88, 159–171.

Schimel, J.P., Jackson, L.E. and Firestone, M.K. (1989) Spatial and temporal effects on plant microbial competition for inorganic nitrogen in a California annual grassland. *Soil Biology and Biochemistry* 21, 1059–1066.

Schlesinger, W.H. (1977) Carbon balance in terrestrial detritus. *Annual Review of Ecology Systematics* 8, 51–81.

Shields, J.A. and Paul, E.A. (1973) Decomposition of 14C labelled plant material under field conditions. *Canadian Journal of Soil Science* 53, 297–306.

Shukla, M.K., Lal, R. and Ebinger, M. (2006) Determining soil quality indicators by factor analysis. *Soil and Tillage Research*, 87, 194–204.

Six, J., Elliott, E.T., Paustian, K. and Doran, J.W. (1998) Aggregation and soil organic matter accumulation in cultivated and native grassland soils. *Soil Science Society of America Journal* 62, 1367–1377.

Steffens, M., Kölbl, A., Giese, M., Hoffmann, C., Totsche, K.U., Beuer, L. and Kögel-Knabner, I. (2009) Spatial varialbility of topsoils and vegetation in a grazed steppe ecosystem in Inner Mongolia (P.R. China). *Journal of Plant Nutrition and Soil Science*, 172, 78–90.

Swift, M.J., Heal, O.W. and Anderson, J.M. (1979) *Decomposition in Terrestrial Ecosystems.* Blackwell, Oxford, UK, pp. 372.

Xiang, S.-R., Doyle, A., Holden, P.A. and Schimel, J.P. (2008) Drying and rewetting effects on C and N mineralization and microbial activity in surface and subsurface California grassland soils. *Soil Biology and Biochemistry* 40, 2281–2289.

8 Managing C and N in Grassland Systems: the Adaptive Cycle Theory Perspective

Roland Bol, Robert M. Dunn and Emma S. Pilgrim

8.1 Introduction

Agricultural land is one of the largest terrestrial biomes on the planet, covering 28–37% of the earth's land surface (Foley *et al.*, 2005; Porter *et al.*, 2009), and it is expected to further expand over the next decade, driven by an increased demand for both food and bioenergy (Steinfeld and Wassenaar, 2007). This increase in the level of food production must be carried out in a sustainable and equitable manner (Godfray *et al.*, 2010). No longer can we give priority to meeting contemporary human needs at the expense of future requirements (Millenium Ecosystem Assessment, 2005a,b; McIntyre *et al.*, 2009). It is widely accepted that, as the inputs of nutrients to pasture are increased, the efficiency with which they are used is generally decreased and this can be detrimental to the environment (Scholefield *et al.*, 2006). However, it is essential that we learn how to increase food production whilst minimizing impacts on other intermediate and final ecosystem services, so these conflicts and trade-offs can be managed more effectively (Foley *et al.*, 2005; Raudsepp-Hearne *et al.*, 2010; Glendining *et al.*, 2009; Burney *et al.*, 2010). Therefore, tackling such looming food security challenges will require multifaceted and cross-disciplinary approaches. Furthermore, attempts to increased agricultural production will, to some extent, depend on inorganic N and P fertilizers, which are anticipated to get increasingly scarce (P) or costly (N).

8.2 Carbon and Nitrogen Cycling in Grasslands

Grasslands themselves are defined as ecosystems where the dominant vegetation component comprises herbaceous species, with less than 10% tree cover (Jones and Donnelly, 2004). Grasslands may contain up to 343 billion tonnes of C, which is a substantial amount of the world's soil organic C, and nearly twice that which is stored in forests worldwide (FAO, 2010; Rumpel, Chapter 7, this volume). Grassland C and N dynamics directly affect plant and soil quality, and, in grazed systems, also meat and milk quality, as well as the resulting emanating pollution. Through their N_2O, CH_4 and CO_2 greenhouse emissions, grassland systems attain influence on the stability and fate of our future climate. Past poor land management has led to the depletion of grassland soil C stocks, but the implementation of improved management practices, e.g. fertilization, irrigation, sowing of favourable forage legumes and grasses, has locally led to enhanced C uptake, restoring C stocks and sequestering atmospheric CO_2 in grassland soils. Sequestering C makes grasslands more resilient to climate variation and climate change, also increasing soil fertility, hence reducing reliance on external N inputs, creating better balance and greater biodiversity (FAO, 2010). The C cycle in grasslands, like other ecosystems (Neff *et al.*, 2002), is intimately process-linked 'up or downstream' with other nutrient dynamics, particularly N (Grandy and Neff, 2008; Roberts *et al.*, 2008; Bol *et al.*, 2009). For the mechanistic interpretation of transformation in one cycle, analogous information about the other is therefore essential (Bol *et al.*, 2006). Therefore, in a sense any (mitigation) dilemma faced by C in agricultural systems, e.g. 'Shall we hoard it or use it?' (Janzen, 2005), could implicitly be a problem for other nutrients as well. Clearly, continued development of biogeochemical tools that can help improve (simultaneous) understanding of these cycles, e.g. isotope tracing and remote measurement techniques (Boegh *et al.*, 2004; Bol *et al.*, 2006; Amelung *et al.*, 2008), are thus of global environmental importance. Faster computing speeds are now enabling improvements in modelling, especially at field or micro scale, to include more mathematical complexity in the dynamic system simulations of C and N processes (Lu *et al.*, 2006; Ptashnyk *et al.*, 2010). Overall, current modelling trends are clearly towards integrated simulations of multi-nutrient cycles, also incorporating crop, atmosphere, microbial biomass, soil water and other components (e.g. Wu *et al.*, 2007). Simulated C and N dynamics of agricultural systems in current models are commonly composed of algorithms for the decomposition of organic matter (OM) combined with plant uptake, mineralization, or immobilization and may also include denitrification and nitrification, input of plant litter, manure and root exudates. The core algorithms of C and N dynamics are being driven by the C decomposition of OM pools, which may vary between models (e.g. GLEAMS, MAGEC, RZWQM, SPACSYS, CENTURY/DAYCENT, see the review by Shepherd *et al.*, 2011; see also Parsons *et al.*, Chapter 1, this volume). Generally, however, these pools are split into a number of fast decomposition pools (e.g. fresh litter or readily degradable compounds) with a higher C:N ratio and pools with slow(er) decomposition containing more stable material (e.g. humus) with a lower C:N ratio. A recent review of 30 models found that a majority of models (e.g. MOTOR, UK-DNDC, DAYCENT, RZWQM; see the review by Shepherd *et al.*, 2011) do link the C and N cycles by the ratio in which they are found in their organic matter component, thus knowing how much C gives the amount of N and vice versa (Shepherd *et al.*, 2011). For example, in the CENTURY model all the N pools and processes have the same structure as the C pools and processes, with prescribed C:N ratios by pool type. The N rates entering or leaving the pools are adjusted such that the C:N ratio of each pool remains fixed, except for the C:N ratio of the metabolic pool, which is allowed to vary. External N inputs are via atmospheric deposition and fertilizer applications or N_2 fixation (Del Grosso *et al.*, 2009).

8.3 Links between Management Type and C and N Patchiness in Grasslands

Among the main reasons for the intensive use of grasslands are the high natural soil fertility and productive capacity of such lands, which make a substantial contribution to the world's livelihoods and food production (FAO, 2010). However, intensive management has grossly perturbed nutrient cycling, causing system disruption of the key biogeochemical processes. Such disruptions include the modification of stoichiometric controls that drive and enable the system to cascade towards a 'natural' nutrient equilibrium (Sinsabaugh *et al.*, 2009; Taylor and Townsend, 2010). The result is the creation of a potentially very leaky system in which nutrients are lost both into water sources, e.g. nitrate (NO_3^-) leaching, and into the air, e.g. ammonia (NH_3) emissions. Similarly, increased emissions of CO_2, methane (CH_4) and nitrous oxide (N_2O) enhance the greenhouse effect and, in the case of N_2O, destroy stratospheric ozone (O_3) (Scholefield *et al.*, 2006). High nutrient inputs may affect the community structure both above and below ground, subsequently impacting on nutrient cycling capabilities (Pilgrim *et al.*, 2010). Grassland systems with higher N inputs have faster N processing rates compared with those of a lower N input, with the consequence that there are greater rates of loss of N from the soil to water and the atmosphere (Scholefield *et al.*, 2006). The spatial distribution and rate of supply of plant-available nutrients in the soil are the most important factors in determining the productivity, as well as sustainability, of managed grassland systems. We know that the soil community plays a pivotal role in nutrient dynamics, particularly in low-input extensive agricultural grasslands, where it is vital for nutrient acquisition by plants (Pilgrim *et al.*, 2010). The supply of nutrients determines both soil fertility and yield and, to a large extent, the potential for deleterious environmental impact due to losses to water and air (Scholefield *et al.*, 2006). However, due to the 'patchiness' on temporal and spatial scales of nutrients, supply and demand are often mismatched (Fig. 8.1), impairing the efficiency of nutrient capture by the plant. Moreover,

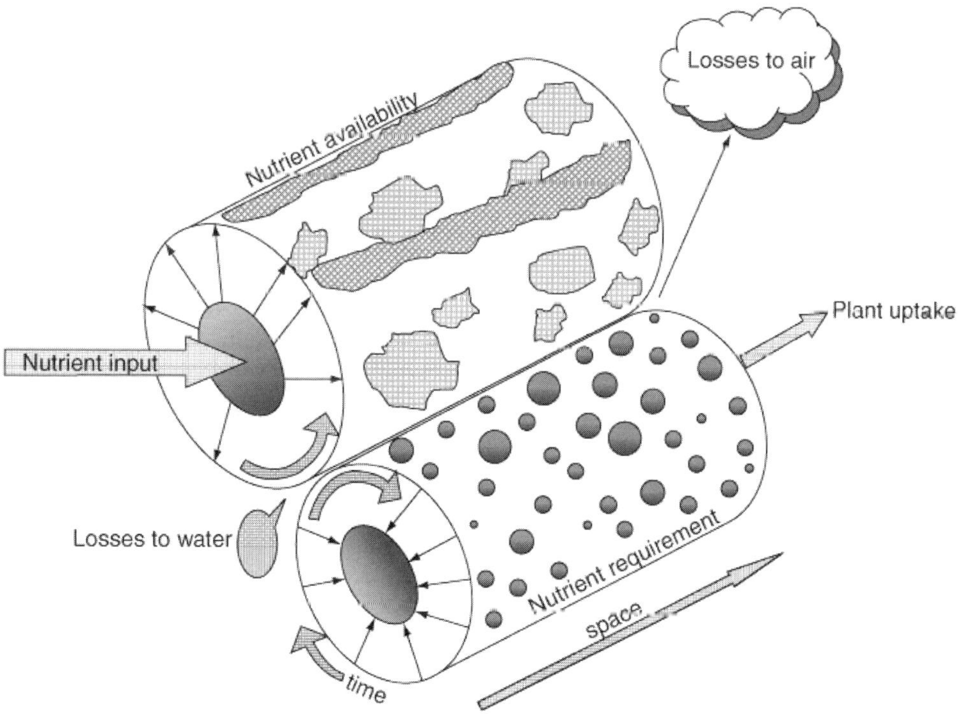

Fig. 8.1. 'Mangle' model of nutrient use efficiency by the plant. The top roller controls soil supply of plant available nutrient, giving rise to spatially (lengthwise) and temporally (circumferential) variable patches of nutrients at different concentrations. This drives the bottom roller (plant uptake) which has holes or sinks for nutrients, which are again spatially and temporally variable and of different strengths. Nutrient use efficiency is determined by the degree of coincidence between supply (patches) and requirements (holes), such that nutrients are lost by being squeezed out of the mangle where coincidence is lacking, upwards to the atmosphere, and downwards to the aqueous environment. Reproduced with permission from Scholefield *et al.* (2006).

the plant must compete with soil microorganisms for nutrients and physicochemical processes that either immobilize nutrients or transport them out of the system (Scholefield *et al.*, 2006; Harrison *et al.*, 2007, 2008). In the next few paragraphs we will discuss how this 'patchiness' of the nutrient supply and demand differs and affects four types of managed grassland systems, which differ in their extent of management levels.

8.4 The Adaptive Cycle Theory Perspective

The early, primary succession stages of any colonization event by plants of a new substrate devoid of soil will lead to a slow accumulation of undecomposed organic matter. Examples of such primary succession are well documented in the ecological literature, and form the basis of many ecological theories of adaptation to, and alteration of, the local environment. The initial community (**Umin**; Fig. 8.2) associated with any vegetated piece of land will inevitably change as the ecosystem matures, with complex links between the fertility of the site, rates of losses and gains in nutrient availabilities and the taxa forming the community. Long-lasting, mature communities come to exist that are dominated by a small number of plant species that competitively exclude many other plant species. In a semi-natural grassland (**Gmin**; Fig. 8.2), biodiversity – here simply defined as the extent of the variety of plant species and their functional characteristics present – will thus, to a large extent, drive and control, through plant 'competition' and 'niche partioning', the level of 'patchiness' (Fig. 8.1) or 'clumping' of resource, C, N and associated nutrients in these grasslands. In these grassland systems where management interventions are few and low-key, this happens mainly through differential effects that vary the input that plant litters and roots have on soil C and nutrient status. In more intensively managed grassland systems (**Gmin →Gmax**; Fig. 8.2), species numbers are

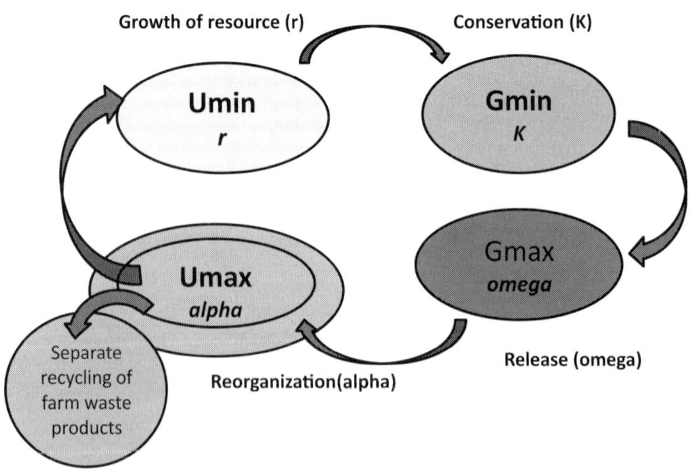

Fig. 8.2. Adaptive cycle of grazed and ungrazed grassland production systems. Umin, grassland systems that are (virtually) ungrazed (no or very little additional inputs); Gmin, low intensity grassland grazing systems (e.g. semi-natural grasslands, hay meadows); Gmax, intensively grazed grasslands (used for beef production or dairy) in which animals are housed for part of the year; Umax, ungrazed grasslands: part of very intensively managed farm animal production systems in which animals are housed all year round. Animal waste products are then (ultimately) mostly re-'used' in subsequent linked production cycles (e.g. for biogas production) which may be off-site. Farm waste products may also be used for controlled 'uniform' spreading to large areas of grasslands or arable land near the farm animal production site.

being increasingly limited to the most productive, ultimately leading to 'single' species pastures. Productivity, at the same time, is increasingly maintained by inorganic and organic fertilizer inputs, providing a spatially 'homogeneous' input of N(PK) or CN(PK). This provides the first level of decoupling of C and N inputs in grassland systems, but they are applied in a managed homogeneous manner (e.g. fertilizer spreading). A further level of decoupling of C and N exists in grazed grasslands (**Gmax**; Figure 8.2), as already discussed by Parsons *et al.* (Chapter 1, this volume), and Rumpel (Chapter 7, this volume), through the activities and behaviours of grazing animals.

The outcome of these activities is that, in any grazed grasslands, the distributions of C and N inputs become spatially heterogeneous. Hence, within these grasslands, C, N and nutrients are spatially spread very patchily, increasing the likelihood of a mismatch between nutrient availability and nutrient requirements by plants and microbes (Fig. 8.1). The mismatch is further enhanced by the fact that both the compositional inputs and spatial distribution of the two excretal returns to grazed pastures by grazing animals, i.e. urine (mainly N) and dung (mainly C), differ (Bol *et al.* 2000, 2004; Dungait *et al.* 2010). In many intensive livestock farming regions, large volumes of liquid cattle slurry, rather than solid manure, have to be disposed of to land. Both options provide opportunities for nutrient losses, particularly of N via NH_3 volatilization and denitrification (Scholefield *et al.*, 2006). Furthermore, as grassland systems get more intensive and animals are housed for increasing periods of the year indoors to aid the control of animal production, there will be more farm-waste products (slurry, manures, dirty water), that have to be stored for varying periods of time, depending on the conditions and legality of spreading these materials back on the land. In short, there is an increasing temporal (and also spatial) discrepancy between animal waste production and its application to grassland systems. Nevertheless, animal wastes are generally homogeneously applied at a specific rate to large areas of the farm, counteracting the increasing patchiness that occurs in more intensive farming systems as more animals are stocked per area of land (compared with less intensive, low-stocking-rate grassland systems used for conservation farming or cutting for silage and hay). All these factors mean that, in the grazed grassland, the overall incorporation of nutrients from herbages into milk or meat by ruminants is inefficient. Globally, N efficiency is 20% for pigs, 34% for poultry, 40% for dairy and just 5% for beef production, and C efficiency, at ~30% (see Parsons *et al.*, Chapter 1, this volume), is not much better.

Only management practices are able to alter the level of nutrient input, the proportion of the year the animals are grazing or fed indoors, the type and quality of diet fed, the type of animal, and the methods of manure storage and disposal to land (Scholefield *et al.*, 2006). Though it is unlikely or, in fact, impossible that the efficiency of plant uptake of available nutrients is ever likely to be 100%, it is a widely held belief that nutrient inefficiency could be improved by employing management techniques to reduce nutrient losses. Clearly, the UK, like many countries, has the need to increase agricultural productivity, hence several new production systems have been identified. To provide some context, it is important to understand that agricultural activities cover about three quarters of the UK land area and produce just less than two thirds of the available food in the UK. However, the UK is not self-sufficient for agricultural production, not even for those products that can be grown in the UK. Indeed, the levels of self-sufficiency have declined from just over 80% in 1991 to less than 60% in 2009. Although, there is acceptance that the UK will never again be 100% self-sufficient as in the 1930s, there is a future need to maintain and increase sustainable agricultural outputs. The best indicator for productivity in the UK is farm income (Defra, 2009). However, between 1998 and 2011, the income for farmers in the UK fell by 8.7%, to an income of less than £21,000 per full time person (well below the UK average income). There has been no significant increase in agricultural land over the past few years. Indeed, dairy and beef herds fell by ~3% in 2009. Dairy farmers are leaving the profession or changing to other ways of farming, leading to potential issues on milk production, not helped by the squeeze on milk prices limiting profits in the current distribution structure – prices that are too largely controlled by supermarkets. Reductions in overall production costs may be possible in larger production systems, due to economies of scale and increased integration of the components of production and waste disposal at the farm scale (**U max + separated farm waste recycling**). One such system, zero grazing, may enable the establishment of dairy farms up to one or two orders of magnitude larger than those of the current 'average' dairy farm (79 cows per unit); more in line with some of the dairy-farming systems in the USA, which can have up to 20,000 cows per unit. In zero-grazing systems, animals are housed all year round and are fed indoors with a mixture of concentrates and fresh cut (not silage) grass. As farm resources would be controlled, with respect to spreading, recycling or disposal, this would allow better integration of nutrient availability versus nutrient requirements in such grassland systems. This would help improve nutrient efficiency while avoiding the pitfalls of the mismatch of supply and demand (Fig. 8.1) that can occur in grazing systems in which animals are housed for parts of the year. There is currently much debate on the largest versions of this type of grazing system; dealing with issues of animal welfare, slurry disposal and the need for important concentrates, this discussion will be ongoing for some time. It should, however, be noted that individual smaller farms are already successfully employing the zero-grazing system in the UK, although some are still being put off by the high levels of investment required for the farm equipment needed to undertake this way of farming.

Clearly, for any one locale, two or more alternative stable states, with seemingly similar nutrient budgets, may come to exist. Interactions between members of the community will influence which state comes into being, and affect how long it will persist. One stable state will switch to another by virtue of a shift in the population size of a species (or guild of species) that has a profound influence on the ecosystem (May, 1977) or, as demonstrated, through a change in management strategies for grassland systems (e.g. **Gmin** to **Gmax** or **Gmax** to **Umax**). Within terrestrial ecosystems, this change could be due to predation pressures from an outbreaking species, or ecological release from other trophic interactions (Carson and Root, 2000). A re-organization of the community then takes place in which species that were formerly dominant will become rare and vice versa. Also, even the most carefully chosen agricultural cultivars will not be as capable of adapting to changes in the local environment as the native vegetation located within the area of land being farmed. This will partly be due to the lack of genotypic variability within the cultivars, and the lack of an evolutionary history between the plant and the local environment, especially where a plant is being moved to an area outside of its native geographical range. Re-colonization of the cleared land by the native species or ecotypes will inevitably occur, possibly to the detriment of important agricultural outcomes such as primary production or nutrient use efficiency (NUE). The resilience of the agriculturally desirable community to change will differ from one location to another, according to the varying conditions of the site at the point when a switch was made from the natural to the agriculturally desired community, and the history of the complex of factors that influence the ecosystem.

Scheffer *et al.* (2001) have suggested that, instead of focusing on the prevention of shifts within the ecosystem, from a desired to an undesired state, more emphasis should be placed on the natural resilience of the community. Measurements of important factors such as nutrient stocks, soil properties and biomass of long-lived organisms will allow steps to be put in place that will prevent the slow degradation of the ecosystem's resilience, allow stochastic events such as droughts, floods and invasion by novel species to be rebuffed, and allow the desired agriculturally useful ecosystem to persist, whilst still being environmentally benign and socially acceptable. Indeed, there are various theories that look at the resilience and critical transitions in nature and society; most of these look in a deductive manner at the emergent patterns in complex systems (Scheffer, 2009). However, one theory, the adaptive cycle theory, looks at the issue from an inductive perspective, i.e. rather than showing the implications of a particular mechanism, this theory aims to summarize consistent patterns that may be recognized in the dynamics of systems, ranging from ecosystems to societies and companies.

The adaptive cycle concept of the heuristic 'Panarchy' framework may thus help us to develop a holistic view (Fig. 8.2) for grassland systems and their management structures, in order to achieve the aims of increasing agricultural production whilst minimizing detrimental effects on the environment. Panarchy is a term devised to describe evolving hierarchical systems with multiple interrelated elements (for details see www.resalliance.org; Gunderson and Holling, 2001). Panarchy is the structure in which such systems, including those of nature (e.g. forests) and of humans (e.g. capitalism), as well as combined human–natural systems (e.g. institutions or groups that govern natural resource use, such as the farming community), are interlinked in continual adaptive cycles of growth, accumulation, restructuring and renewal, across a range of scales. Its essential focus is to rationalize the interplay between change and persistence, some of which is predictable and some of which is unpredictable.

The case we put forward is that if increasing productivity is associated with the improved control and reduction of the 'patchiness' of nutrient resources available to plants, then input/output/losses in grasslands may be further optimized. Ultimately, this would be achieved though a bifurcation of grass/animal production and the animal waste (recycling) components of grassland production systems. This transition would occur in the backloop of the adaptive management, from *omega* to *alpha* (Fig. 8.2), a phase of reorganization leading to renewal and periods that create opportunities for innovation (www.resalliance.org). Interestingly, this transition occurs at the hiatus between research focusing on intensive grassland production and that focusing on animal production. Researchers in these two fields may not always be working in close collaboration, and may miss taking advantage of these opportunities for innovation of the two production systems. A similar schism between (conservation) ecology and grassland production research occurs on the foreloop (Fig. 8.2), from *r* to *K*, which is the slow, incremental phase of growth and accumulation. During the slow sequence from exploitation to conservation, connectedness and stability increase and a capital of nutrients and biomass (in ecosystems) is slowly accumulated and sequestered. Competitive processes lead to a few species becoming dominant, with diversity retained in residual pockets preserved in a patchy landscape. While the accumulated capital is sequestered for the growing, maturing ecosystem, it also represents a gradual increase in the potential for other kinds of ecosystems and futures. Within this part of the foreloop the grassland of system is being developed, more by anthropogenic influence through management strategies than via natural ecosystem development, for strong (productivity) performance. This, in a sense, is building a deep basin of attraction (stability and efficiency), but it also implies rigidity in the management structure and its ability to adapt (Scheffer, 2009). It is only when a critical 'crisis' occurs and a massive transformation of the system is required that a rapid reorganization of the system will take place (backloop).

We are aware that, in the real world, minimum acceptable standards have to be reached in a range of agricultural outputs (e.g. primary production, greenhouse gas production, dirty water, loss of soil fertility, erosion of soil particulates and loss of biodiversity) before any realistic viewpoint may be taken as to whether such a transition is truly favourable. Only then can one manipulate the system to improve on these minimum standards. At this stage, the evaluation of the adaptive management cycle is mostly only conceptual (see Fig. 8.2). We shied away from implementing any firm set rules. Indeed, the current model of the adaptive cycle is only meant to stimulate thought. It focuses attention upon processes of destruction and reorganization, which are often neglected in favour of growth and conservation (in traditional ecology).

Agricultural land has always come from reclaimed natural landscapes, and been altered to suit the farming practice being employed. As such, early stages of agriculture could be thought of as acting as the driver, in causing a shift from one state to another. No manager would want their grassland to switch to another state that would offer less to the production of foodstuffs for humans, and modern farming practices have evolved to try to maintain favourable farming conditions on the land. However, contemporary farming needs to meet multiple objectives. Land managers will be constrained in the options available to them, by social, environmental and economic drivers. A system of management therefore has to be put in place that can deal with multiple objectives that may shift over time, and will also need to be flexible enough to meet freshly emerging objectives should legislation or other influencing factors force a change upon the management strategy employed. Gunderson and Holling (2001) have suggested the use of panarchy as a suitable management system, being a term devised to describe evolving hierarchical systems with multiple interrelated elements. Such a management system will clearly allow multiple objectives (e.g. increasing agricultural production together with optimization of other ecosystem services, as well as increased resilience to climate change) to be met, with the ability to rapidly adapt to any change in any part of the grassland system. Hence, we believe in, and recommend, further exploration of the applicability of the adaptive cycle and panarchy concepts in future grassland studies, as a worthwhile endeavour.

References

Amelung, W., Brodowski, S., Sandhage-Hofmann, A. and Bol, R. (2008) Combining biomarker with stable isotope analysis for assessing the transformations and turnover of soil organic matter. *Advances in Agronomy*, 100, 155–250.

Boegh, E., Thorsen, M., Butts, M.B., Hansen, S., Christiansen, J.S., Abrahamsen, P., Hasager, C.B., Jensen, N.O., van der Keur, P., Refsgaard, J.C., Schelde, K., Soegaard, H. and Thomsen, A. (2004) Incorporating remote sensing data in physically based distributed agro-hydrological modelling. *Journal of Hydrology* 287, 279–299.

Bol, R., Amelung, W., Friedrich, C. and Ostle, N. (2000) Tracing dung-derived carbon in temperate grassland using C-13 natural abundance measurements. *Soil Biology and Biochemistry* 32, 1337–1343.

Bol, R., Petersen, S.O., Christofides, C., Dittert, K. and Hansens, M.N. (2004) Short-term N_2O, CO_2, NH_3 fluxes, and N/C transfers in a Danish grass-clover pasture after simulated urine deposition in autumn. *Journal of Plant Nutrition and Soil Science* 167, 568–576.

Bol, R., Eriksen, J., Smith, P., Garnett, M.H., Coleman, K. and Christensen, B.T. (2006) The natural abundance of ^{13}C, ^{15}N, ^{34}S and ^{14}C in archived (1923–2000) plant and soil samples from the Askov long-term experiments on animal manure and mineral fertilizer. *Rapid Communications in Mass Spectrometry* 19, 3216–3226.

Bol, R., Poirier, N., Balesdent, J. and Gleixner, G. (2009) Molecular turnover time of soil organic matter in particle-size fractions of an arable soil. *Rapid Communications in Mass Spectrometry* 23, 2551–2558.

Burney, J.A., Davis, S.J. and Lobell, D.B. (2010) Greenhouse gas mitigation by agricultural intensification. *Proceedings of the National Academy of Sciences USA* 107, 12052–12057.

Carson, W.P. and Root, R.B. (2000) Herbivory and plant species coexistence: community regulation by an outbreaking phytophagous insect. *Ecological Monographs* 70, 73–99.

Defra (2009) *Agriculture in the UK* (http://www.defra.gov.uk/evidence/statistics/foodfarm/general/auk/latest/documents/AUK-2009.pdf)

Del Grosso, S.J., Ojima, D.S., Parton, W.J., Stehfes, E., Heistemann, M., DeAngelo, B. and Rose, S. (2009) Global scale DAYCENT model analysis of greenhouse gas emissions and mitigation strategies for cropped soils. *Global and Planetary Change* 67, 44–50.

Dungait, J.A.J., Bol, R., Lopez-Capel, E., Bull, I.D., Chadwick, D., Amelung, W., Granger, S.J. and Manning, D.A.C. (2010) Applications of stable isotope ratio mass spectrometry in cattle dung carbon cycling studies. *Rapid Communications in Mass Spectrometry* 24, 495–500.

FAO (2010) *Challenges and Opportunities for Carbon Sequestration in Grassland System*. Technical report on grassland management and climate change mitigation, compiled by Conant, R.T. Food and Agriculture Organization, Rome.

Foley, J.A., Defries, R., Asner, G.P., Barford, C., Bonan, G., Carpenter, S.R., Chapin, F.S., Coe, M.T., Daily, G.C., Gibbs, H.K., Helkowski, J.H., Holloway, T., Howard, E.A., Kucharik, C.J., Monfreda, C., Patz, J.A., Prentice, I.C., Ramankutty, N. and Snyder, P.K. (2005) Global consequences of land use. *Science* 309, 570–574.

Glendining, M.J., Dailey, A.G., Williams, A.G., van Evert, F.K., Goulding, K.W.T. and Whitmore, A.P. (2009) Is it possible to increase the sustainability of arable and ruminant agriculture by reducing inputs? *Agricultural Systems* 99, 117–125.

Godfray, H.C.J., Beddington, J.R., Crute, I.R., Haddad, L., Lawrence, D., Muir, J.F., Pretty, J., Robinson, S., Thomas, S.M. and Toulmin, C. (2010) Food security: the challenge of feeding 9 billion people. *Science* 327, 812–818.

Grandy, A.S. and Neff, J.C. (2008) Molecular C dynamics downstream: the biochemical decomposition sequence and its impact on soil organic matter structure and its function. *Science of the Total Environment* 404, 297–307.

Gunderson, L. and Holling, C.S. (2001) *Panarchy: Understanding Transformations in Systems of Humans and Nature*, Island Press, Washington DC, USA.

Harrison, K.A., Bol, R. and Bardgett, R.D. (2007) Preferences for different nitrogen forms by coexisting plant species and soil microbes. *Ecology* 88, 989–999.

Harrison, K.A., Bol, R. and Bardgett, R.D. (2008) Do plant species with different growth strategies vary in their ability to compete with soil microbes for chemical forms of nitrogen? *Soil Biology and Biochemistry* 40, 228–237.

Janzen, H.H. (2005) The soil carbon dilemma: Shall we hoard it or use it ? *Soil Biology and Biochemistry* 38, 419–424.

Jones, M.B. and Donnelly, A. (2004) Carbon sequestration in temperate grassland ecosystems and the influence of management, climate and elevated CO_2. *New Phytologist* 164, 423–439.

Lu, Y.Z., Li, B.G. and Cui, Y. (2006) Micro-scale spatial variance of soil nutrients under different plant communities. *Scientia Agricultura Sinica* 39, 1581–1588.

May, R.M. (1977) Thresholds and breakpoints in ecosystems with a multiplicity of stable states. *Nature* 269, 471–477.

McIntyre, B.D., Herren, H.R., Wakhungu, J. and Watson, R.T. (2009) *Agriculture at a Crossroads: International Assessment of Agricultural Science and Technology for Development: Vol. IV North America and Europe*. Island Press, Washington, DC.

Millenium Ecosystem Assessment (2005a) *Ecosystems and Human Well-being: Current States and Trends*. Island Press, Washington, DC.

Millenium Ecosystem Assessment (2005b) *Ecosystems and Human Well-Being: Biodiversity Synthesis*. World Resources Institute, Washington, DC.

Neff, J.C., Townsend, A.R., Gleixner, G., Lehman, S.J., Turnbull, J. and Bowman, W.D. (2002) Variable effects of nitrogen additions on the stability and turnover of soil carbon. *Nature* 419, 915–917.

Pilgrim, E.S., Macleod, C.J.A., Blackwell, M.S.A., Bol, R., Hogan, D.V., Chadwick, D.R., Cardenas, L., Misselbrook, T.M., Haygarth, P.M., Hobbs, P., Jarvis, S., Hodgson, C., Murray, P.J. and Firbank, L.G. (2010) Interactions among agricultural production and other ecosystem services delivered from European Grasslands. *Advances in Agronomy* 109, 117–154.

Porter, J., Costanza, R., Sandhu, H., Sigsgaard, L. and Wratten, S. (2009) The value of producing food, energy, and ecosystem services within an agro-ecosystem. *Ambio* 38, 186–193.

Ptashnyk, M., Roose, T. and Kirk, G.J.D. (2010) Diffusion of strongly sorbed solutes in soil: a dual-porosity model allowing for slow access to sorption sites and time-dependent sorption reactions. *European Journal of Soil Science* 61, 108–119.

Raudsepp-Hearne, C., Peterson, G.D. and Bennett, E.M. (2010) Ecosystem service bundles for analyzing tradeoffs in diverse landscapes. *Proceedings of the National Academy of Sciences USA* 107, 5242–5247.

Roberts, P., Bol, R. and Jones, D.L. (2008) Free amino sugar reactions in soil in relation to soil carbon and nitrogen cycling. *Soil Biology and Biochemistry* 39, 3081–3092.

Scheffer, M. (2009) *Critical Transitions in Nature and Society*. Princeton University Press, Princeton, NJ.

Scheffer, M., Carpenter, S., Foley, J.A., Folke, C. and Walker, B. (2001) Catastrophic shifts in ecosystems. *Nature* 413, 591–596.

Scholefield, D., Jewkes, E. and Bol, R. (2006) Nutrient cycling budgets in managed pastures. In: Marschner, P. and Rengel, Z. (eds) *Nutrient Cycling in Terrestrial Ecosystems* (Soil Biology; Vol. 10), Springer, Berlin, pp. 215–246.

Shepherd, S., Wu, L., Chadwick, D. and Bol, R. (2011) A review of quantitative tools for assessing the diffuse pollution response to farmer adaptations and mitigation methods under climate change. *Advances in Agronomy* 112, 1–54.

Sinsabaugh, T.L., Hill, B.H. and Shah, J.J.F. (2009) Ecoenzymatic stoichiometry of microbial organic nutrient acquisition in soil and sediment. *Nature* 462, 795–798.

Steinfeld, H. and Wassenaar, T. (2007) The role of livestock production in carbon and nitrogen cycles. *Annual Review of Environment and Resources* 32, 271–294.

Taylor, P.G. and Townsend, A.R. (2010) Stoichiometric control of organic-nitrate relationships from soils to the sea. *Nature* 464, 1178–1181.

Wu, L., McGechan, M.B., McRoberts, N., Baddeley, J.A. and Watson, C.A. (2007) SPACSYS: integration of a 3D root architecture component to carbon, nitrogen and water cycling–model description. *Ecological Modelling* 200, 343–359.

9 Managing Mineral N Leaching in Grassland Systems

Stewart F. Ledgard, J. Luo and R.M. Monaghan

9.1 Introduction

The productivity of grassland systems is highly dependent on the availability of nitrogen (N) in soil. However, grazed grasslands are relatively inefficient in converting the N that is cycling through the soil/plant system into animal products. This conversion efficiency of N consumed by grazing animals into product (e.g. meat, milk and fibre) is typically about 10–30% compared with about 50–80% for arable crops (e.g. Di and Cameron, 2002). N surplus to animal requirements is excreted in localized patches at very high N loading rates in excess of short–medium term N uptake by pasture, thereby representing hot-spots for N loss.

This chapter discusses the main drivers and extent of N losses by leaching of mineral N (inorganic N as ammonium, nitrite and nitrate) and the effect of management practices on reducing N leaching from grasslands, with emphasis on grazed systems.

9.2 Drivers and Extent of N Leaching

Leaching of N refers to the movement of N in drainage water from the biologically active surface soil to below the root zone of grassland plants. Over time and with increasing drainage, the leached N progresses to groundwater where it can feed into surface waterways. Nitrate in groundwater can be of concern for effects on drinking-water quality, while excess mineral N in surface water can affect biodiversity and the health of invertebrates and fish (e.g. Rabalais, 2002).

The extent of N leaching is a product of the surplus of mineral N in soil (largely dependent on N inputs), the mobility and retention of N forms in soil (affected by soil properties), and the level of drainage (determined mainly by rainfall). The effects of climate and soil physical properties on N leaching have been previously reviewed (e.g. Di and Cameron, 2002). In contrast to the effects of soil physical properties, there is a poor understanding of the effects of soil organic matter or carbon (C) status on N leaching. Research is required to better understand soil C and N sequestration in organic matter and the potential to manipulate it and reduce N leaching.

In temperate areas, leaching occurs predominantly over winter. At this time, the removal of soil mineral N by plant uptake or microbial immobilization is low, soil moisture status is often above field capacity, and rainfall leads to net drainage.

N leaching in sheep- and cattle-grazed pastures generally ranges from ~2–50 kg N/ha/year, while intensive dairy systems with high N inputs can leach up to 200 kg N/ha/year (Fig. 9.1). The wide variation in N leaching at a given level of N input in Fig. 9.1 reflects the variable influences of soil, climate and management factors on the leaching process and provides some indication of the potential for management practices to reduce N leaching.

9.2.1 N Inputs

N surplus in grassland soils is largely determined by the magnitude of N inputs exceeding N removal by plant uptake and microbial immobilization. Thus the main drivers of the potential for mineral N surplus in soil and leaching in grassland systems are external N inputs, which occur predominantly from legume N_2 fixation, N fertilizer and brought-in feed, and from recycling N in animal excreta.

Legume N_2 fixation

In mixed legume/grass pastures, legumes have a key role in providing N inputs from fixation of atmospheric N_2 via symbiotic rhizobia in their root nodules (Lüscher *et al.*, Chapter 11, this volume; Fig. 9.1). Fixation of atmospheric N_2 is determined by a feedback regulation mechanism, so that it decreases as soil mineral N levels increase (e.g. Ledgard *et al.*, 2009). This acts to limit typical levels of N_2 fixation in grazed pastures, enhance overall N efficiency and reduce potential for loss.

The contribution of fixed N to the soil mineral N surplus and N leaching via the mineralization pathway is temperature dependent, thereby coinciding with grass N demand and consequently fixed N is used relatively efficiently. However, fixed N returned to soil via animal excreta provides significant loss potential, as described in more detail later.

N fertilizer inputs

Most grassland studies have shown that N fertilizer is used efficiently by grasses and that direct leaching losses are generally low (unless high rates are applied in winter). Moderate–high annual applications of fertilizer N are typically applied in many split additions through the year and can be linked to the pattern of grass N demand and responsiveness.

A recent review of grazed pasture systems comparing N inputs from clover

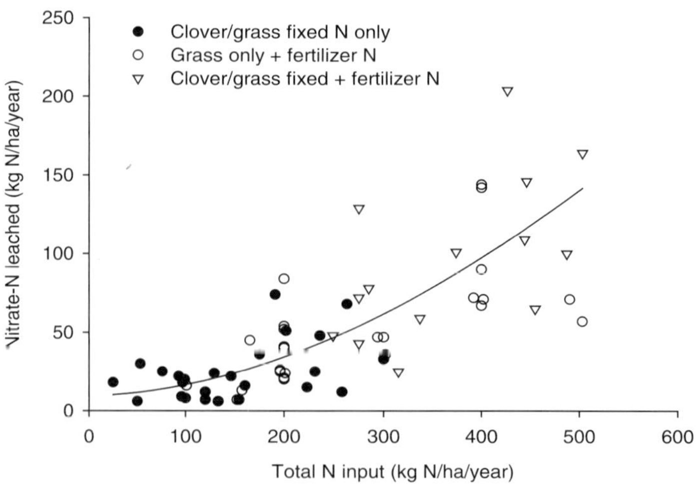

Fig. 9.1. Nitrate leaching from grazed pasture systems as affected by total N input. Data are a summary of studies in Europe and New Zealand (from Ledgard *et al.*, 2009).

N_2 fixation or fertilizer N concluded that N leaching losses were similar under similar total N inputs (Ledgard *et al.*, 2009; also, see Fig. 9.1). This was attributed to the over-riding dominance of animal excreta-N as the main determinant of N leaching losses in grazed pasture systems.

Excreta N inputs

N in animal excreta is returned in localized patches at high loading rates equivalent to 500–1500 kg N/ha. Urine N is rapidly converted to ammonium and nitrate N and can lead to very high N leaching losses, particularly when deposited in autumn or winter. In contrast, the slow release of N from dung results in relatively low N leaching losses.

Figure 9.1 shows that, as fertilizer-N inputs increase, there is an exponential increase in N leaching. While part of this can be due to increased direct leaching of fertilizer N, research has indicated that most of it is associated with higher dietary N concentrations exacerbating N excretion in urine at high rates and higher urine-N leaching. For example, in a farm system receiving 400 kg fertilizer-N/ha/year, the direct leaching of fertilizer N was estimated at about 25 kg N/ha/year, while urine-N leaching was about 75 kg N/ha/year higher than in a nil N farm system comparison (Ledgard *et al.*, 1999). Similar farmlet research and modelling indicate that, if the extra N-boosted pasture had instead been provided as maize silage, it would have increased N leaching by only approximately 5 kg N/ha/year (or 15 kg N/ha/year equivalent after accounting for N leaching from land used to grow the maize silage) (based on Ledgard *et al.*, 2006). The low N concentration in maize silage results in most of the N from feed being converted into milk or excreted in dung, with relatively little in urine.

9.3 Management of N Leaching

Loss of N by leaching from grasslands can be reduced by practices that decrease the N inputs and/or increase the efficiency of N use in the system.

Reducing inputs of fertilizer N can lead to significant reductions in N leaching. For example, Ledgard *et al.* (1999) showed that changing a dairy system from using 200 kg fertilizer-N/ha/year to nil N (both being clover/grass pastures) decreased N leaching by 50% and reduced milk production per ha by 20%. However, with the increasing global production and demand for food, there is a need to maintain or increase production from our grasslands. Thus, emphasis is increasingly placed on approaches that can increase efficiency without incurring production penalties.

In grazed systems, this drive for greater N efficiency needs to target increasing the efficiency of excreta-N recycling, particularly that of urine N. This can occur by manipulating the total amount excreted, the N rate in urine patches, the timing of N excreta deposition to pasture (avoiding high loss risk periods), the form of mineral N retained in the soil and/or the utilization by plants and animals. These aspects will be covered in the following sections by systematically considering mitigation options at soil, plant, animal and farm management levels (summarized in Table 9.1).

9.3.1 Soil N management

Soil N process modifiers have the potential to manipulate specific N transformations in soil, thereby reducing the amount of N in forms susceptible to N leaching. Examples are nitrification inhibitors and compounds that can be used strategically to increase immobilization of mineral N into soil organic N. Nitrogen inhibitors can reduce nitrate N in soil during high-risk periods for losses and potentially increase overall N efficiency. Their greatest advantage in a farm system context occurs when the saved N coincides with a corresponding reduction in N inputs (such as N fertilizer).

The application of a nitrification inhibitor (e.g. dicyandiamide, DCD) in autumn

Table 9.1. Current and potential N mitigation options that act on soil, plant, animal or farm management processes and an indication of their relative effectiveness in reducing N leaching in grazed pastures.

Route of action	Current options and likely reduction in N leaching			Potential options and possible reduction in N leaching		
	Low	Medium	High	Low	Medium	High
Soil		Nitrification inhibitor			Strategic C addition; Constructed wetland or denitrification wall	
Plant	High sugar grass			Greater root activity; High tannins	Inhibitory root exudates	Low N grass
Animal	Higher productivity per animal		Change animal type	Use of lipids or ionophores	Strategic supplementation of diuretic or N modifier	
Management	Optimal N fertilizer management	Lower-N suppl. feed; Strategic winter stand-off	Winter housing and manure management		Targeted mitigation of high N areas	

and winter can reduce N leaching from fertilizer N or urine N by retaining it in the relatively immobile ammonium form. DCD broadcast on soil is used commercially in New Zealand to reduce N losses from pastures grazed by dairy cows, and N leaching can be reduced by 20–50% (e.g. Monaghan *et al.*, 2009).

Another option for reducing surplus mineral N in soil (including urine patches) at high-risk periods is the application of C to immobilize N. Shepherd *et al.* (2010) found that sucrose addition immobilized urine N and decreased N leaching by up to 66%. However, this required very high rates of application and also resulted in significant reduction in pasture production, limiting any practical application of this mitigation option.

When mineral N has leached below the root zone of pastures there may still be opportunities to remove it before it reaches surface waterways. Two options are wetlands and denitrification walls, where C, such as in sawdust, is added in a trench (e.g. Schipper and Vojvodicvukovic, 1998), both having the effect of intercepting the hydrological flow of drainage water to surface waterways. While both options have shown potential to remove nitrate by denitrification, they are highly dependent on the fit to the landscape and the ability to intercept a significant component of the drainage water flows. Additionally, they can lead to N pollution swapping with leached nitrate being converted to N_2O.

9.3.2 Plant mitigation options

Plant characteristics can be selected to increase recovery of mineral N in soil or to reduce risk of N loss by altering N cycling in soil or in animals. Crush *et al.* (2007) showed that grass selected for root proliferation reduced leaching of urine N during high-risk periods.

Use of plants with low N concentration or with specific constituents can influence N cycling. Grasses high in sugar, tannins or saponins can increase N utilization by animals and increase N excretion in the low-risk dung N forms relative to urine N (e.g. Miller *et al.*, 2001). Additionally, tannins or saponins in soil can enhance N immobilization, with potential to decrease N leaching.

Subbarao *et al.* (2006) identified release of natural nitrification inhibitory substances from the roots of *Brachiaria* species, which highlights the potential for plant breeding or genetic modification to incorporate N mitigation traits in key temperate grassland species.

While plants potentially play a central, dominating role in N cycling, none of the above plant characteristics have been tested for N-leaching reduction in grazed systems. It is likely that these effects will be small, except for low-N plants. The N concentration in pasture usually exceeds that of animal requirements and there is considerable potential for reduced N leaching from the use of grasses selected or bred for low N concentration and that won't limit animal production. Thus, research is required on the development of low-N productive grasses.

9.3.3 Animal mitigation options

Differences in animal characteristics can potentially influence the amount of N incorporated in animal tissues and can affect the spread of urinary N. Animals with superior growth rates will convert more N into product and some dairy breeds have a higher N concentration in their milk than others. Similarly, other practices that increase feed conversion efficiency into product, such as increased per-animal productivity and reduced animal replacement rate, will reduce N leaching per unit of animal production. However, net benefits in reducing N leaching are likely to be small (e.g. <10%).

In mixed livestock systems for meat and fibre production, the species of animals differ in their spread of urinary N and therefore in the loading rate of N in urine patches. Hoogendoorn *et al.* (2008) measured twofold lower N leaching from sheep or deer than from beef cows, with the same level of feed and N intake per hectare. This was attributed to the lower concentration of N in sheep and deer urine and greater spread, resulting in lower excretion N loading rates. The potential exists within animal species to achieve greater urine spread through selection or breeding for greater urination frequency but this remains to be tested.

Significant potential also exists to supplement animals directly with compounds that increase N efficiency by improving animal N utilization, increasing urinary N spread or indirectly manipulating soil N processes. The current increase in research focus on reducing animal methane emissions has renewed interest in compounds that influence rumen processes, such as lipids, ionophores and plant extracts, that can also increase utilization of N from feeds (e.g. Hart *et al.*, 2008). However, benefits for reducing N leaching are likely to be small and may not persist over time due to rumen adaptation.

Supplementing animals with salt has been shown to increase water intake, resulting in a diuretic effect with increased frequency of urination, reduced urinary N concentration and increased spread of urine N (Ledgard *et al.*, 2007). This leads to increased recovery by pasture and decreased potential for N leaching. Animals can also be supplemented with N process modifiers that are excreted in urine in an unaltered form and inhibit soil N processes. Ledgard *et al.* (2008) demonstrated that urine patches can be strategically targeted by oral delivery of DCD to animals, with more than 90% of the modifier excreted in the urine, thereby decreasing the quantity needed compared with using broadcast application to soils.

9.3.4 Farm management options

Some of the plant and animal mitigation attributes can be achieved via management practices. For example, farmers can supplement (or substitute) the pasture diet of animals with low-protein-feed sources, down to a level where protein is no longer in excess of animal requirements. This is achieved widely on farms with the use of maize silage, which has a low N concentration (~1.1% N) compared with pasture (~3.0–4.0% N).

One of the most effective management practices for reducing N leaching in areas where animals are grazed all year round is the removal of animals from pasture during late-autumn/winter (e.g. onto stand-off pads or into housing systems). This avoids direct excreta deposition to pasture during high-risk months. The excreta can be collected from the stand-off pads or housing systems, stored and uniformly applied to pasture during periods of low leaching risk. Grazing system studies in New Zealand and USA have shown a 20% decrease in N leaching using winter stand-off systems for 18 h/day (i.e. 6 h grazing/day) (Ledgard *et al.*, 2006) or a 50% decrease from using full winter housing (e.g. Rotz *et al.*, 2009). However, in housed systems the collection, storage and efficient use of manure are critical to retain the large overall reduction in N leaching, but these practices can be associated with corresponding increases in ammonia emissions (i.e. N pollution swapping).

Other farm management practices can also be used to target periods of low or high leaching risk. For example, increasing production through strategic application of fertilizer and manure N (which increases plant N concentration and potential losses) could be restricted to low-risk spring and summer periods. Low-N supplementary feed would most effectively be used in high-risk autumn/winter periods. Ultimately, these practices would be linked to the use of decision support tools that account for N supply from all N sources and align recommended N additions from fertilizer and manure to account for plant N demand, thereby reducing N loss risk. Potentially this could be associated with reduced N inputs, aligned to spatial identification of high N areas and even to avoiding recent urine patches using sensor technology. Similarly, on irrigated soils, the use of precision application for increased water-use efficiency can avoid over-watering and decrease N leaching.

With the increased demand for limiting the effects of agriculture on the environment, the use of mitigation practices is likely to be an integral part of farming in the future. However, the mitigation practices are also likely to be more strategically targeted, both temporally and spatially, for increased cost-effectiveness. Additionally, as agriculture intensifies, there will be a need for the use of multiple mitigations. This will require recognition of where synergies can occur, since in most cases the benefits of multiple mitigations will not be additive, e.g. practices to increase urine-N efficiency in winter, such as the use of diuretics or inhibitors, will be ineffective in systems with winter housing of animals.

9.4 Importance of Whole-system Analyses

Agricultural Life Cycle Assessment (LCA) studies have shown that environmental emissions associated with food and fibre products from grasslands are predominantly from the farm-related stages of the life cycle (e.g. over 90% for N loss to waterways). However, these farm-related stages often include a number of components that need to be accounted for in considering total N losses. For example, dairy studies with supplementation of brought-in maize silage have shown increased milk production with minimal increase in N leaching from grazed pasture, but these ignored the N leaching from the area used for production of the maize silage. When that is accounted for, in land converted out of pasture, there can be a small overall increase in the total (whole-system) amount of N leached per kg of milk produced (e.g. Ledgard *et al.*, 2006). Further analyses have indicated that, if the maize silage production was associated with N mitigation practices (e.g. limited tillage, optimized N fertilizer application, yield optimization and subsequent cover-cropping), it could have reduced whole-system N leaching by up to 30%. Practices to reduce N leaching in grassland/crop rotations have not been discussed in this chapter but were described in detail elsewhere (e.g. Eriksen *et al.*, 2008).

Future dairy farming is likely to include an increasing proportion of very high-producing systems with crop integration and some level of animal housing. Such systems are highly managed and therefore have potential for the integration of a range of mitigation practices to achieve high N use efficiency and low N leaching/kg milk produced. Ultimately, these systems should be based on the use of whole-system analyses and LCA to ensure low N leaching losses and to minimize other environmental emissions (including ammonia, nitrous oxide and total greenhouse gases). In the Netherlands, the De Marke dairy farm examined the potential environmental effects of integrating a range of improved practices, relative to the current, standard, partial housing systems. A combination of higher cow efficiency (more milk/cow, lower replacement rate), greater relative area of maize silage cropping, efficient manure use (optimized storage, injection and timing of application) with reduced N fertilizer use, and greater housing (8 h/day during lactation) resulted in estimated N leaching of 46 kg N/ha/year, compared with 82 kg N/ha/year for standard systems (Aarts *et al.*, 2000; Rotz *et al.*, 2005).

In grassland-only farm systems there are fewer management options to reduce N leaching, and high N efficiency must focus on reducing urinary N losses. This is likely to be associated with a combination of factors, such as optimizing legumes in pasture, use of efficient animal types (e.g. minimizing use of beef cows), using current mitigations where practical, such as strategic winter stand-off and nitrification inhibitor use, strategic timing of operations (e.g. combining restricted N fertilizer use in spring/summer and supplementary low-N feed use in autumn/winter), as well as potential new soil, plant and animal options (Table 9.1) as their efficacy is proven. The greatest benefits will occur from the use of a farm system redesign approach that integrates multiple mitigation practices while accounting for N feedbacks and all N sources that determine total N leaching.

9.5 Conclusions and Implications

Management of N leaching requires knowledge of the key drivers of N cycling and loss and the recognition of the significance of animal urine N as the main source of N leaching in grazed grassland systems. There is a wide range of N mitigation options that focus on decreasing N inputs and/or increasing N efficiency. These may target N processes within the soil, plant or animal, and the current research focus is on the strategic use of N mitigations to increase cost-effectiveness. In practice, the use of N mitigations to increase N efficiency is likely to be of only small–moderate benefit in the longer term because the N retained from reduced leaching remains within the system and will have flow-on effects. However, this benefit can potentially be larger where mitigations account for retained N and coincide with a decrease in N inputs, such as using less N fertilizer or via the natural N-feedback mechanism with legume N_2 fixation.

The importance of N mitigations may vary depending on whether a farm is in an N-sensitive catchment, where there may be a limit placed by regulatory authorities on the amount of N leached per hectare, or in a non-sensitive catchment, in which case the emphasis is more on reducing N leaching per kg of product, for high N efficiency. With future land-use intensification there will be more interest in using multiple mitigations to reduce N leaching and this will require identification of synergies so that different mitigations don't simply target the same N loss process and have little added benefit. The design and optimization of new systems require the use of LCA and system analyses to identify opportunities for reducing several environmental emissions and to ensure that N leaching is decreased on a whole farm system basis.

References

Aarts, H.F.M., Habekotte, G.J. and van Keulen, H. (2000) Nitrogen (N) management in the 'De Marke' dairy farming system. *Nutrient Cycling in Agroecosystems* 56, 231–240.

Crush, J.R., Easton, H.S., Waller, J.E., Hume, D.E. and Faville, M.J. (2007) Genotype variation in root distribution patterns, nitrate interception, and moisture stress response of a perennial ryegrass (*Lolium perenne* L.) mapping population. *Grass and Forage Science* 62, 265–273.

Di, H.J. and Cameron, K.C. (2002) Nitrate leaching in temperate agroecosystems: sources, factors and mitigation strategies. *Nutrient Cycling in Agroecosystems* 46, 237–256.

Eriksen, J., Askegaard, M. and Søegaard, K. (2008) Residual effect and nitrate leaching in grass-arable rotations: Effect of grassland proportion, sward type and fertilizer history. *Soil Use and Management* 24, 373–382.

Hart, K.J., Yañez-Ruiz, D.R., Duval, S.M., McEwan, N.R. and Newbold, C.J. (2008) Plant extracts to manipulate rumen fermentation. *Animal Feed Science and Technology* 147, 8–35.

Hoogendoorn, C.J., Betteridge, K., Ledgard, S.F., Costall, D.A., Park, Z.A. and Theobald, P.W. (2008) Can grazing species be used to manipulate the amount of nitrogen leaching to groundwater? In: *Multifunctional Grasslands in a Changing World*, Vol. I. Guangdong People's Publishing House, Guangzhou, China, pp. 830–831.

Ledgard, S.F., Penno, J.W. and Sprosen, M.S. (1999) Nitrogen inputs and losses from clover/grass pastures grazed by dairy cows, as affected by nitrogen fertiliser application. *Journal of Agricultural Science, (Cambridge)* 132, 215–225.

Ledgard, S., Sprosen, M., Judge, A., Lindsey, S., Jensen, R., Clark, D. and Luo, J. (2006) Nitrogen leaching as affected by dairy intensification and mitigation practices in the Resource Efficient Dairying (RED) trial. In: Currie, L.D. and Hanly, J.A. (eds) *Implementing Sustainable Nutrient Management Strategies in Agriculture*. Occasional Report No. 19. Fertilizer and Lime Research Centre, Massey University, Palmerston North, New Zealand, pp. 263–268.

Ledgard, S.F., Welten, B., Menneer, J.C., Betteridge, K., Crush, J.R. and Barton, M.D. (2007) New nitrogen mitigation technologies for evaluation in the Lake Taupo catchment. *Proceedings of the New Zealand Grassland Association* 69, 117–121.

Ledgard, S.F., Menneer, J.C., Dexter, M.M., Kear, M.J. Lindsey, S. Peters, J.S. and Pacheco, D. (2008) A novel concept to reduce nitrogen losses from grazed pastures by administering soil nitrogen process inhibitors to ruminant animals: a study using sheep. *Agriculture, Ecosystems and Environment* 125, 148–158.

Ledgard, S.F., Schils, R., Eriksen, J. and Luo, J. (2009) Environmental impacts of grazed clover/grass pastures. *Irish Journal of Agricultural Research* 48, 209–226.

Miller, L.A., Moorby, J.M., Davies, D.R., Humphreys, M.O., Scollan, N.D., Macrea, J.C. and Theodorou, K.M. (2001) Increased concentration of water-soluble carbohydrate in perennial ryegrass (*Lolium perenne* L.): Milk production from late-lactation dairy cows. *Grass and Forage Science* 56, 389–394.

Monaghan, R.M., Smith, L.C. and Ledgard, S.F. (2009) The effectiveness of a granular formulation of dicyandiamide (DCD) in limiting nitrate leaching from a grazed dairy pasture. *New Zealand Journal of Agricultural Research* 52, 145–159.

Rabalais, N.N. (2002) Nitrogen in aquatic ecosystems. *Ambio* 31, 102–112.

Rotz, C.A., Taube, F., Russelle, M.P., Oenema, J., Sanderson, M.A. and Wachendorf, M. (2005) Whole-farm perspectives of nutrient flows in grassland agriculture. *Crop Science* 45, 2139–2159.

Rotz, C.A., Soder, K.J., Skimmer, R.H., Dell, C.J., Kleinman, P.J., Schmidt, J.P. and Bryant, R.B. (2009) Grazing can reduce the environmental impact of dairy production systems. *Forage and Grazinglands* doi:10.1094/FG-2009-0916-01-RS.

Schipper, L. and Vojvodicvukovic, M. (1998) Nitrate removal from groundwater using a denitrification wall amended with sawdust - field trial. *Journal of Environmental Quality* 27, 664–668.

Shepherd, M., Menneer, J., Ledgard, S. and Sarathchandra, U. (2010) Application of carbon additives to reduce nitrogen leaching from cattle urine patches on pasture. *New Zealand Journal of Agricultural Research* 53, 263–280.

Subbarao, G.V., Ishikawa, T., Ito, O., Nakahara, K., Wang, H.Y. and Berry, W.I. (2006) A bioluminescence assay to detect nitrification inhibitors released from plant roots: a case study with *Brachiaria humidicola*. *Plant and Soil* 288, 101–112.

10 Greenhouse Gas Emissions from European Grasslands and Mitigation Options

Nina Buchmann

10.1 Introduction

Grasslands are a substantial part of the agricultural landscape. They cover about 70% of the world's agricultural area and – depending on the definitions used – between 22% (Soussana *et al.*, 2007) and 35% (Ciais *et al.*, 2010) of the EU-25 land area. However, our understanding of their biogeochemistry, particularly their greenhouse gas emissions (GHG; CO_2, N_2O, CH_4) still has major gaps, not only for N_2O and CH_4, but also for CO_2 and – most importantly – their interactions with each other and with the water cycle. In this chapter, different measurement techniques for GHG budgets will be introduced and the most important GHG fluxes, their interactions and feedbacks as well as the carbon sequestration potential of grasslands will be discussed. Moreover, several mitigation options will be addressed and future research perspectives proposed.

10.2 Relevance of Grasslands in the European Greenhouse Gas Budget

Grasslands strongly contribute to the European GHG budget, particularly in terms of N_2O emissions from soils (Flechard *et al.*, 2007; Skiba *et al.*, 2009) and CH_4 emissions from livestock (Soussana *et al.*, 2007). Their CH_4 and N_2O emissions almost compensate their rather large soil C sequestration (Schulze *et al.*, 2009). Actually, their soil C sequestration is even higher than that in forests but, due to non-CO_2 GHG emissions, grasslands are considered almost GHG neutral (small GHG sink of -14 ± 18 g Ceq/m^2/year), with a large uncertainty though (Schulze *et al.*, 2009). Similar results were obtained from Ciais *et al.* (2010), who estimated that CH_4 and N_2O emissions offset C sequestration in European grasslands by about 70–80%. Thus, grasslands play a critical role in the European GHG balance and thus also in mitigation efforts.

10.3 Greenhouse Gas Emissions from Grasslands

GHG emissions from grasslands can be measured using various techniques, ranging from soil and leaf chambers (e.g. Flechard *et al.*, 2007) to eddy-covariance (EC) flux measurements (Baldocchi, 2003), both combined with various up-scaling techniques, dependent on the spatial scale of interest (plot, ecosystem, continent or globe). Both techniques have proven their usefulness, and ability to give new insights of biogeochemical cycles and GHG fluxes, in the last decade (e.g. Valentini *et al.*, 2000; Ciais *et al.*, 2005; Soussana *et al.*, 2007; Flechard *et al.*, 2007 and references therein). Particularly high-resolution, EC measurements of net fluxes at the ecosystem scale have increased tremendously over the past few years, although mainly focusing on CO_2 exchange and forest ecosystems (e.g. Aubinet *et al.*, 2000; Baldocchi, 2003, 2008). Non-CO_2 flux measurements over grassland are still scarce, although the number of sites is slowly but steadily increasing (Ciais *et al.*, 2010).

Moreover, in the last couple of years, new cutting-edge technologies for CH_4 and N_2O flux measurements have become available, providing even further opportunities to measure the full suite of GHG fluxes (CO_2, CH_4, N_2O and H_2O_{vapour}). While CO_2 and H_2O_{vapour} fluxes are typically measured with infra-red gas analysers (IRGAs; Aubinet *et al.*, 2000), CH_4 and N_2O fluxes have been typically measured using manually operated or automated chambers combined with gas chromatography or photo-acoustic multi gas analysers (Flechard *et al.*, 2007; Skiba *et al.*, 2009). However, these techniques will slowly but steadily be replaced by high-precision, high-temporal resolution concentration measurements with infra-red laser absorption spectroscopy (Eugster *et al.*, 2007; Ammann *et al.*, 2009).

10.3.1 CO_2 fluxes and carbon sequestration

Using the eddy-covariance method, the net ecosystem CO_2 exchange (NEE) fluxes of many grassland sites have been measured throughout Europe (Ciais *et al.*, 2010). NEE is the small net flux of two large opposing gross fluxes, i.e. photosynthesis and ecosystem respiration. Due to sign conventions, NEE is positive if the ecosystem is a net C source and releases CO_2 to the atmosphere, but negative if the ecosystem is a net C sink and takes up CO_2 from the atmosphere. The CO_2 net flux is thus strongly dependent on environmental conditions (e.g. radiation, temperature, soil moisture availability) affecting plants and soils, but also on vegetation composition and plant physiology (e.g. slow/fast (re) growing plants) and grassland management (e.g. cutting regime, fertilizer applications, irrigation). Therefore, NEE shows pronounced temporal patterns, at diurnal as well as seasonal scales. Typically, CO_2 losses occur during the night due to ecosystem respiration (Fig. 10.1; fluxes with positive signs), while CO_2 uptake occurs during daytime hours (Fig. 10.1; fluxes with negative signs). Abrupt changes seen in these patterns in the figure are due to management (grass cuts; e.g. mid-April, mid-May, end of June, end of August, end of September). The winter months of

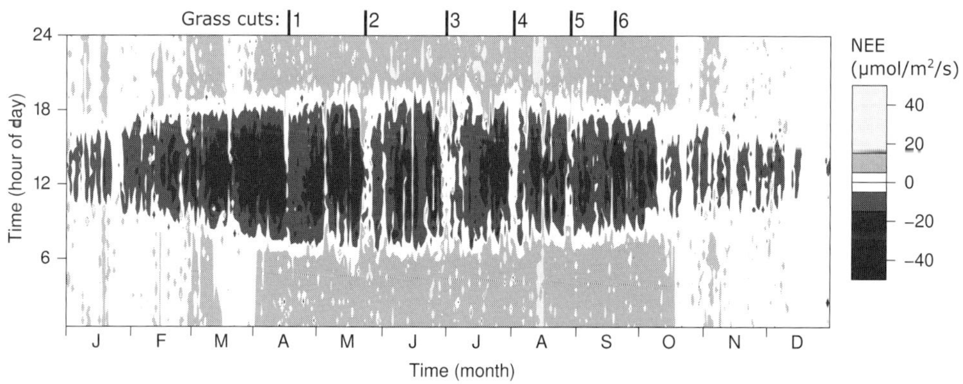

Fig. 10.1. Net ecosystem CO_2 exchange (in μmol $CO_2/m^2/s$) during the year 2007, for an intensively managed grassland in Switzerland. The site, Chamau, is located at 400 m above sea level, in the pre-alpine zone, and is cut six to seven times a year (courtesy of Matthias Zeeman; based on Zeeman *et al.*, 2010).

2006/2007 at the low-land site providing the data for Fig. 10.1 were very mild, so substantial photosynthesis occurred as early as January. After accounting for harvest exports and manure imports of C, this site was a C sink over the entire study year (Zeeman *et al.*, 2010). Partitioning these measured net fluxes into the two gross fluxes is typically done using temperature or light relationships (e.g. Reichstein *et al.*, 2005), but chamber and stable isotope measurements are also used.

Overall, management regimes control the C budgets of grassland sites. Manure applications often represent a large C input into grassland systems (Soussana *et al.*, 2007), sometimes even off-setting harvests almost entirely (Zeeman *et al.*, 2010). On the other hand, extensively managed grasslands have been shown to be rather C sources than C sinks, mainly due to low below-ground C inputs (Ammann *et al.*, 2009). Carbon sequestration in grasslands can only take place in soils, since plants are either cut or grazed. Thus, understanding soil respiration fluxes as well as turnover rates of total soil C stocks or different soil organic matter fractions is of utmost importance if C sequestration potentials are to be estimated under a warmer climate (Davidson and Janssens, 2006).

10.3.2 CH_4 fluxes

Soil CH_4 fluxes strongly depend on environmental conditions. Since CH_4-producing microorganisms are adapted to anaerobic, often wet conditions, highest CH_4 losses are measured from wet, undrained soils, e.g. highly organic soils or peatlands (Mosier, 1998). Typically, soil CH_4 effluxes are highly sporadic, and show no pronounced diurnal pattern as seen, for example, for soil CO_2 fluxes. Furthermore, they interact with management, particularly manure or slurry applications (IPCC, 2007b). On the other hand, soils can also act as CH_4 sinks, as described first for tropical soils (Seiler *et al.*, 1984), later also for temperate prairie soil (Mosier *et al.*, 1991). The magnitude of this sink strongly depends on soil moisture and (nitrogen) management regime (Goulding *et al.*, 1996; Chan and Parkin, 2001), with considerable CH_4 sinks under low moisture conditions and low N fertilizer applications.

However, CH_4 emissions from enteric fermentation in ruminants are by far the largest contributors to total emissions in agriculture, accounting for 27% of global agricultural emissions (IPCC, 2007a). An additional 7% originate from manure treatment. While livestock management, i.e. grazing versus in house as well as feed composition, influences these overall large fluxes, any management changes will strongly impact on animal welfare as well as livestock productivity and thus are heavily debated (Beauchemin *et al.*, 2008).

10.3.3 N_2O fluxes

Agricultural lands, fertilized with organic and inorganic fertilizers (globally 190 million t N and 90 million t N, respectively), lose about 50% of this N due to low plant N use efficiency and inadequate fertilizer practices (IPCC, 2007a). As a consequence, they constitute the largest/only source of N_2O (Robertson *et al.*, 2000), mainly from denitrification and nitrification processes ('Hole in the pipe-model'; Oenema *et al.*, 2005). N_2O fluxes generally increase with N fertilizer application rates, particularly under moist/wet conditions. However, some studies also report N_2O sinks in dry soils (e.g. Ryden, 1981; Neftel *et al.*, 2007), although most of the studies might consider such fluxes as 'background noise'. Since N_2O fluxes are highly sporadic (after fertilization events or heavy rains; Flechard *et al.*, 2007) and do not follow a typical diurnal pattern as known from CO_2 fluxes, continuously measuring N_2O fluxes in the field will provide the most reliable information. Measuring soil N_2O campaign-wise poses the great risk of missing the most relevant periods and peaks (although risk decreases in the long-term, >15 years; N. Brüggemann, 2010, personal communication).

10.3.4 Volatile organic carbon emissions

Volatile organic carbon (VOC) molecules are principal constituents of the plant's defence system, either directly (as e.g. antibiotic, antioxidative compounds, solvents of resin; Loreto and Schnitzler, 2010) or indirectly ('cry for help'; Dicke and Baldwin, 2010), helping plants to adapt or respond to abiotic and biotic stresses. However, although many laboratory experiments confirm the ecological and atmospheric importance of mono-terpene and sesqui-terpene emissions (e.g. Penuelas *et al.*, 1995; Sharkey *et al.*, 2008), little information is available about their seasonal or regional emission patterns. Ciais *et al.* (2010) note the fact that no VOC emission data are available for grasslands and thus conclude that, if model estimates for European grasslands were in the right order of magnitude (50 g C/m^2/year; Lathière *et al.*, 2006), then the net biome productivity (NBP) sink estimate would approach zero.

10.3.5 Greenhouse gas budgets

In order to estimate GHG budgets of grasslands, one needs not only to consider EC measurements of net ecosystem CO_2 exchange (NEE; as done by Gilmanov *et al.*, 2007, 2010), but one also has to account for several inputs (such as fertilizer, manure or slurry applications) as well as outputs (export due to harvest cuts, grazing, fire, volatile organic carbon and methane emissions, erosion, dissolved carbon losses). Since many of these auxiliary import and export fluxes are very difficult to measure, estimates of the GHG budget of grasslands are only available for a few sites (e.g. Ciais *et al.*, 2008). These estimates are generally based on inventory data as well as modelling efforts, rarely on EC measurements (Table 10.1), and often ignore some budget components that are considered small and thus negligible (e.g. fire emissions, erosion losses). While there are a couple of estimates available for the NBP of grasslands (ranging from 36 to 104 g Ceq/m^2/year), estimates for non-CO_2 fluxes, and therefore GHG budgets, are rare (Table 10.1). The most recent compilation of the European grassland GHG balance (Ciais *et al.*, 2010) reports 14–18 g Ceq/m^2/year for European grasslands and points out that the dominance of intensively managed measurement sites probably contributes to an overestimation of the European grassland C sink.

10.4 Mitigation Options

Climate change, due to anthropogenic GHG emissions (IPCC, 2007a), will have pronounced impacts on terrestrial ecosystems, both on managed and unmanaged systems, often with interactions and feedbacks on the climate system (Ciais *et al.*, 2005). International climate change negotiations

Table 10.1 Greenhouse gas budgets of European grasslands.

Source	Methods	NBP	CH_4	N_2O	Total GHG budget
Janssens *et al., 2003*	Stat, Mod	66	–	–	–
Smith et al., 2005	Stat, Mod	36	–	–	–
Schulze *et al., 2009*	EC, Stat, Mod	57	43 (combined)		14
Ciais *et al.*, 2010	EC, Stat, Mod	57–74	48.5	32–52	14–18
Soussana *et al., 2007*	EC	104	32	14	85

Methods used to estimate GHG budgets are given (EC, Eddy-covariance technique to measure net ecosystem GHG exchange; Stat, National/International Statistics; Mod, Modelling efforts). Net biome productivity (NBP) denotes carbon sequestration in g C/m^2/year, CH_4 and N_2O denote CH_4 and N_2O flux in g Ceq/m^2/year, respectively. Total GHG budget is given in g Ceq/m^2/year. Estimates in Janssens *et al.* (2003) are for the European continent, in Smith *et al.* (2005) for EU25 plus Norway and Switzerland, in Schulze *et al.* (2009) as well as in Ciais *et al.* (2010) for EU-25 (CH_4 and N_2O fluxes were recalculated with Global Warming potentials of 8.36 and 126, respectively). In contrast, estimates in Soussana *et al.* (2007) are for nine grassland sites.
–, not available.

try, on the one hand, to find solutions to reduce GHG emissions globally and, on the other hand, to increase sinks for GHG. Different options are available for different sectors at different scales, reaching from a fundamental change in the global energy system to local changes in agricultural management (Stern, 2006; Soussana *et al.*, 2004; Smith *et al.*, 2007).

While GHG emissions from agriculture, mainly CH_4 and N_2O, contribute about 14% to the total global anthropogenic emissions (Stern, 2006), Schulze *et al.* (2009) and Ciais *et al.* (2010) calculated that CH_4 and N_2O emissions almost completely compensate the biological CO_2 sink at the scale of the European continent/ EU-25. A wide array of mitigation options in agriculture is therefore currently being considered (IPCC, 2007b). These options mainly address the reduction or avoidance of GHG emissions, through fossil fuel substitution, optimized fertilizer regimes, enhanced reductions of GHG emissions through specific management practices (e.g. no-tillage, conservation tillage), so they most often focus on the conservation of soil C pools (Lal, 2004). However, the net mitigation effect, according to IPCC (2007b), is not always unequivocal: while fertilization increases productivity in grasslands, thus increases the CO_2 removal, it might also increase N_2O losses to the atmosphere (IPCC, 2007b), but there is little scientific evidence supporting the effect (one star out of three). The same seems to be true for mitigation through species introduction (especially legumes), where controversial effects on N_2O emissions are reported (i.e. low agreement in the literature; IPCC, 2007b and references therein).

In total, IPCC (2007b) considers mitigation options in agriculture, i.e. arable land and grasslands, in the order of magnitude of 6 Gt CO_2 eq/year, mainly by increasing soil C stocks (89%). However, reductions of CH_4 emissions (9%) and of soil N_2O emissions (2%) are also mentioned. McKinsey (2009) calculated the global GHG abatement curves for the agricultural sector (summing up to 4.6 Gt CO_2 eq/year) and concluded that among reductions in CO_2 emissions (representing approximately 75% of total reductions in GHG emissions), reductions in N_2O emissions (approximately 25%) would not only be cost-neutral but actually provide financial benefits to the farmers. However, these GHG abatement curves are based on technical potentials. Taking into account what is realistically achievable (about 10% of the biological potential; Smith *et al.*, 2005), these technical potentials for N_2O mitigation might be as small as < 0.1 Gt CO_2 eq/year.

10.5 Future Research Perspectives

Despite the relevance of grasslands for the global GHG budget and the promising options for climate change mitigation, our understanding of biogeochemical cycles relevant for different gases within the whole suite of important GHG (CO_2, N_2O, CH_4) still has major gaps, not only for N_2O and CH_4, but also for CO_2 and – most importantly – their interactions with each other and with the water cycle. To illustrate these shortcomings: while recent publications indicate that drought affects photosynthesis much more than it affects ecosystem respiration (Ciais *et al.*, 2005; Schwalm *et al.*, 2010), its effects on soil C stocks and microbial soil C turnover are less clear (Davidson and Janssens, 2006). Atmospheric CH_4 concentrations have been stable for nearly one decade (1999–2006) but they started to increase again in 2007 (Dlugokencky *et al.*, 2009). The reasons for these opposing trends are not yet resolved. The global N_2O budget still encounters missing (sink) budget terms (IPCC, 2007a), despite great advancements in science. For example, forest soils have just been shown to act as small but persistent N_2O sinks under drought conditions (Goldberg and Gebauer, 2009), similar to the observations made in a very few studies from grassland soils (Ryden, 1981; Verchot *et al.*, 1999; Flechard *et al.*, 2007; Neftel *et al.*, 2007), contradicting the common view of soils being mainly N_2O sources (Robertson *et al.*,

2000; Schulze *et al.*, 2009). Thus, in order to develop powerful strategies to mitigate climate change, we need a better mechanistic understanding of the biogeochemical cycles in terrestrial ecosystems, particularly grasslands.

Such mechanistic assessment cannot be done with phenomenological observations only, but demands detailed process studies. Different components of the system under study respond very differently to changes in management (or climate), as illustrated above with the response to drought of plants versus microorganisms. Plant species within a given ecosystem react very differently to environmental change (e.g. Thuiller *et al.*, 2005; Gilgen *et al.*, 2010). Time-lags in the response of ecosystems to external drivers might obscure the picture (e.g. Knohl *et al.*, 2005; Bahn *et al.*, 2009). Thus, partitioning the observed net response of the ecosystem into the individual contributions of all the compartments/players involved, taking into account the tight coupling of carbon, nitrogen (N) and water cycles (e.g. Ciais *et al.*, 2005; Sutton *et al.*, 2008), is urgently needed. Management (e.g. manure applications) controls C dynamics in grasslands (Zeeman *et al.*, 2010), but priming effects (see review: Kuzyakov *et al.*, 2000) might counterbalance C uptake of intensively managed grasslands. If the future avenue in climate change mitigation is to use 'green carbon' (TEEB, 2009), then controlled, comprehensive ecosystem-scale experiments are needed (cf. Högberg *et al.*, 2001; Gilgen and Buchman, 2009), where the system is manipulated, taking into account all GHG species, at the same site, at the same time and at high temporal resolution. Only when the response of the most important GHG fluxes to the environment are better understood, on the basis of mechanistic partitioning, including interactions and feedbacks, will the 'evidence' (according to IPCC, 2007b) and the success of mitigation options increase.

Uncertainty arises when only single GHG measurements are carried out, with different methodologies, at sites affected by former land use. Considerable variation in the results may then arise from spatial heterogeneity and differences in experimental set-up and methodology (Gottschalk *et al.*, 2007). Thus, for the future need to introduce adequate land management policies, to improve and to advocate reliable mitigation options, preferentially creating win–win situations, a critical assessment of mitigation strategies needs to include the whole suite of important GHG, and maybe even biodiversity-related mitigation options. Moreover, lateral fluxes need to be taken into account (Schulze *et al.*, 2009; Ciais *et al.*, 2010).

Further avenues might even include mitigation through species introduction (especially legumes). However, we need much closer links between two scientific communities: biogeochemistry/earth system sciences and biodiversity/ecology. Currently, knowledge transfer of scientific achievements between these two disciplines is limited, creating unnecessary time-lags until results are known/applied in the other discipline. It is only recently that first efforts have been taken to overcome this 'traditional' gap, e.g. by Sutton and colleagues (Sutton *et al.*, 2010).

In summary, the assessment of the whole suite of GHG fluxes, including lateral fluxes, supported by auxiliary measurements to account for the tight coupling of biogeochemical cycles and their interactions with the biology of the organisms involved, will provide the mechanistic process understanding we need at the system level to develop the best mitigation strategies for grassland management in the future.

References

Ammann, C., Spirig, C., Leifeld, J. and Neftel, A. (2009) Assessment of the nitrogen and carbon budget of two managed temperate grassland fields. *Agriculture, Ecosystems and Environment* 133, 150–162.

Aubinet, M., Grelle, A., Ibrom, A., Rannik, U., Moncrieff, J., Foken, T., *et al.* (2000) Estimates of the annual net carbon and water exchange of europeran forests: the EUROFLUX methodology. *Advances in Ecological* 182, 452–460

Bahn, M., Schmitt, M., Siegwolf, R., Richter, A. and Brüggemann, N. (2009) Does photosynthesis affect grassland soil-respired CO_2 and its carbon isotope composition on a diurnal timescale? *New Phytologist* 182, 452–460

Baldocchi, D.D. (2003) Assessing the eddy covariance technique for evaluating carbon dioxide exchange rates of ecosystems: past, present and future. *Global Change Biology* 9, 479–492.

Baldocchi, D.D. (2008) Breathing of the terrestrial biosphere: lessons learned from a global network of carbon dioxide flux measurement systems. *Australian Journal of Botany* 56, 1–26.

Beauchemin, K.A., Kreuzer, M., O'Mara, F. and McAllister, T.A. (2008) Nutritional management for enteric methane abatement: a review. *Australian Journal of Experimental Agriculture* 48, 21–27

Chan, A.S.K. and Parkin, T.B. (2001) Effect of land use on methane flux from soil. *Journal of Environmental Quality* 30, 786–797.

Ciais, P., Reichstein, M., Viovy, N., Granier, A., Ogée, J., Allard, V., Buchmann, N., *et al.* (2005). Europe-wide reduction in primary productivity caused by the heat and drought in 2003. *Nature* 437, 529-533.

Ciais, P., Borges, A.V., Abril, G., Meybeck, M., Folberth, G., Hauglustaine, D. and Janssens, I.A. (2008) The impact of lateral carbon fluxes on the European carbon balance. *Biogeosciences* 5, 1259–1271.

Ciais, P., Soussana, J.F., Vuichard, N., Luyssaert, S., Don, A., Janssens, I.A., Piao, S.L., Dechow, R., Lathiere, J., Maignan, F., Wattenbach, M., Smith, P., Ammann, C., Freibauer, A. and Schulze, E.D. (2010) The greenhouse gas balance of European grasslands. *Biogeosciences Discussion* 7, 5997–6050.

Davidson, E.A. and Janssens, I.A. (2006) Temperature sensitivity of soil carbon decomposition and feedbacks to climate change. *Nature* 440, 165-173.

Dicke, M. and Baldwin, I.T. (2010) The evolutionary context for herbivore-induced plant volatiles: beyond the 'cry for help'. *Trends in Plant Sciences* 15, 167–175.

Dlugokencky, E.J., Bruhwiler, L., White, J.W.C., Emmons, L.K., Novelli, P.C., Montzka, S.A., *et al.* (2009) Observational constraints on recent increases in the atmospheric CH_4 burden. *Geophysical Research Letters* 36, L18803.

Eugster, W., Zeyer, K., Zeeman, M., Michna, P., Zingg, A., Buchmann, N. and Emmenegger, L. (2007) Methodical study of nitrous oxide eddy covariance measurements using quantum cascade laser spectrometery over a Swiss forest. *Biogeosciences* 4, 927–939.

Flechard, C.R., Ambus, P., Skiba, U., Rees, R.M., Hensen, A., van Amstel, A., *et al.* (2007) Effects of climate and management intensity on nitrous oxide emissions in grassland systems across Europe. *Agriculture, Ecosystems and Environment* 121, 135–152.

Gilgen, A.K. and Buchmann, N. (2009) Response of temperate grasslands at different altitudes to simulated summer drought differed but scaled with annual precipitation. *Biogeosciences* 6, 2525–2539.

Gilgen, A.K., Signarbieux, C., Feller, U. and Buchmann, N. (2010) Competitive advantage of *Rumex obtusifolius* L. might increase in intensively managed temperate grasslands under drier climate. *Agriculture, Ecosystems and Environment* 135, 15-23.

Gilmanov, T., Soussana, J.F., Aires, L., Allard, V., Ammann, C., Balzarolo, M., *et al.* (2007) Partitioning European grassland net ecosystem CO_2 exchange into gross primary productivity and ecosystem respiration using light response function analysis. *Agriculture, Ecosystems and Environment* 121, 93–120.

Gilmanov, T.G., Aires, L., Barcza, Z., Baron, V.S., Belelli, L., Beringer, J., *et al.* (2010) Productivity, respiration, and light response parameters of world grassland and agroecosystems derived from flux-tower measurements. *Rangeland Ecology and Management* 63, 16–39.

Goldberg, S.D. and Gebauer, G. (2009) Drought turns a Central European Norway spruce forest soil from an N_2O source to a transient N_2 sink. *Global Change Biology* 15, 850–860.

Gottschalk, P., Wattenbach, M., Neftel, A., Fuhrer, J., Jones, M., *et al.* (2007) The role of measurement uncertainties for the simulation of grassland net ecosystem exchange (NEE) in Europe. *Agriculture, Ecosystems and Environment* 121, 175–185.

Goulding, K.W.T., Willison, T.W., Webster, C.P. and Powlson, D.S. (1996) Methane fluxes in aerobic soils. *Environmental Monitoring and Assessessment* 42, 175–187.

Högberg, P., Nordgren, A., Buchmann, N., Taylor, A.F.S., Ekblad, A., Högberg, M.N., *et al.* (2001) Large-scale forest girdling shows that current photosynthesis drives soil respiration. *Nature* 411, 789–792.

IPCC (2007a) *Climate Change 2007: The Physical Science Basis. Contribution of Working Group I to the 4AR of the Intergovernmental Panel on Climate Change.* Solomon, S. et al. (eds.), Cambridge University Press, UK.

IPCC (2007b) *Climate Change 2007: Mitigation. Contribution of Working Group III to the 4AR of the Intergovernmental Panel on Climate Change.* Metz, B. et al. (eds), Cambridge University Press, UK

Janssens, I.A., Freibauer, A., Ciais, P., Smith, P., Nabuurs, G.J., Folberth, G., Schlamadinger, B., Hutjes, R.W.A., Ceulemans, R., Schulze, E.D., Valentini, R. and Dolman, A.J. (2003) Europe's terrestrial biosphere absorbs 7 to 12% of European anthropogenic CO_2 emissions. *Science* 300, 1538–1542.

Knohl, A., Werner, R.A., Brand, W.A. and Buchmann, N. (2005) Short-term variations in $\delta^{13}C$ of ecosystem respiration reveals link between assimilation and respiration in a deciduous forest. *Oecologia* 142, 70–82.

Kuzyakov, Y., Friedel, J.K. and Stahr, K. (2000) Review of mechanisms and quantification of priming effects. *Soil Biology and Biochemistry* 32, 1485–1498.

Lal, R. (2004) Soil carbon sequestration impacts on global climate change and food security. *Science* 304, 1623–1627.

Lathière, J., Hauglustaine, D.A., Friend, A.D., De Noblet-Ducoudré, N., Viovy, N. and Folberth, G.A. (2006) Impact of climate variability and land use changes on global biogenic volatile organic compound emissions. *Atmospheric Chemistry and Physics* 6, 2129–2146.

Loreto, F. and Schnitzler, J.P. (2010) Abiotic stresses and induced BVOCs. *Trends in Plant Sciences* 15, 154–166.

McKinsey (2009) Pathways to a low carbon economy. Version 2 of the global greenhouse gas abatement cost curve. www.mckinsey.com/globalGHGcostcurve.

Mosier, A.R. (1998) Soil processes and global change. *Biology and Fertility of Soils* 27, 221–229.

Mosier, A., Schimel, D., Valentine, D., Bronson, K. and Parton, W. (1991) Methane and nitrous-oxide fluxes in native, fertilized and cultivated grasslands. *Nature* 350, 330–332

Neftel, A., Flechard, C., Ammann, C., Conen, F., Emmenegger, L. and Zeyer, K. (2007) Experimental assessment of N_2O back-ground fluxes in grassland systems. *Tellus* 59B, 470–482.

Oenema, O., Wrage, N., Velthof, G.L., van Groenigen, J.W., Dolfing, J. and Kuikman, P.J. (2005) Trends in global nitrous oxide emissions from animal production systems. *Nutrient Cycling in Agroecosystems* 72, 51–65.

Penuelas, J., Llusia, J. and Estiarte, M. (1995) Terpenoids: a plant language. *Trends in Ecology and Evolution* 10, 289.

Reichstein, M., Falge, E., Baldocchi, D., Papale, D., Valentini, R., Aubinet, M., *et al.* (2005) On the separation of net ecosystem exchange into assimilation and ecosystem respiration: review and improved algorithm. *Global Change Biology* 11, 1–16.

Robertson, G.P., Paul, E.A. and Harwood, R.R. (2000) Forcing of the atmosphere contributions of individual gases to the radiative greenhouse. *Science* 289, 1922–1925.

Ryden, J.C. (1981) N_2O exchange between a grassland soil and the atmosphere. *Nature* 292, 235–237.

Schulze, E.D., Luyssaert, S., Ciais, P., Freibauer, A., Janssens, I.A., *et al.* (2009) Importance of methane and nitrous oxide for Europe's terrestrial greenhouse-gas balance. *Nature Geosciences* 2, 842–850.

Schwalm, C., Williams, C., Schaefer, K., Arneth, A., Bonal, D., Buchmann, N., *et al.* (2010) Assimilation exceeds respiration sensitivity to drought: A FLUXNET synthesis. *Global Change Biology* 16, 657–670.

Seiler, W., Conrad, R. and Scharffe, D. (1984) Field studies of methane emission from termite nests into the atmosphere and measurements of methane uptake by tropical soils. *Journal of Atmospheric Chemistry* 1, 171–186.

Sharkey T.D., Wiberley A.E. and Donohue, A.R. (2008) Isoprene emission from plants: why and how. *Annals of Botany* 101, 5–18.

Skiba, U., Drewer, J., Tang, Y.S., van Dijk, N., Helfter, C., Nemitz, E., *et al.* (2009) Biosphere–atmosphere exchange of reactive nitrogen and greenhouse gases at the NitroEurope core flux measurement sites: Measurement strategy and first data sets. *Agriculture, Ecosystems and Environment* 133, 139–149.

Smith, J., Smith, P., Wattenbach, M., Zaehle, S., Hiederer, R., Jones, R.J.A., Montanarella, L., Rounsevell, M.D.A., Reginster, I. and Ewert, F. (2005) Projected changes in mineral soil carbon of European croplands and grasslands, 1990–2080. *Global Change Biology* 11, 2141–2152.

Smith, P., Martino, D., Cai, Z., Gwary, D., Janzen, H.H., Kumar, P. *et al.* (2007) Greenhouse gas mitigation in agriculture. *Philosophical Transactions of the Royal Society of London, B*, 363, 789–813.

Soussana, J.F., Loiseau, P., Viuchard, N., Ceschia, E., Balesdent, J., Chevallier, T. and Arrouays, D. (2004) Carbon cycling and sequestration opportunities in temperate grasslands. *Soil Use and Management* 20, 219-230.

Soussana, J.F., Allard, V., Pilegaard, K., Ambus, P., Amman, C., Campbell, C., *et al.* (2007) Full accounting of the greenhouse gas (CO_2, N_2O, CH_4) budget of nine European grassland sites. *Agriculture, Ecosystems and Environment* 121, 121–134.

Stern, Sir N. (2006) *The Economics of Climate Change.* Cambridge University Press, Cambridge, UK

Sutton, M.A., Simpson, D., Levy, P.E., Smith, R.I., Reis, S., van Oijen, M. and de Vries, W. (2008) Uncertainties in the relationship between atmospheric nitrogen deposition and forest carbon sequestration. *Global Change Biology* 14, 2057–2063.

Sutton, M.A., Oenema, O., Erisman, J.W., Grennfelt, P., Beier, C., Billen, G., *et al.* (2010) *Managing the European Nitrogen Problem.* Centre for Ecology and Hydrology (CEH) and the Partnership for European Environmental Research (PEER).

TEEB (2009) *The Economics of Ecosystems and Biodiversity for National and International Policy Makers – Summary: Responding to the Value of Nature.* Banson Production, Cambridge, UK.

Thuiller, W., Lavorel, S., Araújo, M.B., Sykes, M.T. and Prentice, I.C. (2005) Climate change threats to plant diversity in Europe. *Proceedings of the National Academy of Sciences USA* 102, 8245–8250.

Valentini, R., Matteucci, G., Dolman, A.J., Schulze, E.D., Rebmann, C., Moors, E.J., *et al.* (2000) Respiration as the main determinant of carbon balance in European forests. *Nature* 404, 861–865.

Verchot, L.V., Davidson, E.A., Cattfinio, J.H., Ackerman, I.L., Erickson, H.E. and Keller, M. (1999) Land use change and biogeo-chemical controls of nitrogen oxide emissions from soils in eastern Amazonia. *Global Biogeochemical Cycles* 13, 31–46.

Zeeman, M.J., Hiller, R., Gilgen, A.K., Michna, P., Plüss, P., Buchmann, N. and Eugster, W. (2010) Management, not climate, controls net CO_2 fluxes and carbon budgets of three grasslands along an elevational gradient in Switzerland. *Agricultural and Forest Meteorology* 150, 519–530.

11 Role and Impacts of Legumes in Grasslands for High Productivity and N Gain from Symbiotic N_2 Fixation

Andreas Lüscher, Jean-François Soussana and Olivier Huguenin-Elie

11.1 Introduction

Under fertile agricultural conditions, monocultures of selected highly productive grass species produce high forage yields when strongly fertilized with nitrogen (N) (Frame, 1991; Daepp *et al.*, 2000). However, synthesis of mineral N fertilizer is highly energy demanding and the application of large quantities of N fertilizer is costly and associated with negative environmental impacts (Jarvis *et al.*, 1995). While symbiotic N_2 fixation from photosynthetic carbon is 'greenhouse gas neutral', greenhouse gas emissions of 3.3 and 8.6 kg CO_2-equivalent (global warming potential over 100 years, according to IPCC in 2007) are calculated per kg of urea N and ammonium nitrate N, respectively, for production and transport (Ecoinvent Centre, 2008). Thus, substitution of fertilizer N by an improved exploitation of symbiotic N_2 fixation in agricultural grasslands would be an important contribution to sustainable and resource-efficient agricultural systems.

11.2 Functional Diversity for High Grassland Productivity

Experiments in relatively species-rich and nutrient-poor grassland model systems showed that biomass production was enhanced in species-rich swards compared with the average monoculture yield (Hooper *et al.*, 2005; overyielding, Trenbath, 1974). Plant communities with higher species richness (species number) are expected to (i) utilize available resources better, due to species niche complementarity; and (ii) have a higher probability of showing positive interspecific interactions; and they may also contain highly productive species that dominate the community (selection effect). Symbiotic N_2 fixation by legumes is considered a key mechanism behind the functional diversity effects observed in grassland systems. Biomass productivity also benefits from diversity in nutrient-rich, intensively managed grasslands (Kirwan *et al.*, 2007; Nyfeler *et al.*, 2009) and, thus, mixtures, rather than heavily fertilized grass monocultures, could improve the provision of ecosystem services. Optimizing the use of mixtures therefore has critical importance for worldwide forage production and sustainable agricultural practice.

To be the most efficient system at a given level of growth resources, a mixture has to be better than the best-performing monoculture (transgressive overyielding, Trenbath, 1974) and not just better than the average monoculture. In a meta-analysis of 44 biodiversity experiments that manipulated plant species richness, Cardinale *et al.* (2007) found that the most diverse

 Grassland Productivity and Ecosystem Services (eds G. Lemaire, J. Hodgson and A. Chabbi)

mixtures on average achieved only 88% of the biomass of the most productive monocultures. Transgressive overyielding occurred in only 12% of the experiments and it took on average about 5 years before transgressive overyielding occurred. In general agricultural practice, the low rate of occurrence of transgressive overyielding may be an important reason for the lack of adoption of mixtures, especially under high input conditions.

Which strategy (monocultures, simple or complex mixtures) is most appropriate to improve resource efficiency and productivity in intensive, fertile grassland systems? The positive diversity–productivity relationship that was observed in biodiversity experiments suggested that herbage production could be maximized by maximizing biodiversity. However, there are several reasons to question this strategy. First, in experiments, mixtures are generally assembled by randomly allocating species from a pool of species with highly differing yield potentials. This may result in a selection effect: diverse communities are more likely to contain and become dominated by highly productive species. This may give the diverse mixtures a yield advantage. However, in practical agriculture, the a priori knowledge of the best-performing species under given growth conditions affects how mixtures are assembled. The best-performing species is selected for a monoculture followed by the second-best-performing species for a bi-species mixture, and so on. As a result, in agricultural systems, more diverse mixtures suffer from containing more species with low yield potential than less diverse mixtures. Second, with increasing species number, experiments showed a decrease in yield gain per species added, i.e. a log–linear relationship between species number (log scale) and yield (linear scale). This may be related to functional redundancy of the new species added, compared with the (increasing) number of species that were already in the mixture. Third, in fertile conditions, the yield of energy- and protein-rich forage strongly increases with an increase in the frequency of defoliation (combined with an adapted increase in fertilization) (Herrmann *et al.*, 2005), but these management conditions also dramatically decrease plant species richness (Gough *et al.*, 2000). Thus, there seems to be a trade-off between high productivity due to intensive management and high species richness. Fourth, optimizing mixed grassland communities requires knowledge not only of the effects of species richness, but also of the effects of species evenness (the second component of species diversity). It seems obvious that species have to grow in significant abundance to modify biomass production of the community. Thus, an efficient strategy may be to mix a low number of species with comparable competitive abilities, resulting in equilibrated mixtures (i.e. with high evenness of species relative abundances), rather than to mix many species with highly differing competitive abilities. So far, there has been only a small number of experiments that address the effects of relative species abundance (evenness) on the whole plant-community production (Gibson *et al.*, 1999). Here, we hypothesize that the best strategy to profit from functional diversity under fertile, intensive grassland conditions would be to mix a relatively low number of selected key species. To develop productive mixtures with high-diversity effects, these species should fulfil the following prerequisites: (i) be highly productive, (ii) be functionally different and (iii) allow mixtures with equilibrated relative species proportions.

11.3 Transgressive Overyielding with Grass–Clover Mixtures: Clover Proportion Matters

A pan-European experiment carried out at 32 sites in 17 countries tested whether high yields and large diversity effects can be achieved, under typical agricultural conditions with mixtures of only four forage plant species, by maximizing evenness, and by selecting a priori highly productive species that are functionally different. Functional diversity was maximized by selecting four species belonging to four different functional groups. Functional groups were defined as the factorial combination of the plant's ability to symbiotically fix N_2 (grasses versus legumes) and the plant's development strategy (fast establishing versus slow establishing but persistent) (Kirwan *et al.*, 2007; Nyfeler *et al.*, 2009). The mixture with the highest evenness (25% of each of the four species) showed the largest diversity effect. Compared with the mean value for the four species in monoculture (overyielding), the increased yield of this mixture was 47% when averaged over all sites (Kirwan *et al.*, 2007). Most remarkably, there was significant transgressive overyielding for most of the mixtures (higher yield than best monoculture yield) in the first harvest year (Kirwan *et al.*, 2007). Preliminary analysis of the data from the mid-European sites indicated that this result persisted during the 3 years of the experiment, with transgressive overyielding being 6%, 20% and 16% in years 1, 2 and 3, respectively (Lüscher *et al.*, 2008). Functional diversity effects were mainly related to the plant's ability to symbiotically fix N_2 (between the functional groups grasses and legumes). Functional diversity effects related to the plant's development strategy (between the functional groups fast and slow developing) were clearly smaller but still significant (Nyfeler *et al.*, 2009).

According to the strong grass–clover functional diversity effect, performance of the mixture depended strongly on its grass:clover composition. At the Swiss site, the maximum yield of the mixtures was reached at a clover percentage in the sward between 36 and 70%, depending on the year and the N fertilizer treatment (Nyfeler *et al.*, 2009). More importantly, at the low (50 kg N/ha/year) and the moderate (150 kg N/ha/year) N level, the yield of mixtures was higher or at least as high as that of the highest yielding monoculture over a wide range of about 15 to 90% clover in the sward (Nyfeler *et al.*, 2009). All these findings clearly demonstrate that grass–clover mixtures have significant advantages when compared with their monocultures that are managed at the same cutting frequency and fertilizer input. Thereby, it does not matter whether grass–clover mixtures are compared with the clover or grass monocultures: they were better than any of the monocultures. These advantages of the mixtures were surprisingly robust: they occurred over the whole gradient of climatic conditions from the Mediterranean to the Sub-Arctic, and they occurred in every experimental year and over a wide range of clover proportions in the mixture.

11.4 Large N Gain from Symbiotic N_2 Fixation: Clover Proportion Matters

Comparing the yield of heavily N-fertilized pure grass swards with the yield of moderately fertilized grass–clover mixtures illustrates the potential N fertilizer savings associated with grass–clover mixtures. At the Swiss site of the pan-European experiment, grass–clover mixtures fertilized with 50 or 150 kg of N/ha/year attained yields in the range of the heavily fertilized (450 kg of N/ha/year) monocultures of the highly productive grass species *Lolium perenne* (14.3 t dry matter (DM)/ha/year) and *Dactylis glomerata* (17.2 t DM/ha/year) if the mixture's clover percentage was between 30 and 80% (Nyfeler *et al.*, 2009). At 50 and 150 kg of N/ha/year, symbiotic N_2 fixation was the main source of N for the grass–clover mixtures across a wide range of clover proportions (Nyfeler *et al.*, 2011) . It reached values of about 300 kg of N/ha/year, thus agreeing with results from earlier studies under productive conditions (e.g. Zanetti *et al.*, 1997). These high amounts demonstrate that additional access to N from the atmosphere can be expected, and is the most important component of grass–clover mixtures' yield advantage in comparison with grass monocultures. This cannot, however, explain the mixtures' yield advantage compared with the legume monocultures.

Most interestingly, the amount of N from symbiosis was maximized not in pure legume stands but in mixtures with 80 and 61% legumes (in years 1 and 2, respectively). Stimulation of symbiotic N_2 fixation of the legumes (per cent legume N derived from symbiosis) when grown in mixture was so strong that a legume percentage in the mixture of only 60 and 37% (in years 1 and 2, respectively) was sufficient to attain the same amount of N from symbiosis as in pure legume stands (Nyfeler *et al.*, 2011). In conclusion, optimal grass–legume mixtures can fix more N_2 from the atmosphere than legume monocultures. This raises the question of how N input by symbiotic N_2 fixation is regulated at the scale of legume plants and the whole sward.

11.5 Feedback Regulation of Symbiotic N_2 Fixation: Effect of Environmental Conditions

Here, we show how available empirical evidence fits a conceptual model in which symbiotic N_2 fixation, at both whole sward level and the level of an individual legume plants is regulated to close the gap between availability of non-symbiotic N sources (mainly available N in the soil) and N demand (Soussana and Hartwig, 1996; Hartwig, 1998; Soussana and Tallec, 2010). Environmental triggers that affect plant growth ultimately alter the demand for nutrients, including N. This concept suggests that generally high productivity that results in high N demand is not a trade-off but a stimulator for high symbiotic N_2 fixation. We briefly discuss here the role of accompanying grasses, N availability and atmospheric CO_2 concentration in feedback regulation of symbiotic N_2 fixation.

The stimulatory effect of accompanying grasses on symbiotic N_2 fixation activity of clover (per cent N derived from symbiosis), reported above, fits this conceptual model well. Symbiotic N_2 fixation of the clover plants was very high in grass-dominated swards at the low and moderate N fertilizer levels (50 and 150 kg of N/ha/year). In these conditions, N availability for the legumes was very low, with most of the N from non-symbiotic sources being taken up by the competitive grass component and very little by the legume component of the mixture (Nyfeler *et al.*, 2011). In legume-dominated swards (> 60% legumes), however, the legumes' activity in N_2 fixation was down-regulated. This was most probably due to two reasons: firstly, the legumes got access to N from non-symbiotic N sources due to the low abundance of the competitive grasses in the sward and, secondly, the N demand of the whole sward was reduced since these clover-dominated swards were less productive than well-balanced grass–clover mixtures (Nyfeler *et al.*, 2011). Both these processes narrowed the gap between N demand and availability of N from non-symbiotic sources in the legume-dominated swards.

Down-regulation of the N_2 fixation activity of clover was also observed if high amounts of fertilizer N were added to the sward (Zanetti *et al.*, 1996, 1997; Nyfeler *et al.*, 2011), strongly increasing availability of non-symbiotic N in the soil. Though it may reduce symbiotic N_2 fixation, mineral N supply does not alter the above-ground net primary productivity of legumes grown as monocultures (Soussana *et al.*, 1995; Hebeisen *et al.*, 1997; Tallec *et al.*, 2008). Thus, feedback regulation is due to increased availability of N from non-symbiotic sources that replaces symbiotically fixed N. In mixed swards, however, an increased availability of mineral N from fertilizer is accompanied by an increased growth rate and competitive ability of grasses, resulting in a decreased growth rate of clover plants (Soussana *et al.*, 1995; Faurie *et al.*, 1996; Tallec *et al.*, 2008); this reduction in N demand of clover leads to a further narrowing of the gap between N demand and availability for clover. As a consequence, at the whole sward level, gain of symbiotically fixed N_2 at high levels of N fertilizer is reduced due to the combination of less clover growth and lower fixation activity.

Elevated atmospheric CO_2 concentrations stimulate photosynthesis, leading to increased plant productivity and modified water and nutrient cycling (Kimball *et al.*, 2002). Experiments under optimal conditions show that doubling atmospheric CO_2 concentration increases leaf photosynthesis by 30 to 50% (Ainsworth and Long, 2005) and canopy photosynthesis by 30% (Casella and Soussana, 1997). This demonstrates that, in ecosystems where no parallel increase in N input occurs, an increase of the C:N ratio of the ecosystem happens when atmospheric CO_2 concentration increases. Many results confirm that N is the major limiting factor in the yield response of grasslands to elevated CO_2 concentrations. Thus, elevated CO_2 concentrations resulted in a decrease in the N nutrition index of grasses (Soussana *et al.*, 1996; Zanetti *et al.*, 1997). Legumes, with their access to the unlimited N source of the atmosphere, have the potential to close such an increased gap between N demand and N availability of the ecosystem. Indeed, in fertilized grasslands, legumes benefit more from elevated atmospheric CO_2 concentrations than do non-fixing species (Hebeisen *et al.*, 1997; Lüscher *et al.*, 1998). This results in a significant increase in symbiotic N_2 fixation due to increased legume proportions in the sward and to increased per cent N derived from symbiosis of the legume plant (Soussana and Hartwig, 1996; Zanetti *et al.*, 1997). An experiment with N_2-fixing and non-fixing lucerne (*Medicago sativa*) showed that the uptake of mineral N from the soil did not increase under elevated atmospheric CO_2 concentrations (Lüscher *et al.*, 2000). The additional harvested N under high atmospheric CO_2 concentrations was derived solely from increased activity of symbiotic N_2 fixation (reviewed in Lüscher *et al.* 2004; Soussana and Lüscher, 2007).

11.6 Conclusions

The results reviewed demonstrate that, in fertile grassland systems, exploitation of functional diversity can be successfully achieved from a modest increase in the number of productive species that differ in their functional traits and especially in mixtures with balanced relative abundance of species. A main component of functional diversity effects in grassland systems is the gain of large amounts of additional N from the atmosphere thanks to the presence of legume species. However, access to atmospheric N_2 is not the only important component of functional diversity because it alone cannot explain the highly significant yield advantage of mixtures compared with legume monocultures. Exploitation of all these diversity effects could reduce both environmental and economic costs. It remains to be examined whether a further increase in species richness will lead to a further increase in mixture performance.

The results confirm the heuristic power of the conceptual model that regulation of symbiotic N_2 fixation in legume-based systems is tightly coupled to the N demand of the system. Interestingly, exploitation of additional N from the atmosphere, by symbiotic N_2 fixation, was stimulated by the grass component of the mixture, leading to mutual grass–legume interactions. This demonstrates that the most efficient strategy to benefit from legumes, for high productivity and N gain from the atmosphere, is to grow them in association with non-N_2-fixing species.

Effects of functional diversity in fertile grasslands were consistent over a broad range of environmental conditions and persisted for 3 years; symbiotic N_2 fixation was further stimulated under the elevated atmospheric CO_2 conditions expected in the future. This suggests a relatively broad applicability of the grass–clover mixtures for sustainable and resource-efficient grassland systems. Knowledge gaps remain about the benefits of legumes under more marginal growth-limiting conditions. In such conditions, benefits may be smaller due to growth limitation reducing N demand and thus leading to down-regulation of symbiotic N_2 fixation.

A strong decrease in legume percentage in the swards in the last year of the pan-European experiment confirms difficulties in maintaining a reasonable abundance of legumes in mixtures (Guckert and Hay, 2001). Developing management practices to establish and maintain balanced mixtures optimized for exploitation of functional diversity is a challenge for future research and development. Combining species or genotypes in mixtures that have comparable competitive abilities would result in more balanced mixtures and combining species with high combining ability would result in better resource exploitation through niche differentiation (Lüscher and Jacquard, 1991).

There is no trade-off between high productivity and high gains of symbiotically fixed N_2 because they are positively linked through N demand involved in the regulation of symbiotic N_2 fixation. Research is needed on the link between legumes, their symbiotic N_2 fixation and other ecosystem services, and this book can contribute to this. Due to the stable C:N ratio of soil organic matter, C-sequestration into the soil ultimately also means N sequestration. This suggests a positive relationship between N gain from symbiotic N_2 fixation and C sequestration (Rumpel, Chapter 7, this volume). Down-regulation of symbiotic N_2 fixation, seen when N demand does not exceed N availability in the soil, suggests less N losses in legume-based systems than in heavily N fertilized systems without legumes (Ledgard *et al.*, Chapter 9, this volume).

Acknowledgements

We are grateful to J. Connolly, J. Finn, J. Nösberger and M. Suter for helpful discussions.

References

Ainsworth, E.A. and Long, S.P. (2005) What have we learned from 15 years of free-air CO_2 enrichment (FACE)? A meta-analytic review of the responses of photosynthesis. *New Phytologist* 165, 351–371.

Cardinale, B.J., Wrigh, J.P., Cadotte, M.W., Carroll, I.T., Hector, A., Srivastava, D.S., Loreau, M. and Weis, J.J. (2007) Impacts of plant diversity on biomass production increase through time because of species complementarity. *Proceedings of the National Academy of Sciences USA* 104, 18123–18128.

Casella, E. and Soussana J.-F. (1997) Long-term effect of CO_2 enrichment and temperature increase on the carbon balance of temperate grass sward. *Journal of Experimental Botany* 48, 1309–1321.

Daepp, M., Suter, D., Almeida, J.P.F., Isopp, H., Hartwig, U.A., Frehner, M., Blum, H., Nösberger, J. and Lüscher, A. (2000) Yield response of *Lolium perenne* swards to free air CO_2 enrichment increased over six years in a high N input system on fertile soil. *Global Change Biology* 6, 805–816.

Ecoinvent Centre (2008) *Ecoinvent Data - The Life Cycle Inventory Data, V2.1.* Swiss Centre for Life Cycle Inventories, Dübendorf, ISBN 3-905594-38-2. Available at: www.ecoinvent.org (accessed January 15, 2010).

Faurie, O., Soussana, J.-F. and Sinoquet, H. (1996) Radiation interception, partitioning and use in grass–clover mixtures. *Annals of Botany* 77, 35–45.

Frame, J. (1991) Herbage production and quality of a range of secondary grass species at five rates of fertilizer nitrogen application. *Grass and Forage Science* 46, 139–151.

Gibson, D.J., Connolly, J., Hartnett, D.C. and Weidenhamer J.D. (1999) Designs for greenhouse studies of interactions between plants. *Journal of Ecology* 87, 1–16.

Gough, L., Osenberg, C.W., Gross, K.L. and Collins, S.L. (2000) Fertilization effects on species density and primary productivity in herbaceous plant communities. *Oikos* 89, 428–439.

Guckert, A. and Hay, R.K.M. (2001) The overwintering, spring growth, and yield in mixed species swards, of white clover in Europe. *Annals of Botany* 88, 667–668.

Hartwig, U.A. (1998) The regulation of symbiotic N_2 fixation: a conceptual model of N feedback from the ecosystem to the gene expression level. *Perspectives in Plant Ecology, Evolution and Systematics* 1, 92–120.

Hebeisen, T., Lüscher, A., Zanetti, S., Fischer, B.U., Hartwig, U.A., Frehner, M., Hendrey, G.R., Blum, H. and Nösberger, J. (1997) Growth response of *Trifolium repens* L and *Lolium perenne* L. as monocultures and bi-species mixture to free air CO_2 enrichment and management. *Global Change Biology* 3, 149–160.

Herrmann, A., Kelm, M., Kornher, A. and Taube, F. (2005) Performance of grassland under different cutting regimes as affected by sward composition, nitrogen input, soil conditions and weather - a simulation study. *European Journal of Agronomy* 22, 141–158.

Hooper, D.U., Chapin, F.S., Ewel, J.J., Hector, A., Inchausti, P., Lavorel, S., Lawton, J.H., Lodge, D.M., Loreau, M., Naeem, S., Schmid, B., Setala, H., Symstad, A.J., Vandermeer, J. and Wardle, D.A. (2005) Effects of biodiversity on ecosystem functioning: a consensus of current knowledge. *Ecological Monographs* 75, 3–35.

Jarvis, S.C., Scholefield, D. and Pain, B.F. (1995) Nitrogen cycling in grazing systems. In: Bacon P.E. (ed.) *Nitrogen Fertilization in the Environment*, Marcel Dekker, New York, pp. 381–419.

Kimball, B.A., Kobyashi, K. and Bindi, M. (2002) Responses of agricultural crops to free-air CO_2 enrichment. *Advances in Agronomy* 77, 293–368.

Kirwan, L., Lüscher, A., Sebastià, M.T., Finn, J.A., Collins, R.P., Porqueddu, C., Helgadottir, A., Baadshaug, O.H., Brophy, C., Coran, C., Dalmannsdóttir, S., Delgado, I., Elgersma, A., Fothergill, M., Frankow-Lindberg, B.E., Golinski, P., Grieu, P., Gustavsson, A.M., Höglind, M., Huguenin-Elie, O., Iliadis, C., Jørgensen, M., Kadziuliene, Z., Karyotis, T., Lunnan, T., Malengier, M., Maltoni, S., Meyer, V., Nyfeler, D., Nykanen-Kurki, P., Parente, J., Smit, H.J., Thumm, U. and Connolly, J. (2007) Evenness drives consistent diversity effects in an intensive grassland system across 28 European sites. *Journal of Ecology* 95, 530–539.

Lüscher, A. and Jacquard, P. (1991) Coevolution between interspecific plant competitors. *Trends in Ecology and Evolution* 6, 355–358.

Lüscher, A., Hendrey, G.R. and Nösberger, J. (1998) Long-term responsiveness to free air CO_2 enrichment of functional types, species and genotypes of plants from fertile permanent grassland. *Oecologia* 113, 37–45.

Lüscher, A., Hartwig, U.A., Suter, D. and Nösberger, J. (2000) Direct evidence that symbiotic N_2 fixation in fertile grassland is an important trait for a strong response of plants to elevated atmospheric CO_2. *Global Change Biology* 6, 655–662.

Lüscher, A., Daepp, M., Blum, H., Hartwig, U.A. and Nösberger, J. (2004) Fertile temperate grassland under elevated atmospheric CO_2 - role of feed-back mechanisms and availability of growth resources. *European Journal of Agronomy* 21, 379–398.

Lüscher, A., Finn, J.A., Connolly, J., Sebastià, M.T., Collins, R.P., Fothergill, M., Porqueddu, C., Brophy, C., Huguenin-Elie, O., Kirwan, L., Nyfeler, D. and Helgadóttir, Á. (2008) Benefits of sward diversity for agricultural grasslands. *Biodiversity* 9, 29–32.

Nyfeler, D., Huguenin-Elie, O., Suter, M., Frossard, E., Connolly, J. and Lüscher, A. (2009) Strong mixture effects among four species in fertilized agricultural grassland led to persistent and consistent transgressive overyielding. *Journal of Applied Ecology* 46, 683–691.

Nyfeler, D., Huguenin-Elie, O., Suter, M., Frossard, E. and Luscher, A. (2011) Grass–legume mixtures can yield more nitrogen than legume pure stands due to mutual stimulation of nitrogen uptake from symbiotic/non-symbiotic sources. *Agriculture, Ecosystems and Environment* 140, 155–163.

Soussana, J.-F. and Hartwig, U.A. (1996) The effects of elevated CO_2 on symbiotic N_2 fixation. A link between the C and N cycles in grassland ecosystems. *Plant and Soil* 187, 321–332.

Soussana, J.-F. and Lüscher, A. (2007) Temperate grasslands and global atmospheric change: a review. *Grass and Forage Science* 62, 127–134.

Soussana, J.-F. and Tallec, T. (2010) Can we understand and predict the regulation of biological N_2 fixation in grassland ecosystems? *Nutrient Cycling in Agroecosystems* 88, 197–213.

Soussana, J.-F., Vertès, F. and Arregui, M.C. (1995) The regulation of clover shoot growing points density and morphology during clover decline in mixed swards. *European Journal of Agronomy* 4, 205–215.

Soussana, J.-F., Casella, E. and Loiseau, P. (1996) Long-term effects of CO_2 enrichment and temperature increase on a temperate grass sward. II. Plant nitrogen budgets and root fraction. *Plant and Soil* 182, 101–114.

Tallec, T., Diquelou, S., Lemauviel, S., Cliquet, J.B., Lesuffleur, F. and Ourry, A. (2008) Nitrogen:sulphur ratio alters competition between *Trifolium repens* and *Lolium perenne* under cutting: Production and competitive abilities. *European Journal of Agronomy* 29, 94–101.

Trenbath, B.R. (1974) Biomass productivity of mixtures. *Advances in Agronomy* 26, 177–210.

Zanetti, S., Hartwig, U.A., Lüscher, A., Hebeisen, T., Frehner, M., Fischer, B.U., Hendrey, G.R., Blum, H. and Nösberger, J. (1996) Stimulation of symbiotic N_2 fixation in *Trifolium repens* L. under elevated atmospheric pCO_2 in a grassland ecosystem. *Plant Physiology* 112, 575–583.

Zanetti, S., Hartwig, U.A., van Kessel, C., Lüscher, A., Hebeisen, T., Frehner, M., Fischer, B.U., Hendrey, G.R., Blum, H. and Nösberger, J. (1997) Does nitrogen nutrition restrict the CO_2 response of fertile grassland lacking legumes? *Oecologia* 112, 17–25.

12 Efficiency of Phosphorus Cycling in Different Grassland Systems

Richard W. McDowell and Peter J.A. Kleinman

12.1 Introduction

It is estimated that current known global reserves of phosphorus (P) will be exhausted in about 50–100 years, with peak P production anticipated in the next several decades (Cordell *et al.*, 2009; Vaccari, 2009). Moreover, excessive past and present use has promoted P loss from land to water and accelerated freshwater eutrophication (Carpenter *et al.*, 1998). This in turn has led to the decline of freshwater for recreation and as a fishery or source of potable water. Economically, there is a considerable cost to both treating water and mitigating P loss from soils enriched with P (Wilson and Carpenter, 1999). For example, due to eutrophication, the cost to treat water for public supply and the loss of biodiversity and amenity in England and Wales is estimated at £100 million annually, while the cost to remediate this damage is about £55 million (Pretty *et al.*, 2003). These facts stress that the efficient use of P is important now and set to increase in the future.

Grassland systems that support livestock farming, account for a considerable proportion of the earth's productive land use (~30%; Steinfeld *et al.*, 2006), but vary greatly in their intensity. Efficiency can be viewed from both an environmental and economic standpoint. The system selected usually reflects economic return, unless regulation restricting P losses from land to water enforce management changes or an alternative system to be adopted. A wide range of environmental benefits are often associated with grazed systems, largely on the premise that, when managed well, such systems mimic natural grassland ecosystems and therefore recycle P more efficiently than do confined operations (Rook *et al.*, 2005; Tallowin *et al.*, 2005). However, inefficiencies exist for people managing grazing systems, who must balance nutrient management with other farm priorities, such as herd health, profitability, quality of life, farm management objectives and socio-economic pressures. Furthermore, there is a growing body of work that describes adverse impacts of grazing systems on soil and water quality (e.g. Stout *et al.*, 2000; McDowell, 2008).

In this chapter we will examine the cycling and loss of P within different grassland systems, focusing on those applicable to dairy farming. Our objective is to shed light on the merits of different systems in different regions and how losses within each system could be managed to make the most efficient use of P.

12.2 Dairy Farming Systems

Many modern pasture-based dairy systems borrow from New Zealand's dairy industry.

In New Zealand nearly all dairy farms are operated as grazed systems using a combination of cool season grasses and legumes designed to yield high-quality forage under the prevailing climate (either with or without irrigation) and soils. The nitrogen-fixing capacity of legumes and nearly year-round growth have resulted in a highly profitable system that also requires little P to maintain a soil-test P concentration optimal for pasture growth (20–40 mg Olsen P/kg). However, the success of New-Zealand-style systems in other parts of the world has been mixed, particularly given difficulties in maintaining high pasture productivity. Often, if forage production has been poor in summer or winter, supplementary feed is required either in the paddock or in the barnyard. Feeds vary from concentrates to total mixed rations (TMRs) to fodder crops. Alternatively, graziers may adapt by including warm-season grasses (e.g. Bermuda grass) or drought-tolerant legumes (e.g. subterranean clover) in their pastures, to improve yield and quality during critical periods.

In contrast to grazed systems, total confinement systems house and milk cows in barns, and tend to be larger (from 100s to 10,000 cows), to take advantage of better economies of scale (Muller, 1992). Milk yield, on a per cow basis, is greater in confinement systems than in grazed systems due to less variable feed quality/quantity, less energy expended in walking and the maintenance of body temperature, and different animal genetics. With the evolution of TMRs, a nutritionally balanced diet can be tailored to the cow. However, P is often added to the TMR as an insurance policy against poor health. Any P fed in excess of that required by the cow is not absorbed in the cow's gastrointestinal tract and is excreted in the faeces and increases the risk of P loss in manure spread on land (Dou *et al.*, 2003).

Combining grazed and confined systems, the hybrid dairy relies upon a range of pasture uses, from grazing during the growing season and confinement during the winter to grazing at night, when it's cooler, and feeding stored grain during the day. Many operators of hybrid dairies also use a partial TMR that can correct the nutritional imbalances associated with variable pasture quality and quantity (Bargo *et al.*, 2002). Recently, in New Zealand's heavily pasture-based systems, the use of housing has increased in areas with climatic extremes (Clark *et al.*, 2007). Fortunately, as these farms originated from grazed systems, the quantity of land available for spreading of manure tends to be greater, and in better balance with offtakes, than seen in many hybrid dairies that originated as confined operations.

12.3 Economic Efficiency

Many studies have compared the costs and profitability of different dairy systems. The most profitable system depends upon the climate and resources available (e.g. soil, fertilizer and feed). However, where conditions allow (good spring to autumn pasture growth), intensively managed grazed systems can make more efficient use of assets, labour, supplementary feed (if required) and fertilizer than confined systems (Dartt *et al.*, 1999). Despite producing less milk than confined systems (3–5%), the lower cost of production (about 10–20%) means that intensively managed grazed systems are therefore more profitable (Cunningham, 1993; Elbehri and Ford, 1995). However, if pasture production drops in these systems, for example due to drought, then confined systems may become more profitable by sustaining production.

While climatic variability may be better buffered in the short term by confined compared with grazed systems, confined systems are still subject to profit fluctuations, influenced instead by the price of the raw materials that comprise TMRs and concentrate. Furthermore, policy changes, such as those that led to the diversion of maize to ethanol production in the USA, can dramatically increase the cost of feed (Coyle, 2007). Depending on regional supply chains and demand, this can mean that the cost to produce a litre of milk is the same as the cost of feed (e.g. concentrate; Parker *et al.*, 1992), making the operation unprofitable. However, the cost to produce milk has generally been about half that of concentrate in the USA (Kleinman and Soder, 2008), yielding good economic return when used in hybrid dairies (Soder and Rotz, 2001).

12.4 Environmental Efficiency

12.4.1 Phosphorus cycling

Figure 12.1 shows stylized versions of P cycles within grazed and confined dairy operations. Both systems import fertilizer and/or feed to produce product. The magnitude of P inputs is determined by a combination of factors, including soil chemical and physical properties, the history and intensity of land use, and the ability of the producer to afford and/or recognize the need for P fertilizer (Sibbesen and Runge-Metzger, 1995). In most low-intensity grassland systems, P fertilizer is the only input, while larger proportions of imported P derive from P in supplementary feed and/or manure as production intensity increases. As a rule of thumb, intensification results in greater P losses, as components of the farm system become saturated with P and become leaky. A typical example is the increase in P loss in runoff with soil-test P concentration (McDowell and Sharpley, 2001). For example, Williams and Haynes (1992) measured P in soil, plants and animals from a flood-irrigated trial in New Zealand. The total gains and losses of P for unfertilized pasture were compared with those of pasture that had received 188 or 376 kg P/ha/year as superphosphate for 38 years (Table 12.1). Of the fertilizer P applied (up to 1329 kg P/ha), total soil P increased by 780 kg P/ha, which, over the 38 years of the study, would equate to an annual increase

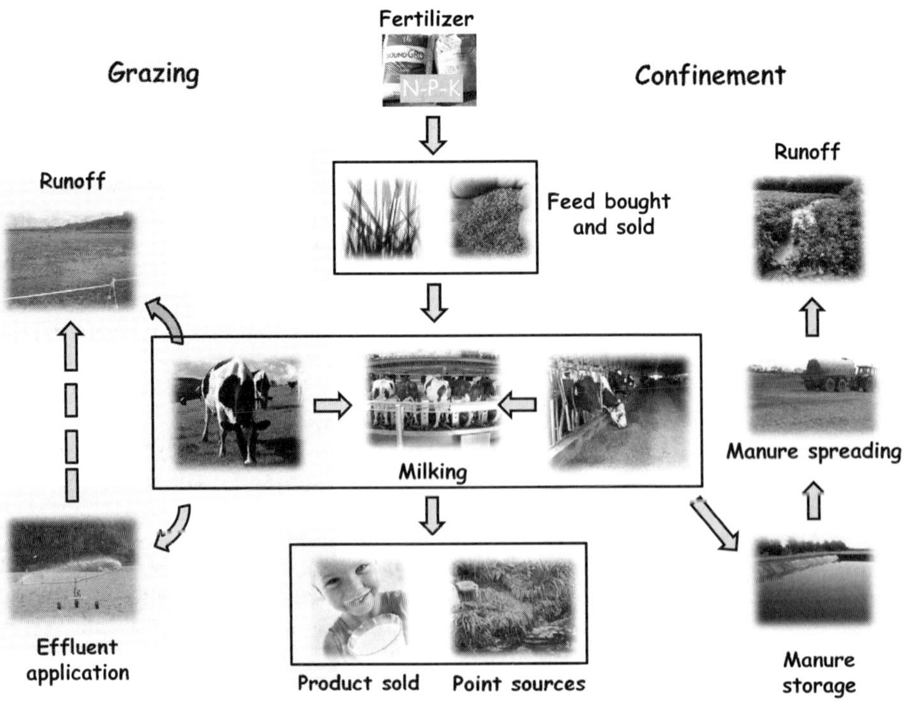

Fig. 12.1. The P cycles within grazed and confined dairy systems.

Table 12.1. Total losses and gains of phosphorus over 38 years for the 188 and 376 kg superphosphate/ha treatments (kg P/ha) (taken from Williams and Haynes, 1992).

Measurement	Wilderness	188 kg P/ha/year	376 kg P/ha/year
Fertilizer P applied	0	664	1329
Total soil (0–20 cm depth)	2020	2360	2800
Change in soil P	–	340	780
P removed in animal products	–	52	58
P present in roots	–	8	12
P present in ungrazed herbage	–	17	14
Total P accounted for	–	417	864
P potentially lost to water	–	247	465

– Not measured.

of ~20 kg P/ha. More importantly, this large surplus was also accompanied by 247 and 465 kg P/ha that could not be accounted for in produce removed or in the various residual soil and plant pools, respectively. The authors concluded that most of the unaccounted P may have been removed in surface runoff, as suspended soil particles and animal manure during flood irrigation, although some leaching of P may have occurred in the shallow stony soil (Udic Dystrochrept).

Although enrichment of topsoil and potential for P loss in runoff occurs in grazed dairy systems, a wide body of literature documents that the P surplus in a grazed system tends to be low (e.g. Hirstov *et al.*, 2005) and that any increase is due to excessive applications of fertilizer, not effluent/manure P. This is mainly due to the larger size of grazed farms and the deposition and wide distribution of excreta around the farm during grazing. While, both systems share the need to spread excreta on land, the effluent applied to pasture in a grazed system is typically more dilute than manure applied to land from a confined system. In confined systems, manure is also commonly applied according to the N requirement of the crop. However, since the N:P ratio in many manures is small and does not match that required by most pasture or row crops, too much P is usually applied (Smith *et al.*, 1998). Since the crop cannot utilize this added P, soil-test P concentrations increase (Toth *et al.*, 2006). Historically, the build-up of P on dairy farms has been exacerbated by the addition of P in feed. Dou *et al.* (2003) surveyed 612 dairy farms in north-east USA and observed that dietary P concentrations for lactating cows ranged from 3.6 to 7.0 g/kg of feed dry matter. The mean concentration (4.4 g/kg) was about a third greater than that recommended by the National Research Council (2001). The reason for overfeeding P was the widely held perception that high P rations improved reproductive efficiency (Tallam *et al.*, 2005). While it has been shown that severe P deficiency can impair reproduction (Goff, 1998), there is no evidence that feeding P in excess of recommended levels benefits reproduction (Wu and Satter, 2000; Wu *et al.*, 2000). Thus, improved ration management represents a cornerstone of many modern nutrient management programmes aimed at dairies.

Table 12.2 summarizes a few examples of P inputs and losses associated with different systems and their management. It is evident from the wide range of losses that significant opportunities exist in confinement and hybrid systems to decrease imported P in feed and make better use of P generated on the farm. Ghebremichael *et al.* (2007) modelled two hybrid-grazing farms at 1.1 and 1.8 cows/ha, comparing the status quo with a strategy to import less P in feed. The results indicated that there was no drop in production but on-farm P surpluses decreased by 43 to 60%. In theory, hybrid-grazing operations have a high degree of flexibility, for example, they can chose to graze if conditions allow or not if pastures are too wet and runoff likely.

Table 12.2. Examples of P inputs and losses in confined, hybrid and grazed dairy systems.

System	Location	P added (kg P/ha)	P lost (kg P/ha)	Reference
Confined				
Dairy manure to:[a]				
Mixed rotations	Sweden	24	0.20	Granstedt (2000)
Maize – winter	NY	108	8.75	Klausner *et al.* (1976)
Maize	WI	100	6.20	Mueller *et al.* (1984)
Maize – spring	MN	21	0.50	Young and Mutchlerm (1976)
Maize – autumn	MN	55	2.59	
Fescue – dry	AL	104	8.21	Reese *et al.* (1982)
Fescue – slurry	AL	112	13.55	
Fescue	SC	142	1.85	McLeod and Hegg (1984)
Hybrid				
60% grazing[b]	PA	15	1.00	Rotz *et al.* (2002)
100% grazing[b]	PA	11	1.00	
30% grazing[b], 1.1 cows/ha	NY	10	1.90	Ghebremichael *et al.* (2007)
30% grazing[b], 1.8 cows/ha	NY	14	2.20	
Grazed				
Non-effluent block	NZ	38	1.70	Houlbrooke *et al.* (2008)
Effluent block	NZ	31	2.60	
Tile drained, 100N	NZ	50	0.10	Monaghan *et al.* (2000)
Tile drained, 400N	NZ	50	0.40	
Flood irrigated	NZ	25	3.40	Carey *et al.* (2004)
Flood irrigated[c]	Aus	44	9.70	Austin *et al.* (1996)

[a] Dairy manure applied to land and P lost in surface runoff.
[b] Percentage of total farm area grazed by cattle from May to October.
[c] Irrigation occurred within 1 month of superphosphate application.
Location abbreviations are state abbreviations in the USA except for NZ (New Zealand) and Aus (Australia).

In contrast, many confined operations have a large P surplus and hence tend to lose much P to surface water. Similarly, although P surpluses are generally low in grazed systems, grazing during winter, as well as the continual need to land-apply effluent (including at unsuitable times of year), mean that P losses still occur in this system unless conscientious management is practised.

12.4.2 Managing the environmental impact of confined systems

The confinement of dairy cattle emphasizes the importance of manure handling and the potential for point sources within the farm. Point sources include barns, barnyards or feeding pads, silos and the manure storage infrastructure. These areas have been highlighted as significant sources of P lost to surface water and, therefore, targeting them with mitigation strategies is highly cost-effective (Edwards *et al.*, 2008). The combination of limited infiltration and large volumes of excreta make surface runoff from point sources particularly harmful to surface waters. Strategies to minimize surface runoff from barns and milking parlours involve intercepting clean runoff from rainfall away before it reaches these areas and treating any runoff from barns or feedlots with filter strips, constructed wetlands or sediment ponds. Point source best management practices (BMPs) can be tremendously effective in improving downstream water quality. For instance, Sorodes and Normand (1999) recorded a 63% decrease in total P losses to a nearby waterway after routing dairy effluent through a newly constructed wetland. However, the lifespan of filter strips and constructed wetlands can be limited due to sedimentation and channelized flow (Schellinger and Clausen, 1992).

More difficult to manage is the application of stored manure to farm soils. Although many dairy systems have large land bases that enable manure to be applied at rates not in excess of plant uptake, the accumulation of nutrients (especially P) in the soils of certain fields is well documented, particularly fields near to barns (e.g. Aarons *et al.*, 2004). Since P tends to be well sorbed to soil, repeated surface application enriches and stratifies P in the topsoil. Not only will the extra P probably increase soil P concentrations, it also preferentially increases the water extractable P (WEP) fraction, which is strongly related to P in surface runoff (Kleinman *et al.*, 2002). In addition, low-density organic matter fractions in manure are highly susceptible to erosion when manure is broadcast (McDowell and Sharpley, 2002).

As an alternative to surface application, technologies that incorporate manure as a slurry into the soil will act to decrease the potential for P losses in surface runoff (Daverede *et al.*, 2004). However, no systems exist to incorporate semi-solid manures (e.g. bedded pack) into pasture soils, although new technologies geared toward directly incorporating dry manures (<30% moisture) hold promise (Warren *et al.*, 2008). In areas where surface runoff is dominated by rainfall in excess of soil infiltration rates, the use of aeration equipment can significantly decrease surface runoff, which also suggests that aeration may decrease P losses from surface-applied manure (Moore *et al.*, 2005). However, any benefit from aeration in areas where surface runoff occurs via saturation-excess conditions is unlikely (Franklin *et al.*, 2007).

In addition to the method of manure application, rate and timing will influence P loss (Kleinman and Sharpley, 2003). Widely distributing manure across a dairy farm to lower the rate of P applied at any one point can decrease the potential for P loss from individual fields. Furthermore, the availability of nutrients in applied manure to surface runoff tends to decline quickly with time: the maximum availability and potential for loss is soon after application. For instance, Preedy *et al.* (2001) observed that 1.8 kg/ha of P losses in runoff occurred over the 7 days after dairy slurry was surface applied. With time, slurry infiltration, non-surface-runoff-producing rainfall washing manure nutrients into the soil, manure crusting and invertebrate action all serve to protect P from surface runoff (Vadas *et al.*, 2007). Thus, 2 months after application, Mueller *et al.* (1984) reported a 76% decrease in dissolved P concentrations in surface runoff compared with to events soon after application.

Increasingly, due to more specialization and the better profitability of larger systems, confined operations and some hybrid dairy operations are exporting manure at their own cost off-farm, to avoid water quality issues associated with excess nutrients. This action is usually forced by regulations, court orders and neighbours (e.g. Texas Water Resources Institute, 2006). Some areas are even exploring the use of point/non-point source credit trading programmes to leverage resources from wastewater treatment industries to agriculture where, the argument is made, they are more cost-effective (Horan *et al.*, 2002). While manure export and nutrient credit trading offer an alternative to repeated, heavy application of manure nutrients to farm soils above levels necessary for crop production or applying manure at times of high runoff risk, depending where the manure is sent, it can simply shift the problems of P losses to the recipient.

12.4.3 Managing the environmental impact of grazed systems

Management to decrease P losses from grazed pastures focuses on a few key areas. These include: minimizing the direct deposition of dung into waterways with fencing; optimizing stocking rates to suit pasture production; avoiding grazing wet pastures; and the use of fertilizer or effluent to maintain soil-test P concentrations at the optimum for plant growth;

and avoiding areas or times likely to produce surface runoff.

In a paddock that doesn't receive effluent, McDowell *et al.* (2007) found that, for a grazed dairy in New Zealand, approximately 10, 30, 20 and 40% of P losses in surface runoff during a year were attributable to fertilizer, dung, pasture-plant and soil (including treading) components, respectively. The pasture component was attributed to P released during the ripping of pasture and the loss of P from denuded material. To avoid dung confounding the results, Mundy *et al.* (2003) mowed their pasture and were able to show that P losses in flood irrigation doubled once the pasture had been cut. Had the pasture been grazed, McDowell *et al.* (2003a) showed that the influence of treading on P losses increases much more once the pasture has been grazed for more than 24 h. McDowell (2006a) also showed that the potential for P loss from dung would have been initially high but decreased exponentially with time. The same relationship is also true for fertilizer, but the initial potential loss depends upon the water solubility of the fertilizer, which is why some studies have shown less P loss in soils with reactive phosphate rock (water solubility < 1%) applied compared with superphosphate (> 90% water solubility) (McDowell *et al.*, 2003b).

Failure to exclude cattle from streams can result in a disproportionate impact of pasture-fed cattle on water quality. In a 1200 km^2 catchment in the USA facing P-based water quality restrictions, James *et al.* (2007) estimated that 2800 kg of P was annually defecated into pasture streams by dairy cattle, with an additional 5600 kg P defecated within 10 m of the streams. In-stream defecation was equivalent to 12% of the annual P loadings attributed to all agricultural sources. Elsewhere, McDowell and Wilcock (2007) observed enriched concentrations of total P in stream flow associated with trampling and destabilization of the stream bank by stock and other factors. Thus stream exclusions for pastured cattle provide cost-effective reductions in P loadings but require consideration of drinking water source, laneway layout and fence maintenance.

The application of excessive P to maintain soil P concentration beyond the plant optimum, on the basis that it is 'money in the bank', also leads to significant P losses. Once enriched with P, a soil can take a long time to decrease concentrations below environmentally acceptable concentrations. Johnston and Poulton (1976) showed that soils that had received no farmyard manure since 1901 took 73 years to deplete Olsen P concentrations from 63.9 and 69.2 to 12.4 and 11.9 mg/kg in off-takes. Due to stratification of P in topsoil, a quicker strategy to decrease P losses is tillage, which, in a grazed dairy, can occur as part of a farm re-grassing programme, and redistribution of P within the plough layer (Sharpley, 2003).

On the effluent block of grazed dairies, effluent is applied according to the amount of potassium present, generally meaning that the effluent block comprises at least 10% of the farm. A large area means that there is usually insufficient P to maintain the soil-test P concentration, and supplemental fertilizer P is required. However, poor effluent application can result in excessive P losses. A recent strategy has been to increase storage so that effluent applications avoid times of year when the soil is wet and loss via surface runoff or drainage is likely. In addition, the use of low rate application (< 4 mm/h), rather than the traditional 'travelling irrigator' (~120 mm/h), allows for effluent and the P entrained therein to interact and be taken up by the soil, thereby minimizing losses. Houlbrooke *et al.* (2006) showed that a low rate strategy decreased P losses by 67%.

Recently, surface runoff from farm infrastructure such as lanes, tracks and gateways has been highlighted as potentially greater sources of P loss than pasture (Lucci *et al.*, 2010). Examining the loads of P lost from these areas in relation to the pasture during storm events, McDowell and Srinivasan (2009) confirmed that this was the case during summer and autumn, when infiltration-excess conditions occurred, but that pasture still dominated loads on an annual basis. However, the importance of farm infrastructure cannot be discounted, as losses during summer and autumn are when detrimental effects on surface-water quality are most likely (Jarvie *et al.*, 2006). Strategies to mitigate this loss generally involve engineering solutions, such as diverting flow back onto pasture, but the application of steel melter slag to the side of a laneway was shown to decrease P losses by about 95% and losses to a small catchment by about 66% (McDowell, 2007).

12.5 Hybrids: the Way Forward?

The premise for a hybrid system tends to be based on economics, but may also suit environmental goals in certain regions. A hybrid system may enable the farmer to minimize P losses by using confinement at times when the potential for losses is greater, and applying manure to areas where, for instance, surface runoff is unlikely, at other times. For example, decreased P losses may occur if a barn is used to house cows over winter instead of having them graze winter forage crops. This scenario occurs in the cooler southern regions of New Zealand, where pasture growth over winter is poor and maintenance feed is achieved by sowing about 10% of the farm into fodder crops (usually a *Brassica* sp.) that are grazed bit by bit over the winter months. However, grazing cropland in wet, but not frozen, conditions results in severe pugging and erosion (McDowell, 2006b). Soil conditions, together with dung deposition and saturated conditions, promote P loss via surface runoff (McDowell, 2006b). The losses (up to 2 kg P/ha) are much greater than those recorded on ungrazed pasture during this time (~0.5–1.0 kg P/ha) (McDowell and Houlbrooke, 2009). The use of a wintering barn would remove cattle from pasture and the need for winter forage crops, but introduces confined feeding and point-source concerns that must be weighed.

Compared with confined systems, the main environmental benefits of a hybrid system include a larger land area on which to apply manure, thereby avoiding increases in soil-test P, and the better distribution of dung around the farm during grazing. Although it is likely that this would decrease P losses, no data yet exist to confirm this probability.

12.6 Concluding Remarks

The cycling of phosphorus in grassland systems heavily depends upon the interaction between natural resources (e.g. climate, soils, etc.), land use and management. Although the need to make a profit largely drives the use of P on-farm, unless P is used wisely, losses at different stage of the P cycle will occur. It is therefore important to optimize both the system and components of the system to the resources available, so that the best use of a limited P source is achieved. The range of dairy systems varies widely, but common among all is the increased potential for losses and impacts on surface-water quality with increasing intensification. Problems like substantial P surpluses are common in confined systems, leading to soil P enrichment. Problems also exist in grazed systems despite many advocating they possess a better environmental footprint. Solutions to improve the efficiency of P use by minimizing P losses exist in both types of system and range from simple operational changes (e.g. not grazing wet paddocks) to more substantial changes in infrastructure (e.g. manure storage and application systems). By incorporating components of both systems, hybrid dairies may have increased flexibility to choose a suitable option and minimize P losses further, but work is required to confirm this hypothesis. Independent of environmental considerations, the increasing scarcity of P resources places emphasis on the farm manager to utilize these strategies and ensure that P is used efficiently.

References

Aarons, S.R., Melland, A.R. and Gourley, C.J.P. (2004) Nutrient distribution within a dairy farm. Third Australian New Zealand Soil Conference, SuperSoil 2004. Published online: http://www.regional.org.au/au/asssi/supersoil2004/s16/oral/1622_aaronss.htm (accessed 10 April 2010)

Austin, N.R., Prendergast, J.B. and Collins, M.D. (1996) Phosphorus losses in irrigation runoff from fertilized pasture. *Journal of Environmental Quality* 25, 63–68.

Bargo, F., Muller, L.D., Delahoy, J. and Cassidy, T. (2002) Performance of high producing dairy cows with three different feeding systems combining pasture and total mixed rations. *Journal of Dairy Science* 85, 2948–2963.

Carey, P.L., Drewry, J.J., Muirhead, R.W. and Monoghan, R.M. (2004) Potential for nutrient and faecal bacteria losses from a dairy pasture under border-dyke irrigation: a case study. *Proceedings of the New Zealand Grassland Association* 66, 141–149.

Carpenter, S.R., Caraco N.F., Correll D.L., Howarth R.W., Sharpley A.N. and Smith V.H. (1998) Nonpoint pollution of surface waters with phosphorus and nitrogen. *Ecological Applications* 8, 559–568.

Clark, D.A., Caradus, J.R., Monaghan, R.M., Sharp, P. and Thorrold, B.S. (2007) Issues and options for future dairy farming in New Zealand. *New Zealand Journal of Agricultural Research* 50, 203–221.

Cordell, D., Drangert, J.-O. and White, S. (2009) The story of phosphorus: global food security and food for thought. *Global Environmental Change* 19, 292–305.

Coyle, D. (2007) The future of biofuels: a global perspective. *Amber Waves* 5, 24–29.

Cunningham, L.C. (1993) An economic comparison of grazing intensity for lactating dairy cattle in northeastern Pennsylvania. MS thesis, The Pennsylvania State University, University Park.

Dartt, B.A., Lloyd, J.W., Radke, B.R., Black, J.R. and Kaneene, J.B. (1999) A comparison of profitability and economic efficiencies between management intensive grazing and conventionally managed dairies in Michigan. *Journal of Dairy Science* 82, 2412–2420.

Daverede, I.C., Kravchenko, A.N., Hoeft, R.G., Nafziger, E.D., Bullock, D.G., Warren, J.J. and Gonzini, L.C. (2004) Phosphorus runoff from incorporated and surface-applied liquid swine manure and phosphorus fertilizer. *Journal of Environmental Quality* 33, 1535–1544.

Dou, Z., Ferguson, J.D., Fiorini, J., Toth, J.D., Alexander, S.M., Peterson, A.B., Sims, J.T. and Wu, Z. (2003) Phosphorus feeding levels and critical control points on dairy farms. *Journal of Dairy Science* 86, 3787–3795.

Edwards, A.C., Kay, D., McDonald, A.T., Francis, C., Watkins, J., Wilkinson, J.R. and Wyer, M.D. (2008) Farmyards, an overlooked source for highly contaminated runoff. *Journal of Environmental Management* 87, 551–559.

Elbehri, A. and Ford, S.A. (1995) Economic analysis of major dairy forage systems in Pennsylvania: the role of intensive grazing. *Journal of Production Agriculture* 8, 501–506.

Franklin D.H., Cabrera, M.L. West, L.T. Calvert V.H. and Rema, J.A. (2007) Aerating grasslands: effects of runoff and phosphorous losses from applied broiler manure. *Journal of Environmental Quality* 36, 208–215.

Ghebremichael, L., Cerosaletti, P., Veith, T., Hamlett, J. and Gburek, W. (2007) Economic and phosphorus-related effects of precision feeding and forage management at a farm-scale. *Journal of Dairy Science* 90, 3700–3715.

Goff, J.P. (1998) Phosphorus deficiency. In: Howard, J.L. and Smith, R.A. (eds) *Current Veterinary Therapy 4: Food Animal Practice.* W.B. Saunders Co., Philadelphia, PA. pp. 218–220.

Granstedt, A. (2000) Increasing the efficiency of plant nutrient recycling within the agricultural system as a way of reducing the load to the environment – experience from Sweden and Finland. *Agriculture, Ecosystems and Environment* 80, 169–185.

Hirstov, A.N., Hazen, W. and Ellsworth, J.W. (2005) Efficiency of use of imported nitrogen, phosphorus, and potassium and potential for reducing phosphorus imports on Idaho dairy farms. *Journal of Dairy Science* 89, 3702–3712.

Horan, R.D., Shortle, J.S. and Abler, D.G. (2002) Point-nonpoint nutrient trading in the Susquehanna River basin. *Water Resources Research* 38, 1050.

Houlbrooke, D.J., Monaghan, R., Smith, C. and Nicolson, C. (2006) Reducing contaminant losses following application of farm dairy effluent to land using a K-line irrigation system. In: Currie, L.D. and Hanly, J.A. (eds) *Implementing Sustainable Nutrient Management Strategies in Agriculture Occasional Report No. 19.* Fertilizer and Lime Research Centre, Massey University, Palmerston North, New Zealand, pp. 290–300.

Houlbrooke, D.J., Horne, D.J., Hedley, M.J., Snow, V.O. and Hanly, J.A. (2008) Land application of farm dairy effluent to a mole and pipe drained soil: implications for nutrient enrichment of winter-spring drainage. *Australian Journal of Soil Research* 46, 45–52.

James, E., Kleinman, P. Veith, T. Stedman, R. and Sharpley, A (2007) Phosphorus contributions from pastured dairy cattle to streams of the Cannonsville Watershed, New York. *Journal of Soil and Water Conservation* 62, 40–47.

Jarvie, H.P., Neal, C. and Withers, P.J.A. (2006) Sewage-effluent phosphorus: a greater risk to river eutrophication than agricultural phosphorus? *Science of the Total Environment* 360, 246–253.

Johnston, A.E. and Poulton, P.R. (1977) Yields on the exhaustion land and changes in the NPK content of the soils due to cropping and manuring, 1852–1975. In: *Rothamsted Experimental Station Report for 1976*, Part 2 Rothamsted, Harpenden, UK, pp. 53–101.

Kleinman, P.J.A. and Sharpley, A.N. (2003) Effect of broadcast manure on runoff phosphorus concentrations over successive rainfall events. *Journal of Environmental Quality* 32, 1072–1081.

Kleinman, P.J.A. and Soder, K. (2008) The impact of hybrid dairy systems on air, soil and water quality: focus on nitrogen and phosphorus cycling. In: McDowell, R.W. (ed.) *Environmental Impacts of Pasture-Based Farming*. CAB International, Wallingford, UK, pp. 249–276.

Kleinman, P.J.A., Sharpley, A.N., Moyer, B.G. and Elwinger, G.F. (2002) Effect of mineral and manure phosphorus sources on runoff phosphorus. *Journal of Environmental Quality* 31, 2026–2033.

Klausner, S.D., Zwerman, P.J. and Ellis, D.F. (1976) Nitrogen and phosphorus losses from winter disposal of dairy manure. *Journal of Environmental Quality* 5, 47–49.

Lucci, G.M., McDowell, R.W. and Condron, L.M. (2010) Phosphorus and sediment loads/transfer from potential source areas in New Zealand dairy pasture. *Soil Use and Management* 26, 44–52.

McDowell, R.W. (2006a) Contaminant losses in overland flow from cattle, deer and sheep dung. *Water, Air and Soil Pollution* 174, 211–222

McDowell, R.W. (2006b) Phosphorus and sediment loss in a catchment with winter forage grazing of cropland by dairy cattle. *Journal of Environmental Quality* 35, 575–583.

McDowell, R.W. (2007) *Assessment of Altered Steel Melter Slag and P-socks to Remove Phosphorus from Streamflow and Runoff from Lanes*. Report to Environment B.O.P., Land & Environmental Management Group, AgResearch, Invermay Agricultural Centre, Private Bag 50034 Mosgiel, New Zealand. http://www.envbop.govt.nz/TechReports/TechReports-070601-AssessmentAlteredSteelmelterslag.pdf (accessed 21 April 2010).

McDowell, R.W. (ed.) (2008) *Environmental Impacts of Pasture-Based Farming*. CAB International, Wallingford, UK.

McDowell, R.W. and Houlbrooke, D.J. (2009) Management options to decrease phosphorus and sediment losses from irrigated cropland grazed by cattle and sheep. *Soil Use and Management* 25, 224–233.

McDowell, R.W. and Sharpley, A.N. (2001) Approximating phosphorus release from soils to surface runoff and subsurface drainage. *Journal of Environmental Quality* 30, 508–520

McDowell, R.W. and Sharpley. A.N. (2002) Phosphorus transport in overland flow in response to position of manure application. *Journal of Environmental Quality* 31, 217–227.

McDowell, R.W. and Srinivasan, M.S. (2009) Identifying critical source areas for water quality: 2. Validating the approach for phosphorus and sediment losses in grazed headwater catchments. *Journal of Hydrology* 379, 68–80.

McDowell, R.W. and Wilcock, R.J. (2007). Sources of sediment and phosphorus in stream flow of a highly productive dairy farmed catchment. *Journal of Environmental Quality* 36, 540–548.

McDowell, R.W., Drewry, J.J., Muirhead, R.W. and Paton, R.J. (2003a) Cattle treading and phosphorus and sediment loss in overland flow from grazed cropland. *Australian Journal of Soil Research* 41, 1521–1532.

McDowell, R.W., Monaghan, R.M. and Carey, P.L. (2003b) Potential phosphorus losses in overland flow from pastoral soils receiving long-term applications of either superphosphate of reactive phosphate rock. *New Zealand Journal of Agricultural Research* 46, 329–337.

McDowell, R.W., Nash, D.M. and Robertson, F. (2007) Sources of phosphorus lost from a grazed pasture receiving simulated rainfall. *Journal of Environmental Quality* 36, 1281–1288.

McLeod, R.V. and Hegg, R.O. (1984) Pasture runoff water quality from application of inorganic nitrogen sources. *Journal of Environmental Quality* 13, 122–126.

Monaghan, R.M., Paton, R.J., Smith, L.C. and Binet, C. (2000) Nutrient losses in drainage and surface runoff from a cattle-grazed pasture in Southland. *Proceedings of the New Zealand Grassland Association* 62, 99–104.

Moore, P.A. Jr, Formica, S.J., Van Epps, M. and Delaune, P.B. (2005) Effect of pasture renovation on nutrient runoff from pastures fertilized with manure. In: *Proceedings of the 7th International Symposium on Livestock Environment, May 18–20, 2005, Beijing, China.* American Society of Agricultural and Biological Engineers, St. Joseph, Michigan, pp. 301.

Mueller, D.H., Wendt, R.C. and Daniel, T.C. (1984) Phosphorus losses as affected by tillage and manure application. *Soil Science Society of America Journal* 48, 901–905.

Muller, L.D. (1992) Feeding Management Strategies. In: Van Horn, H.H. and Wilcox, C.J. (eds) *Large Dairy Herd Management. Management Services.* American Dairy Science Association, Champaign, IL, pp. 326–335.

Mundy, G.N., Nexhip, K.J., Austin, N.R. and Collins, M.D. (2003) The influence of cutting and grazing on phosphorus and nitrogen in irrigation runoff from perennial pasture. *Australian Journal of Soil Research* 41, 675–685.

National Research Council (2001) *Nutrient Requirements of Dairy Cattle, 7th revision.* National Academy of Science, Washington, DC.

Parker, W.J., Muller, L.D. and Buckmaster, D.R. (1992) Management and economic implications of intensive grazing on dairy farms in the Northeastern States. *Journal of Dairy Science* 75, 2587–2597.

Preedy, N., McTiernan, K., Mathews, R., Heathwaite, L. and Haygarth. P. (2001) Rapid incidental phosphorus transfers from grassland. *Journal of Environmental Quality* 30, 2105–2112.

Pretty, J.N., Mason, C.F., Nedwell, D.B., Hine, R.E., Leaf, S. and Dils, R. (2003) Environmental costs of freshwater eutrophication in England and Wales. *Environmental Science and Technology* 32, 201–208.

Reese, L.E., Hegg, R.O. and Gantt, R.E. (1982) Runoff water quality from dairy pastures in the Piedmont region. *Transactions of the American Society of Agricultural Engineers* 25, 697–701.

Rook, A.J., Tallowin, J.R.B., Rutter, S.M., Orr, R.J., Hopkins, A. and Gibb, M.J. (2005) Behavioural and community ecology: mechanisms driving biodiversity in pasture-based systems. In: Rook, A.J. and Penning, P.D. (eds) *Grazing Management, The Principles and Practice of Grazing, for Profit and Environmental Gain, within Temperate Grassland Systems.* British Grassland Society, Reading, UK. pp. 54–65.

Rotz, C.A., Sharpley, A.N., Satter, L.D., Gburek, W.J. and Sanderson, M.A. (2002) Production and feeding strategies for phosphorus management on dairy farms. *Journal of Dairy Science* 85, 3153.

Schellinger, G.R. and Clausen. J.C. (1992) Vegetative filter treatment of dairy barnyard runoff in cold regions. *Journal of Environmental Quality* 21, 40–45.

Serodes, J.B. and Normand, D. (1999) Phosphorus removal in agricultural wastewater by a recently constructed wetland. *Canadian Journal of Civil Engineering* 26, 305–311.

Sharpley, A.N. (2003) Soil mixing to decrease surface stratification of phosphorus in manured soils. *Journal of Environmental Quality* 32, 1375–1384.

Sibbesen, E. and Runge-Metzger, A. (1995) Phosphorus balance in European agriculture - status and policy options. In: Tiessen, H. (ed.) *Phosphorus in the Global Environment.* John Wiley and Sons, Chichester, UK, pp 43–58.

Smith, K.A., Chalmers, A.G., Chambers, B.J. and Christie, P. (1998) Organic manure phosphorus accumulation, mobility and management. *Soil Use and Management* 14, 154–159.

Soder, K.J. and Rotz, C.A. (2001) Economic and environmental impact of four levels of concentrate supplementation in grazing dairy herds. *Journal of Dairy Science* 84, 2560–2572.

Steinfeld, H., Gerber, P., Wassenaar, T., Castel, V., Rosales, M. and de Haan. C. (2006) *Livestock's Long Shadow: Environmental Issues and Options.* FAO, Rome.

Stout, W.L., Fales, S.L., Muller, L.D., Schnabel, R.R., Elwinger, G.F. and Weaver, S.R. (2000) Assessing the effect of management intensive grazing on water quality in the northeast U.S. *Journal of Soil and Water Conservation* 55, 238–243.

Tallam, S.K., Ealy, A.D., Bryan, K.A. and Wu. Z. (2005) Ovarian activity and reproductive performance of dairy cows fed different amounts of phosphorus. *Journal of Dairy Science* 88, 3609–3618.

Tallowin, J.R.B., Rook, A.J. and Rutter, S.M. (2005) Impact of grazing management on biodiversity of grasslands. *Animal Science* 81, 193–198.

Texas Water Resources Institute (2006) Dairy compost utilization by the public and public entities. http://compost.tamu.edu/index.php (accessed 13 April 2010).

Toth, J.D., Dou, Z., Ferguson, J.D., Galligan, D.T. and Ramberg, C.F. Jr (2006) Nitrogen- vs. phosphorus-based dairy manure applications to field crops: nitrate and phosphorus leaching and soil phosphorus accumulation. *Journal of Environmental Quality* 35, 2302–2312.

Vaccari, D.A. (2009) Phosphorus: a looking crisis. *Scientific American* June, 42–47.

Vadas, P.A., Harmel, R.D. and Kleinman, P.J.A. (2007) Transformations of soil and manure phosphorus after surface application of manure to field plots. *Nutrient Cycling in Agroecosystems* 77, 83–99.

Warren, J. G., Sistani, K. R.Way, T. R. Mays, D. A. and Pote, D. H. (2008) A new method of poultry litter application to perennial pasture: subsurface banding. *Soil Science Society of America Journal* 72, 1831–1837.

Williams, P.M. and Haynes, R.J. (1992) Balance sheet of phosphorus, sulphur and potassium in a long term grazed pasture supplied with superphosphate. *Fertilizer Research* 31, 51–60.

Wilson, M.A. and Carpenter, S.R. (1999) Economic valuation of freshwater ecosystem services in the United States, 1977-1997. *Ecological Applications* 9, 772–783.

Wu, Z. and Satter, L.D. (2000) Milk production and reproductive performance of dairy cows fed two concentrations of phosphorus for two years. *Journal of Dairy Science* 83, 1063.

Wu, Z., Satter, L.D. and Sojo, R. (2000) Milk production, reproductive performance, and fecal excretion of phosphorus by dairy cows fed three amounts of phosphorus. *Journal of Dairy Science* 83, 1028–1041.

Young, R.A. and Mutchlerm, C.K. (1976) Pollution potential of manure spread on frozen ground. *Journal of Environmental Quality* 5, 174–179.

Part III

Grasslands as Regulating Biodiversity Dynamics

Editors: Gilles Lemaire and Abad Chabbi

Biodiversity within grasslands has to be analysed at three levels of organization: plant population and communities, microfauna, and soil microbes; and according to two points of view: patrimonial and functional biodiversity. In Part III we focus particularly on functional biodiversity, which should be involved in the processes analysed within Parts I and II, for providing different ecosystem services. But we must understand that the contribution of grasslands to patrimonial biodiversity is, in itself, an ecosystem service and this topic is developed within some chapters of Part IV. Chapter 13 develops the important feature that grassland biodiversity has to be viewed in relation to climate changes, and that long-term experimental platforms allowing the manipulation of some climate parameters are necessary for understanding and forecasting future trends in the vegetation dynamics of grasslands. The influence of grassland management systems on grassland vegetation dynamics and its biodiversity is analysed within Chapter 14, with an important discussion about antagonisms between grassland intensification and species diversity or the possibilities for restoring biodiversity. Chapter 15 develops a theoretical framework for an approach to the functional biodiversity within grassland, using functional traits of plant species in order to relate more accurately the floristic composition of grasslands to the different ecosystem services, through the effects and responses of functional groups of species. Chapter 16 deals with the analysis of the role and importance of genetic diversity for a better adaptation of plant populations and communities, either in natural or in sown grasslands, with a discussion of the role and impact of inter- versus intra-specific diversity and their implications for an effective plant-breeding strategy. Chapter 17 contributes to understanding the role and impact of soil microbe populations and communities, in the light of the new molecular technologies that allow analysis of the diversity of these microbes and their functional role in C and N cycles. The importance and role of soil animal diversity in grasslands are discussed within Chapter 18, with a strong focus on the effect of grassland management on soil fauna communities, on the responses of soil diversity to successional gradients, and the role of soil ecosystem engineers on grassland ecosystem functioning.

13 Predicting Biodiversity Dynamics of Grasslands under Global Change: the Role of Long-term Manipulations of Climate

Andrew P. Askew, Jason D. Fridley and J. Philip Grime

13.1 Introduction

For many different purposes related to food production, the preservation of ecosystem services and the maintenance of biodiversity, it is essential that we not only devise models of future impacts of climate change on ecosystems but also conduct experiments to test the validity of such models. This article refers to the 'Buxton experiment' – a long-term simulation of changes of climate now expected later this century on an ancient limestone grassland ecosystem situated in the southern Pennines of north Derbyshire, UK. The experiment is now in its 18th year of continuous climate manipulation, and accounts of the results have been published (Grime *et al.*, 2000, 2008; Bates *et al.*, 2005; Fridley *et al.*, 2011). Our objective here is not to review the methods and results in detail but to consider the following questions:

- What has been learned so far?
- Should the experiment continue?
- How can we maximize the value of long-term experiments?

We address these questions in turn below. First, however, it may be helpful to ask to what extent the results of an experiment of this kind can be predicted on theoretical grounds.

13.2 Predictions

13.2.1 General predictions of ecosystem resistance and resilience

Often our initial research objective is to obtain a broad characterization of sensitivity to climate change. We can pursue this objective at all scales from that of an individual organism to that of an entire ecosystem. Across this wide range in scale it is usually helpful to differentiate between two facets of response. The first is resistance (the capacity to withstand changed conditions) and the second is resilience (the ability to recover rapidly and completely from any effects of changed conditions). By extension from the laws of physics, scientists such as Westman (1978) and Leps *et al.* (1982) have successfully applied the concepts of resistance and resilience in ecology, and an experiment conducted across five neighbouring herbaceous communities (MacGillivray *et al.*, 1995) supports the hypothesis that, in vegetation at least, there is a trade-off between resistance and resilience. In consequence, traits such as long lifespan and slow growth confer resistance but are also associated with low resilience. When this theoretical framework is applied to the grassland that is the focus of this article, it is predicted that, in response to climate manipulation, we will observe high resistance and low resilience. At the end of this article we examine the significance of this prediction more closely.

13.2.2 More detailed predictions

For the purposes of classification and mapping of resistance and resilience of vegetation and ecosystems across large areas of landscape, the concepts outlined under the previous heading hold considerable promise. However, they are not appropriate where it is necessary to obtain a much more detailed knowledge of the many physiological and population processes that underlie the sequence of events expected to take place as a changing climate transforms physical, chemical and biotic conditions at specific sites. Figure 13.1 illustrates seven steps through which it is predicted (Grime *et al.*, 2007) that vegetation is likely to pass as it responds to climate change. In the remainder of this article we examine the extent to which the various phenomena predicted in Fig. 13.1 have been observed in response to the climate forcing applied in the 'Buxton experiment'.

13.3 What Has Been Learned So Far?

For the first 5 years of its existence, the Buxton experiment was associated with a sister experiment in which identical climate manipulations were applied to an area of limestone grassland situated at Wytham, near Oxford in southern England. Not only was this second site exposed to a warmer and drier climate, but it occupied an area of relatively productive semi-natural grassland on an abandoned wheat field. A comparison of the two experiments allowed investigation of whether differences in response to the experimental changes in climate were dependent upon the different background climatic conditions or were due to the very different agricultural histories of the two sites. Although both sites were situated on shallow, limestone soils, the Buxton grassland had a long history of sheep grazing without exposure to ploughing or fertilizer application, whereas the soil at Wytham contained fertilizer residues supporting a community of productive herbaceous species.

The results after 5 years (Grime *et al.*, 2000) provided no evidence that differences between the two sites in the magnitude, rate and nature of the changes in vegetation were related to differences in

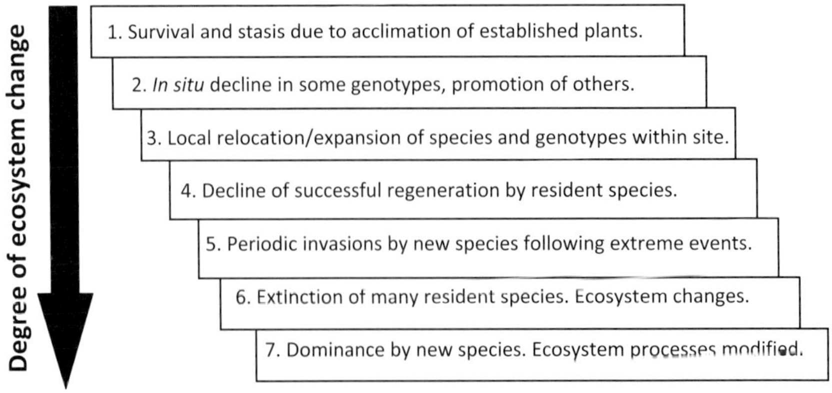

Fig. 13.1. Seven-step model of sequential vegetation responses to climate change (from Grime *et al.*, 2007). A comprehensive address of climate change impacts on grassland vegetation includes investigations of all processes, although the relative importance of each will vary depending on site conditions, landscape history and the magnitude of climate forcing.

the climates experienced during the experiment or before it commenced. The contrast in response appeared to be driven by site productivity: manipulations of rainfall and temperature initiated profound and rapid vegetation transformations at Wytham but little change was recorded at Buxton. These results are consistent with the prediction, outlined earlier in this article, that greatest resistance to climate change is to be expected in unproductive habitats. Here the slow growth, potentially long lifespans of individuals, and the longevity and climatic tolerance of the plant tissues confer resistance but simultaneously restrict the rate of vegetation response.

Although the Wytham experiment ended after 5 years, when funding was discontinued, subsequent recording has confirmed the resistance of the Buxton grassland to climate manipulation (Grime *et al.*, 2008). However, results from the most recent investigations (Fridley *et al.*, 2011) suggest that it would be unwise to attribute this feature exclusively to climatic tolerance and the 'slow dynamics' of the vegetation associated with infertile soils. First we must recognize strong evidence (Fig. 13.2) that in habitats of uneven soil depth, such as those at Buxton, species vulnerable to drought may be able to persist by retreating to local pockets of deeper soil. Secondly, there is increasing evidence that species can be buffered against extinction by local genetic diversity in plant populations. Genetic diversity is conspicuous in ancient calcareous grasslands and has been shown to slow the rate of decline in species richness in experimental plant communities (Booth and Grime, 2003; Fridley *et al.*, 2007; Fridley and Grime, 2010). A current challenge in the work at Buxton is to assess the contributions of genetic diversity and genetic change in component plant populations to the resistance of the vegetation to 18 years of climate manipulation.

Another possible reason for the relative constancy of species composition at Buxton is the failure of the plots to be colonized by new species capable of exploiting the changed climatic conditions.

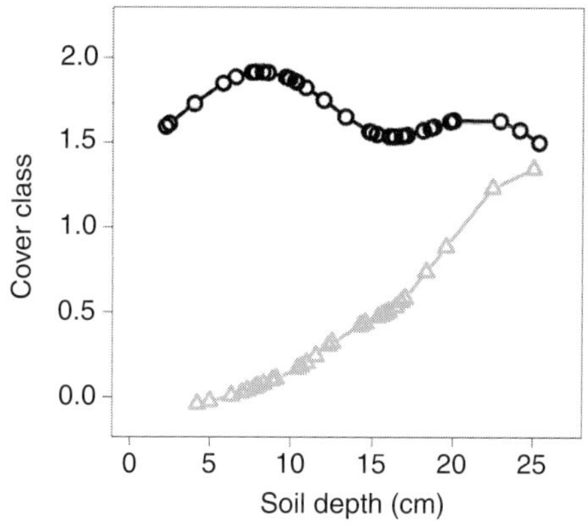

Fig. 13.2. The abundance of the grassland forb *Lotus corniculatus* in response to more than 15 years of summer drought depends strongly on soil conditions that vary at fine spatial scales (< 1 m^2). These curves represent generalized additive models of *Lotus* abundance (100 cm^2 cover) in control (black line, circles) and summer droughted plots (grey line, triangles) with respect to soil depth. Although *Lotus* has responded little to enhanced soil moisture, it has retreated to the deepest microsites in response to chronic drought, which has prevented its local extinction.

This hypothesis is under test in current experiments that simulate invasions, by introducing seeds obtained from southern populations at lower altitudes.

13.4 Should the Experiment Continue?

Our answer to this question is an unequivocal 'Yes'. A recent review (Silvertown *et al.*, 2010) documents the scarcity of long-term experiments in the UK and on a world scale, and explains the need to create more long-term experiments to investigate the causes and consequences of the global changes that are arising from shifting patterns and intensity in the impacts of land use and climate. There are additional reasons, some specific to the Buxton site, why we need to continue the experiment:

1. Many of the small shrubs, herbs, sedges, grasses and bryophytes in the Buxton grassland have long lifespans and display only intermittent success in regeneration. To describe their responses to the climate treatments properly, long-term recording is necessary. This is not only to observe the fate of individuals already established but to document reproductive success or failure, particularly following unusual climatic events.

2. As described in Fig. 13.1, we have much more work to do before it will be possible to establish the role of genetic variation in the persistence of plant populations under changed climatic conditions. An intriguing aspect of the response of the Buxton grassland to the manipulations is the extent to which the capacity of new species to invade could be negated by expansion, within the populations of residents, by those genotypes best attuned to the new conditions.

3. Maintenance of the experiment into the future is also essential if we are to test our understanding of the sequence of events whereby some components of both the plant community and the whole ecosystem may be displaced and a new equilibrium established with the changed climate. Particularly with respect to the soil environment, progress in ecosystem research is likely to depend upon the application of new ideas and the development of techniques, some of which have not yet been invented.

4. During the course of a long-term experiment, phenomena may be encountered that cannot be satisfactorily explained simply by prolonging the experiment. Sometimes it is helpful to initiate new treatments, and, in anticipation of this requirement, three additional untreated plots were included in each replicate block of the Buxton experiment. These so-called 'spare' plots have received the same management (simulated grazing each October) over the course of the experiment and are therefore available for use at any time. One set of spare plots was activated in 2007, as part of an experiment measuring susceptibility to invasion by southern grasses: comparisons of invasion success have been drawn between plots that had experienced 15 years of heating and drought treatment and others that had remained untreated until 2007. Another set of spare plots came into play in 2010, in a quite different experiment examining the role of interspecific competition in the responses of grassland on shallow and deep areas of the experimental plots.

Not all of the problems encountered in long-term experiments can be resolved by inserting new treatments into existing experiments. Often the complexity and irregularities of natural communities and ecosystems defy analysis, and a more controlled experimental system is required. For this reason, adjacent to the main experiment at Buxton, an area is devoted to outdoor, open-top microcosms in which plant communities of controlled composition have been assembled on calcareous soil and subjected to climate manipulation. In this way, treatments can be applied with a precision and a degree of replication that is rarely possible in natural grassland.

13.5 How Can We Maximize the Effectiveness of Long-term Experiments?

An evaluation of previous long-term experiments in the UK (Silvertown *et al.*, 2010) concludes that many were terminated prematurely and, for this reason, most failed to provide adequate tests of our understanding of the trajectories and mechanisms of ecological responses to environmental and anthropogenic change. However, it would be misleading to suggest that longevity by itself guarantees success. On the basis of our experiences at Buxton, we suggest that, to justify the effort and cost of maintaining long-term experiments, radical changes in philosophy and design are necessary. In particular, care must be taken in choosing the locations of these experiments. It is important that the selected communities and ecosystems are representative of types of widespread ecological or economic importance, so that the insights gained have wide relevance and applicability. Wherever possible, it is helpful to use sites for which databases exist not only with respect to the taxonomic identity of component organisms but also to the functional biology of the most abundant and influential plants, animals and microorganisms.

The logistics of long-term experiments may be critical to their success: it is initially exciting to work in remote locations but, if the accommodation for staff and services to support good science during long periods are not available, sustained and appropriate effort may not be achieved. For many purposes, considerable foresight is essential to anticipate the unexpected developments and opportunities that occur during the majority of well-designed, long-term experiments. Already, earlier in this article, we have emphasized the need for flexibility in research tactics, which is possible through the provision of 'spare plots' that from the outset are thoroughly integrated into the experimental design and site management. Similar considerations apply to plot size. Initially, small plots may suffice for monitoring vegetation, but much larger areas may be needed for more sophisticated measurement and sampling as attention shifts towards genetics, ecosystem processes or invasion biology. Sometimes these developments can be addressed by new experiments outside the confines of the original experiment but this is not an option if, as in the Buxton experiment, the requirement is to apply new treatments in the context of ecosystems uniquely conditioned by a particular climate and management history.

13.6 Conclusions

If the contrasted responses of the Buxton and Wytham grasslands to climate manipulation prove to be representative of a consistent difference between productive and unproductive ecosystems, it is possible that this pair of experiments may have provided the basis for a generalization of wide applicability in climate-change research. The results emerging from the extended investigation at Buxton suggest that long-term climate manipulations can also allow investigations into the mechanisms by which communities and ecosystems respond to a changing climate. Some insights arise as a simple consequence of long-term, standardized recording. Increasingly, however, ecologists are adding to the value and power of the original experimental design by introducing new dimensions 'along the way'. Some of these can be located in spare plots that were created as part of the original design and provide the space in which new ideas and techniques can be applied and different problems addressed. In this way, the original purposes of the long-term manipulations are preserved but the objectives and methods can be continually updated.

References

Bates, J.W., Thompson, K. and Grime, J.P. (2005) Effects of long-term climatic change on the bryophytes of a limestone grassland community. *Global Change Biology* 11, 757–769.

Booth, R.E. and Grime, J.P. (2003) Effects of genetic impoverishment on plant community diversity. *Journal of Ecology* 91, 721–730.

Fridley, J.D. and Grime, J.P. (2010) Community- and ecosystem-level consequences of intraspecific genetic diversity in grassland microcosms of varying species diversity. *Ecology* 91, 2272–2283.

Fridley, J.D., Grime, J.P. and Bilton, M. (2007) Genetic identity of interspecific neighbours mediates plant responses to competition and environmental variation in a species-rich grassland. *Journal of Ecology* 95, 908–915.

Fridley J.D., Grime J.P. and Askew, A.P. (2011) Soil heterogeneity buffers community response to climate change in a species-rich grassland. *Global Change Biology* 17, 2002–2011.

Grime, J.P, Brown, V.K., Thompson, K., Masters, G.J., Hillier, S.H, Clarke, I.P., Askew, A.P, Corker, D. and Kielty, J.P. (2000) The response of two contrasted grasslands to simulated climate change. *Science* 289, 762–765.

Grime, J.P, Hodgson, J.G. and Hunt, R. (2007) *Comparative Plant Ecology*, 2nd edn. Castlepoint Press, Colvend, UK.

Grime, J.P., Fridley, J.D., Askew, A.P., Thompson, K., Hodgson, J.G. and Bennett, C.R. (2008) Long-term resistance to simulated climate change in an infertile grassland. *Proceedings of the National Academy of Sciences USA* 105, 10,028–10,032.

Leps, J., Osbornova-Kosinova, J. and Rejmanek, K. (1982) Community stability, complexity, and species life-history strategies. *Vegetatio* 511, 53–63.

MacGillivray, C.W, Grime, J.P. and the ISP team (1995) Testing predictions of the resistance and resilience of vegetation subjected to extreme events. *Functional Ecology* 9, 640–649.

Silvertown, J., Tallowin, J., Stevens, C., Power, S.A., Morgan, V., Emmett, B., Hester, A., Grime, J.P., Morecroft, M., Buxton, R., Poulton, P., Jinks, R. and Bardgett, R. (2010) Environmental myopia: a diagnosis and a remedy. *Trends in Ecology and Evolution* 25, 556–561.

Westman, W.E. (1978) Measuring inertia and resilience of ecosystems. *Bioscience* 28, 705–710.

14 Interactions Between Grassland Management and Species Diversity

Bertrand Dumont and Jerry R.B. Tallowin

14.1 Introduction

Natural and semi-natural grasslands are globally threatened ecosystems (White and Vannaselt, 2001). Semi-natural grasslands are defined here as ecosystems that are managed, by mowing and/or grazing, to prevent natural succession to forests. Semi-natural grasslands have a relatively natural character and notable richness in species, due to the absence of, or very low input of, fertilizers and continuous use as grasslands over a long period. The wide diversity of semi-natural grassland communities across landscapes reflects altitudinal, edaphic and other environmental gradients that have influenced the development of plant species assemblages from regional species pools. Different management practices have then further influenced the community assemblages, through interacting with specific plant ecological strategies to provide community mosaics at a range of scales, all of which contribute to overall landscape biodiversity richness. Although we primarily focus on interactions between grassland management and species richness in this chapter, we recognize that biodiversity is the sum of total biotic variation, from gene to landscape. Natural and semi-natural grasslands make a major contribution to global biodiversity (White and Vannaselt, 2001; Aguiar, 2005). The biological richness of natural and semi-natural grasslands reflects the wide diversity of the plant communities, of distinctly different botanical composition and character, that they encompass, as evidenced by the number of high-nature-value temperate grasslands of Europe (Veen *et al.*, 2009). Loss of their biodiversity through agricultural expansion and intensification and through other human developments erodes the wide range of ecosystem services that natural and semi-natural grasslands provide (Aguiar, 2005; Wade *et al.*, 2008).

14.2 Grassland Intensification and Biodiversity: Antagonism and Restoration Strategies

Agricultural intensification until the latter decades of the 20th century had a singular functional objective: to improve the provisioning service value of agroecosystems. The result of this intensification has been structural homogenization of farmed landscapes (Benton *et al.*, 2003). At the field scale, this has meant drastic simplification of vegetation communities, with reduced botanical diversity, phenological variation and architectural complexity, all of which have had major negative impacts on overall farmland biodiversity (Benton *et al.*, 2003). The use of

 Grassland Productivity and Ecosystem Services (eds G. Lemaire, J. Hodgson and A. Chabbi)

fertilizers to boost grassland productivity, combined with increased frequency of disturbance, by either mowing or grazing, favours a relative few competitive plant species at the expense of stress-tolerant species (Grime, 2002), with consequent loss of overall diversity (Tallowin *et al.*, 2005). Increasing grazing intensity alone on natural and semi-natural grasslands can also damage their biodiversity value, as the introduction of cattle grazing in more arid rangelands in the USA has demonstrated (Fleischner, 1994). In temperate grassland ecosystems, evidence suggests that, in order to maintain high biodiversity value, grazing should remove only a fraction of the forage that is available for consumption. Derivation of optimum stocking rates (carrying capacity) for maintaining high grassland biodiversity is complex. Factors such as scale and abiotic heterogeneity of a site and the size of herbivore or mixture of herbivores that are used will affect biodiversity, as will be discussed later in this chapter. Mown high-nature-value semi-natural grasslands are also sensitive to frequency of disturbance, sustainable management being a single mid- to late-growing season cut, followed, in many cases, by grazing of the regrowth. At the farm scale, intermediate levels of such biodiversity-friendly management rules (e.g. 40% of hay paddocks cut after flowering, threshold sward height of 8 cm at turn-out in grazed paddocks) are assumed to benefit floristic diversity without affecting system production (Jouven and Baumont, 2008).

With the sustainability of intensive farming systems under increasing scrutiny, impetus to develop more multifunctional agroecosystems is likely to grow in the new millenium. For many biodiversity-degraded farmed landscapes, restoration of an increased amount of biodiversity will be required (Wade *et al.*, 2008), but this faces considerable constraints (Bakker and Berendse, 1999; Hodgson *et al.*, 2005). Elevated nutrient availability will be a major factor limiting the restoration of greater plant diversity in temperate grasslands (Janssens *et al.*, 1998; Bakker and Berendse, 1999; Suding *et al.*, 2005). Reducing nitrogen inputs to below the critical load for the existence of natural and semi-natural grassland communities will be slow and costly for most intensively farmed landscapes (Bakker and Berendse, 1999). Modest improvement of grassland plant diversity can, however, be achieved by reducing nitrogen fertilization/availability on grasslands (Baer *et al.*, 2003; Marini *et al.*, 2008). Rapid and/or large changes in species composition are unlikely to occur in many mesotrophic grasslands where nutrient levels remain high even in the absence of fertilizer inputs (Marriott *et al.*, 2004). An impoverished species pool at the landscape scale also impedes restoration of ecological diversity (Poschlod and Bonn, 1998). To achieve rapid but modest improvements in the biodiversity of highly improved agricultural landscapes requires species re-introduction (Walker *et al.*, 2004), but this is unlikely to provide habitats for ecological specialists (Pywell *et al.*, 2007). In order to achieve modest levels of multifunctionality in terms of ecosystem services, the re-introduction of mixtures of competitive legume and non-legume forb (dicotyledonous) species into temperate grasslands seems to provide a rapid and practical compromise (Mortimer *et al.*, 2006; Bradbury *et al.*, 2010; Table 14.1).

14.3 Reducing Grazing Intensity: Benefits and Threats According to Nutrient Availability

Increasing structural heterogeneity at a range of scales, from within field to farm and region, will be critical for the restoration of grassland biodiversity (Wilson *et al.*, 2005). Grazing offers a key management tool with which to manipulate grassland structure and biodiversity value. A key mechanism by which grazing herbivores shape the structure of plant communities is through selective defoliation, which arises as the result of expression of herbivore dietary choices. Persistent selective grazing of some plant species can fundamentally alter plant community boundaries and resultant community mosaic scales (Bakker,

Table 14.1. A selection of competitive plant species with a range of potential values for creating multifunctional grasslands (adapted from Mortimer *et al.*, 2006).

	Total no. of associated invertebrates	Total no. of associated farmland birds	Livestock feed value	Potential soil restructuring value
Lotus corniculatus	93		+++	
Medicago sativa	35		+++	+++
Medicago lupulina	22		+	
Onobrychis viciifolia	26		+	+++
Trifolium dubium	58	9	+	
Trifolium pratense	93	9	+++	++
Achillea millefolium	88	9	++	+
Centaurea nigra	67	9		++
Plantago lanceolata	93	11	++	++
Rumex acetosa	134		+	++
Sanguisorba minor	6–11		++	I

1998; Farnsworth and Anderson, 2001). However, selective grazing effects on interspecific competition within the plant community can be largely negated under high stocking rates/grazing intensity, where only plant species that have strong defences against herbivory remain undisturbed. Analysing vegetation dynamics under contrasting management regimes identified the emergence of plant types that shared common attributes. Identification of these functional types provides a powerful approach to predict vegetation response to environmental factors in agriculturally managed grasslands (see Garnier and Navas, Chapter 15, this volume). As an example, 12 years of extensification in a fertile temperate grassland led to the dominance of tall and late-flowering plants of low specific leaf area, nitrogen content and digestibility, which would have no grazing tolerance but a strong ability to compete for light (Louault *et al.*, 2005). Grazing intensity is thus another key management variable with which to influence the structure, composition and thus biodiversity value of grazed pastures (Fig. 14.1). Lenient stocking rates benefit grassland biodiversity value by creating mosaics of habitats defined by short and tall vegetation patches with contrasting growth forms and competitive interactions (Milchunas *et al.*, 1988). The manipulation of stocking rate therefore offers a potentially important tool for conservation management. How large herbivores affect vegetation diversity in temperate grasslands, however, depends on the background conditions. Plant species richness has been shown to be favoured by a reduction of stocking rate in ecosystems with moderate to low soil fertility (Kruess and Tscharntke, 2002; Louault *et al.*, 2005; Pöyry *et al.*, 2006). Forbs and non-competitive grasses increased in abundance following a reduction of cattle stocking rate in a species-rich upland pasture with low soil nutrient status (Dumont *et al.*, 2009; Fig. 14.1). Conversely, dramatic changes in dominance have been reported in fertile grasslands that were taken out of intensive use for biodiversity conservation purposes (Marriott *et al.*, 2004). Reduction of grazing intensity also enhanced dominance of grasses and decreased plant diversity in two cattle-grazed fertile pastures (Scimone *et al.*, 2007). In these two grasslands, there were clearer shifts in species richness and evenness than in the semi-natural grasslands with which they were compared.

14.4 Matching Type of Livestock to Desired Biodiversity Outcomes

Different livestock species can also have significantly different effects on plant communities (Grant *et al.*, 1996; Loucougaray *et al.*, 2004; Loydi *et al.*, 2010). The number of such studies has, however, probably been limited by methodological

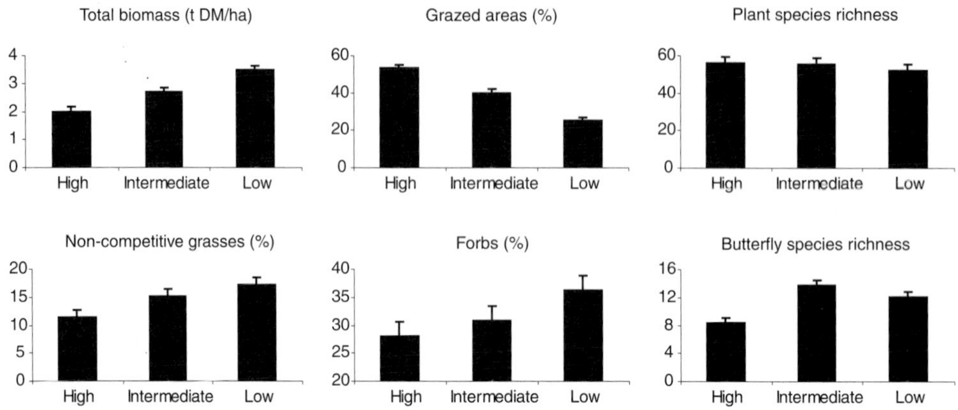

Fig. 14.1. The effect of grazing intensity on sward structural heterogeneity and biodiversity value (adapted from Dumont *et al.*, 2009). All measurements were made during summer at the plot scale in an upland pasture of central France. High grazing intensity: 1.4 LU/ha, with 1 LU = 600 kg live weight; Intermediate grazing intensity: 1.0 LU/ha; Low grazing intensity: 0.6 LU/ha.

obstacles (Bakker, 1998). Information on the processes leading to divergent vegetation dynamics is even more limited. The species of grazer is an important component in determining vegetation dynamics, since different species have different feeding styles (Rook *et al.*, 2004). Cattle are less selective than sheep, due to their larger mouth size, and are more willing to graze taller vegetation. Horses are able to graze closer to the ground than cattle as a result of having incisor teeth in both their upper and lower jaws. The studies by Grant *et al.* (1996) and Warren *et al.* (2002) demonstrated the effect of differences in selectivity between cattle and sheep on sward composition. However, under different climatic and/or edaphic conditions, the difference in impact on the grassland plant community between cattle and sheep may be relatively minor compared with the impact of grazing intensity per se (Stewart and Pullin, 2008; Dumont *et al.*, 2010) or grazing season (Wallace *et al.*, 2008). A reason for this is that both sheep and cattle select for legumes, forbs and short rather than tall vegetative patches, consistent with the usually higher nutritive value of these short vegetative patches (Dumont *et al.*, 2007). Adler *et al.* (2001) coined the term 'patch grazing' to describe this behaviour. Selective grazing by cattle on legumes and forbs and avoidance of reproductive grasses has been shown to result in some degree of intra- and even inter-annual stability of grazing patterns in a fertile grassland that was lightly grazed by cattle (Rossignol *et al.*, 2011). In the same experiment, plots grazed by sheep also showed some level of patch stability between the peak of biomass and the autumn, but no inter-annual stability. Grazing by cattle thus seems more likely to maintain relatively short cover over successive years, and would consequently favour functional vegetation diversity by creating relatively stable open patches, enabling prostrate forbs and legumes to compete with tall grasses. In the long term, this may lead to a clearer effect of grazer species on sward botanical composition, as these patches and under-grazed areas could become dominated by very different plant species (Willms *et al.*, 1988). Cattle grazing has indeed been shown to increase heterogeneity in sub alpine grasslands in the Pyrenees. In contrast, highly selective sheep grazing produced a strong homogenization effect, but also favoured the presence of a specific set of species with a high conservation value (Sebastià *et al.*, 2008). Horses graze short patches even more than cattle (Menard *et al.*, 2002) and should thus also maintain the mosaic of tall and short vegetation stands. In a coastal grassland of western France, horse grazing enhanced the development of rosette, sub-halophyte and halophyte species where the soil was saline, which increased species richness. The greatest effect was, however, observed under mixed grazing by cattle and horses, as cattle limited the strongly competitive grasses that developed on horse latrine areas (Loucougaray *et al.*, 2004). The long-term effects of horse feeding style still need to be investigated in different grassland communities.

In the literature, there are only a small number of direct breed comparisons under controlled conditions that indicate consistent differences in the choice made by sheep or cattle breeds. These comparisons support the hypothesis that the so-called breed effect can largely be explained by differences in body size and the consequent allometric relationships with food intake, digestibility and selectivity (Dumont *et al.*, 2007). The weak amplitude of these differences did not allow clear management implications to be identified when the impacts of traditional and commercial livestock breeds were compared on four grassland communities across Europe (Scimone *et al.*, 2007; WallisDeVries *et al.*, 2007). In addition to the effect of body size, prior experience of biodiverse pastures during early life can affect subsequent diet selection. Sheep that experienced two generally unpalatable plant species during their first year ate more of them, than naive sheep, during the next 2 years (Arnold and Maller, 1977). On a diverse tropical savannah, the diet of goats also varied with the origin and year of birth of the animals, even though they then grazed the same pasture for 4 years (Biquand and Biquand-Guyot, 1992). Early dietary experience can have persistent effects on food consumption, through changes in critical physiological mechanisms. Nine months after initial exposure, sheep reared on low-quality roughage ingested more from it than naive sheep because they had a better nitrogen economy (Distel *et al.*, 1996). However, other experimental evidence suggests that the effects of experience in young animals can be of short duration (Rook *et al.*, 2004). In order to ensure maximum nature conservation benefits are achieved by manipulating the early learning experience of grazers, a better understanding of interactions between animal types and forage types in shaping vegetation communities is required.

14.5 Grazing and the Conservation of Faunal Diversity in Grasslands

Irrespective of changes in plant species richness and evenness, the rapid changes in vegetation structure that result from the application of contrasting stocking rate treatments will influence other aspects of biodiversity. Decreasing stocking rate will allow a higher structural diversity of grassland to develop, which will provide a greater number of ecological niches and buffered microclimates than vegetation under more intensive grazing pressures (Morris, 2000; Pöyry *et al.*, 2006; WallisDeVries *et al.*, 2007). Decreasing stocking rate usually benefits the populations of invertebrate species that are ill adapted to frequent disturbance of their habitat (Fig. 14.1), through allowing undisrupted trophic interactions between plants and herbivorous insects (Kruess and Tscharntke, 2002; Pöyry *et al.*, 2006; Dumont *et al.*, 2009). For example, ungrazed plants that are allowed to flower provide important food resources for nectar and pollen feeding guilds of invertebrates. Where the grazer selects flowers, as in the case of sheep observed by Öckinger *et al.* (2006), this can be particularly detrimental to butterfly populations (Fig. 14.2). The changes in seasonal patterns of utilization affect plant morphology, allowing an increase in flower production and thus a greater potential for species to produce seeds. An empirical recommendation for increasing biodiversity in semi-natural grasslands can be to take off animals from some of the plots during the main flowering period. This had a positive impact on butterfly populations in an upland pasture that was grazed by cattle at a high stocking rate, where butterfly species richness was

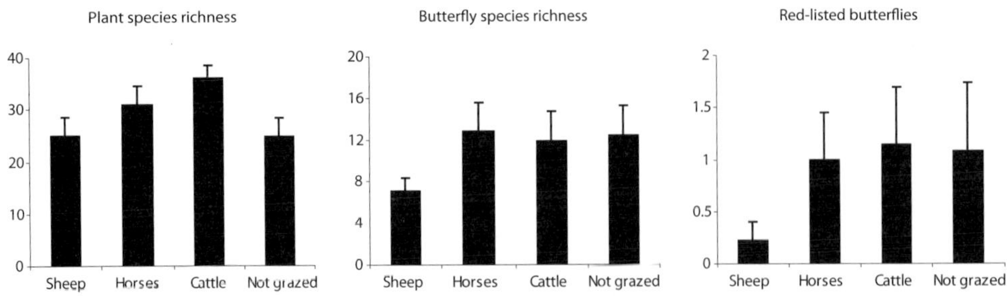

Fig. 14.2. The effect of grazing by different herbivore species on species richness of plants, butterflies in general and rare, 'red-listed' butterflies, in semi-natural grasslands in southern Sweden (Öckinger *et al.*, 2006).

2.5 times higher and butterfly abundance twofold higher than in plots that were continuously grazed at the same stocking rate (Farruggia *et al.*, 2008). This observation supports the application of temporally varying grazing intensities for grassland nature conservation management. With regard to bird dynamics that are driven by both direct and indirect effects of grazing, reconciling ecological and productive performance objectives of a grassland system may also require the implementation of a temporal shift in grazing sequences. The maintenance of wader populations within a grassland agroecosystem was indeed achieved, at intermediate levels of productive performance, by implementing a late summer grazing sequence without any spring grazing and thus no trampling (Sabatier *et al.*, 2010).

14.6 Conclusion

Restoration and maintenance of species richness/biodiversity in grasslands is necessary for multifunctionality. In order to achieve a high biodiversity value, the maximum stocking rates and thus potential output from moderately fertile temperate grasslands will probably have to be strongly reduced. WallisDeVries (1998) proposed the use of only 25% of forage availability for conservation purposes in temperate semi-natural grasslands. Thus the levels of biodiversity/multifunctionality that grassland agroecosystems provide will depend upon the amount of support that society is prepared to invest in them. Management for biodiversity conservation requires the integration of ecological and socio-economic systems at a landscape scale. There is an urgent need to identify complementarities and trade-offs between agri-environmental scheme objectives, to maintain or enhance farmland biodiversity, and those aimed at maintaining other ecosystem services. This has important implications for the economic cost and practicality of implementing environmental management practices in grassland-based livestock farming systems.

References

Adler, P.B., Raff, D.A. and Lauenroth, W.K. (2001) The effect of grazing on the spatial heterogeneity of vegetation. *Oecologia* 128, 465–479.

Aguiar, M.R. (2005) Biodiversity in grasslands: current changes and scenarios for the future. In: Reynolds, S.G. and Frame, J. (eds) *Grasslands: Developments Opportunities Perspectives*, FAO and Science Publishers, Rome.

Arnold, G.W. and Maller, R.A. (1977) Effects of nutritional experience in early and adult life on the performance and dietary habits of sheep. *Applied Animal Ethology* 3, 5–26.

Baer, S.G., Blair, J.M., Collins, S.L. and Knapp A.K. (2003) Soil resources regulate productivity and diversity in newly established tallgrass prairie. *Ecology* 84, 724–735.

Bakker, J.P. (1998) The impact of grazing on plant communities. In: WallisDeVries, M.F., Bakker, J.P. and Van Weiren, S.E. (eds) *Grazing and Conservation Management*, Kluwer Academic Publishers, Dordrecht, pp. 137–184.
Bakker, J.P. and Berendse, F. (1999) Constraints in the restoration of ecological diversity in grassland and heathland communities. *Trends in Ecology and Evolution* 14, 63–68.
Benton, T.G., Vickery, J.A. and Wilson, J.D. (2003) Farmland biodiversity: is habitat heterogeneity the key? *Trends in Ecology and Evolution* 18, 182–188.
Biquand, S. and Biquand-Guyot, V. (1992) The influence of peers, lineage and environment on food selection of the criollo goats (*Capra hircus*). *Applied Animal Behaviour Science* 34, 231–245.
Bradbury, R.B., Stoate, C. and Tallowin, J.R.B. (2010) Lowland farmland bird conservation in the context of wider ecosystem service delivery. *Journal of Applied Ecology* 47, 986–993.
Distel, R.A., Villalba, J.J., Laborde, H.E. and Burgos, M.A. (1996) Persistence of the effects of early experience on consumption of low-quality roughage by sheep. *Journal of Animal Science* 74, 965–968.
Dumont, B., Rook, A.J., Coran, Ch. and Röver, K.U. (2007) Effects of livestock breed and grazing intensity on biodiversity and production in grazing systems. 2. Diet selection. *Grass and Forage Science* 62, 159–171.
Dumont, B., Farruggia, A., Garel, J.P., Bachelard, P., Boitier, E. and Frain, M. (2009) How does grazing intensity influence the diversity of plants and insects in a species-rich upland grassland on basalt soils? *Grass and Forage Science* 64, 92–105.
Dumont, B., Carrère, P., Rossignol, N., Chadoeuf, J., Farruggia, A., Ginane, C. and Louault, F. (2010) Selective grazing, patch stability and vegetation dynamics in a rotationally-grazed pasture. *Grassland Science in Europe* 15, 937–939.
Farnsworth, K.D. and Anderson, A.R.A. (2001) How simple grazing rules can lead to persistent boundaries in vegetation communities. *Oikos* 95, 15–24.
Farruggia, A., Dumont, B., Leroy, T., Duval, C. and Garel, J.P. (2008) Ecological rotation favours biodiversity in beef cattle systems in the French Massif Central. *Grassland Science in Europe* 13, 60–62.
Fleischner, T.L. (1994) Ecological costs of livestock grazing in western North America. *Conservation Biology* 8, 629–644.
Grant, S.A., Torvell, L., Sim, E.M., Small, J.L. and Armstrong, R.H. (1996) Controlled grazing studies on *Nardus* grassland: effects of between-tussock sward height and species of grazer on *Nardus* utilization and floristic composition in two fields of Scotland. *Journal of Applied Ecology* 33, 1053–1064.
Grime, J.P. (2002) *Plant Strategies, Vegetation Processes, and Ecosystem Properties*, 2nd Edn. John Wiley and Sons Ltd, Chichester, UK.
Hodgson, J.G., Montserrat-Marti, G., Tallowin, J.R.B., Thompson, K., Diaz, S., Cabido, M. *et al.* (2005) How much will it cost to save grassland diversity? *Biological Conservation* 122, 263–273.
Janssens, F., Peeters, A., Tallowin, J.R.B., Bakker, J.P., Bekker, R.M., Fillat, F. and Oomes, M.J.M. (1998) Relationship between soil chemical factors and grassland diversity. *Plant and Soil* 202, 69–78.
Jouven, M. and Baumont, R. (2008) Simulating grassland utilization in beef suckler systems to investigate the trade-offs between production and floristic diversity. *Agricultural Systems* 96, 260–272.
Kruess, A. and Tscharntke, T. (2002) Contrasting responses of plant and insect diversity to variation in grazing intensity. *Biological Conservation* 106, 293–302.
Louault, F., Pillar, V.D., Aufrère, J., Garnier, E. and Soussana, J.F. (2005) Plant traits and functional types in response to reduced disturbance in a semi-natural grassland. *Journal of Vegetation Science* 16, 151–160.
Loucougaray, G., Bonis, A. and Bouzillé, J.B. (2004) Effects of grazing by horses and/or cattle on the diversity of coastal grasslands in western France. *Biological Conservation* 116, 59–71.
Loydi, A., Distel, R.A. and Zalba, S.M. (2010) Large herbivore grazing and non-native plant invasions in Montane grasslands of central Argentina. *Natural Areas Journal* 30, 148–155.
Marini, L., Fontana, P., Scotton, M. and Klimek, S. (2008) Vascular plant and Orthoptera diversity in relation to grassland management and landscape composition in the European Alps. *Journal of Applied Ecology* 45, 361–370.
Marriott, C.A., Fothergill, M., Jeangros, B., Scotton, M. and Louault, F. (2004) Long-term impacts of extensification of grassland management on biodiversity and productivity in upland areas. A review. *Agronomie* 24, 447–462.
Menard, C., Duncan, P., Fleurance, G., Georges, J.Y. and Lila, M. (2002) Comparative foraging and nutrition of horses and cattle in European wetlands. *Journal of Applied Ecology* 39, 120–133.
Milchunas, D.G., Sala, O.E. and Lauenroth, W.K. (1988) A generalized model of the effects of grazing by large herbivores on grassland community structure. *American Naturalist* 132, 87–106.

Morris, M.G. (2000) The effects of structure and its dynamics on the ecology and conservation of arthropods in British grasslands. *Biological Conservation* 95, 129–142.
Mortimer, S.R., Kessock-Philip, R., Potts, S.G., Ramsay, A.J., Roberts, S.P.M., Woodcock, B.A.., Hopkins, A., Gundrey, A., Dunn, R., Tallowin, J.R.B., Vickery, J. and Gough, S. (2006) *Review of the Diet and Micro-habitat Values for Wildlife and the Agronomic Potential of Selected Grassland Plant Species*, English Nature Research Report Number 697, Peterborough, UK.
Öckinger, E., Eriksson, A.K. and Smith, H.G. (2006) Effects of grassland abandonment, restoration and management on butterflies and vascular plants. *Biological Conservation* 133, 291–300.
Poschlod, P. and Bonn, S. (1998) Changing dispersal processes in the central European landscape since the last ice age: an explanation for the actual decrease of plant species richness in different habitats? *Acta Botanica Neerlandica* 47, 27–44.
Pöyry, J., Luoto, M., Paukkunen, J., Pykälä, J., Raatikainen, K. and Kuussaari, M. (2006) Different responses of plants and herbivore insects to a gradient of vegetation height: an indicator of the vertebrate grazing intensity and successional age. *Oikos* 115, 401–412.
Pywell, R.F., Bullock, J.M., Tallowin, J.R.B., Walker, K.J., Warman, E.A. and Masters, G. (2007) Enhancing diversity of species-poor grasslands: an experimental assessment of multiple constraints. *Journal of Applied Ecology* 44, 81–94.
Rook, A.J., Dumont, B., Isselstein, J., Osoro, K., WallisDeVries, M.F., Parente, G. and Mills J. (2004) Matching type of livestock to desired biodiversity outcomes in pastures - a review. *Biological Conservation* 119, 137–150.
Rossignol, N., Chadoeuf, J., Carrère, P. and Dumont, B. (2011) A hierarchical model for analyzing the stability of vegetation patterns created by grazing in temperate pastures. *Applied Vegetation Science* 14, 189–199.
Sabatier, R., Doyen, L. and Tichit, M. (2010) Modelling trade-offs between livestock grazing and wader conservation in a grassland agroecosystem. *Ecological Modelling* 221, 1292–1300.
Scimone, M., Rook, A.J., Garel, J.P. and Sahin, N. (2007) Effects of livestock breed and grazing intensity on grazing systems: 3. Effects on diversity of vegetation. *Grass and Forage Science* 62, 172–184.
Sebastià, M.-T., de Bello, F., Puig, L. and Taull, M. (2008) Grazing as a factor structuring grasslands in the Pyrenees. *Applied Vegetation Science* 11, 215–222.
Stewart, G.B. and Pullin, A.S. (2008) The relative importance of grazing stock type and grazing intensity for conservation of mesotrophic 'old meadow' pasture. *Journal for Nature Conservation* 16, 175–185.
Suding, K.N., Collins, S.L., Gough, L., Clark, C., Cleland, E.E., Gross, K.L., Milchunas, D.G. and Pennings, S. (2005) Functional- and abundance-based mechanisms explain diversity loss due to N fertilization. *Proceedings of the National Academy of Sciences USA* 102, 4387–4392.
Tallowin, J.R.B., Smith, R.E.N., Goodyear, J. and Vickery, J.A. (2005) Spatial and structural uniformity of lowland agricultural grassland in England: a context for low biodiversity. *Grass and Forage Science* 60, 225–236.
Veen, P., Jefferson, R., de Smidt, J. and van der Straaten, J. (2009) *Grasslands in Europe of High Nature Value.* KNNV Publishing, Zeist, The Netherlands.
Wade, M.R., Gurr, G. M. and Wratten, S.D. (2008) Ecological restoration of farmland: progress and prospects. *Philosophical Transactions of the Royal Society of London B* 363, 831–847.
Walker, K.J., Stevens, P.A., Stevens, D.P., Mountford, J.O., Manchester, S.J. and Pywell, R.F. (2004) The restoration and re-creation of species-rich lowland grassland on land formerly managed for intensive agriculture in the UK. *Biological Conservation* 119, 1–18.
Wallace, J.M., Wilson, L.M. and Launchbaugh, K.L. (2008) The effect of targeted grazing and biological control on yellow starthistle (*Centaurea solstitialis*) in Canyon grasslands of Idaho. *Rangeland Ecology and Management* 61, 314–320.
WallisDeVries, M.F. (1998) Habitat quality and the performance of large herbivores. In: WallisDeVries, M.F., Bakker, J.P. and Van Weiren, S.E. (eds) *Grazing and Conservation Management*. Kluwer Academic Publishers, Dordrecht, pp. 275–320.
WallisDeVries, M.F., Parkinson, A.E., Dulphy, J.P., Sayer, M. and Diana, E. (2007) Effects of livestock breed and grazing intensity on biodiversity and production in grazing systems. 4. Effects on animal diversity. *Grass and Forage Science* 62, 185–197.
Warren, J., Christal, A. and Wilson, F. (2002) Effects of sowing and management on vegetation succession during grassland habitat restoration. *Agriculture, Ecosystems and Environment* 93, 393–402.
White, R.P. and Vannaselt, W. (2001) Still and silent ecosystems: declining grassland biodiversity. In: *Pilot Analysis of Global Ecosystems: Grassland Ecosystems.* EarthTrends 2001, World Resources Institute, Washington, DC, USA.

Willms, W.D., Dormaar, J.F. and Schaalje, G.B. (1988) Stability of grazed patches on rough fescue grasslands. *Journal of Range Management* 41, 503–508.

Wilson, J.D., Whittingham, M.J. and Bradbury, R.B. (2005) The management of crop structure: a general approach to reversing the impacts of agricultural intensification on birds? *Ibis* 147, 453–463.

15 Assessing the Functional Role of Plant Diversity in Grasslands: a Trait-based Approach

Eric Garnier and Marie-Laure Navas

15.1 Introduction

The objective of this chapter is to provide some insights into the functional role of plant diversity in grasslands. To do so, we will follow a trait-based approach to species functioning, and show how it can be used, in combination with community structure, to infer some properties of ecosystems, and to serve as a basis for grassland management.

We will stick to the idea that a trait is measured at the level of an individual organism (see McGill *et al.*, 2006, for more details; Lavorel *et al.*, 2007), with the following definition: 'any morphological, physiological or phenological feature measurable at the individual level, from the cell to the whole-organism level' (Violle *et al.*, 2007). We will present recent developments in grassland ecology using traits, largely based on the response-effect framework proposed by Lavorel and Garnier (2002) and further refined by Suding *et al.* (2008). A simplified scheme of this framework, incorporating ecosystem services provided by grasslands, is presented in Fig. 15.1: environmental drivers act as filters sorting species according to the value of their traits (so-called 'response traits'), which results in a functional structure of communities depending on the type and strength of these filters. In turn, the functional structure of communities, defined as the value, range and relative abundance of traits, has various impacts on ecosystem properties (via so-called 'effect traits') and services (Diaz *et al.*, 2007b). We examine below how this general framework applies to grasslands.

15.2 Management Practices in Grasslands

If we are to understand quantitative relationships between plant functioning – as assessed here by traits – and environmental drivers, standardization of variables is necessary. This has been done for trait measurements over the last two decades (cf. Garnier *et al.*, 2007, and references therein) but, as stressed elsewhere (Hodgson *et al.*, 2005; McGill *et al.*, 2006; Garnier *et al.*, 2007), much remains to be done in the case of environmental variables.

In grasslands, the defoliation regime (either grazing or mowing for hay making) and the supply of fertilizers appear as the main management practices affecting vegetation structure and functioning (Fig. 15.1). How can these be assessed quantitatively?

Characterizing the disturbance regime in grasslands is complex and requires several variables. Building on the disturbance descriptors developed by White and

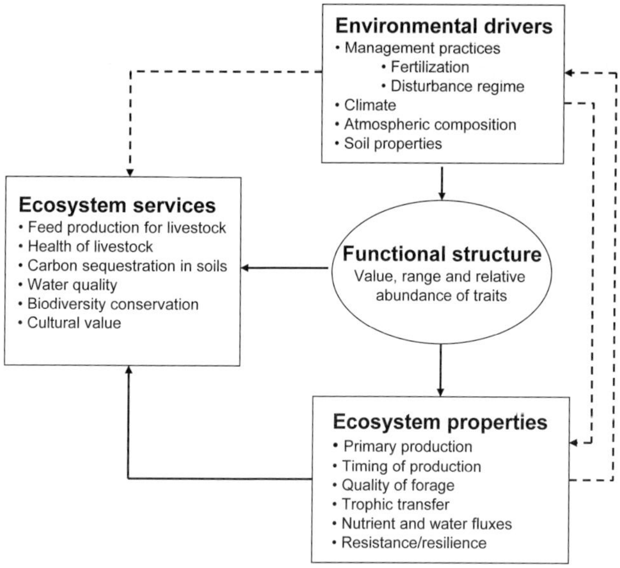

Fig. 15.1. The functional structure of community reflects the differential response of species to environmental drivers, which depend on their trait values, and affects ecosystem properties. Although this scheme is general to any type of ecosystem, examples pertaining to grasslands and dealt with in the present chapter have been selected in the different boxes. Solid arrows indicate relationships addressed in this chapter. Adapted from Diaz *et al.* (2007b).

Pickett (1985), Duru *et al.* (1998) and Kleyer (1999), a particular effort has been devoted to the identification of such variables during the course of a recent European programme (Garnier *et al.*, 2007). Five parameters were used to describe the disturbance regime: (i) pre-treatment regime (how was the field used before its current use); (ii) type of disturbance (grazing, identity of grazer, mowing); (iii) onset of disturbance (Julian day in the year); (iv) return interval of disturbance (year); and (v) intensity of disturbance (percentage biomass removed). The task of quantification across the 11 sites of the programme was conducted adequately for the first three parameters, while the qualification of intensity was fine in most situations. By contrast, the parameterization of the return interval proved delicate, pointing to the need for more work, on the design of indices and their practical estimation, before this aspect of disturbances can be quantified adequately.

Although the assessment of nutrient limitation could theoretically be done through soil measurements, there is currently no simple method to assess the amount of nitrogen or phosphorus a given soil can release and make available to plants. This has led to the development of nutrient indices for nitrogen (NNI; Lemaire and Gastal, 1997) and phosphorus (PNI; Duru and Ducrocq, 1997). NNI (see Table 15.1 for a list of abbreviations used in the text) is based on plant nitrogen concentration determination and directly estimates the nutrient status of the vegetation in relation to the degree of potential growth. The NNI has been successfully applied to diagnose the nitrogen nutritional status of mixed crops (Cruz and Soussana, 1997) as well as native grassland (Duru *et al.*, 1997). PNI is based on simultaneous measurements of nitrogen and phosphorus concentrations of the vegetation. Within sites, PNI is positively correlated with soil Olsen P values, a soil fraction related to available phosphorus for plants (Jouany *et al.*, 2002; Garnier *et al.*, 2007), but this relationship does not seem to hold across sites.

Table 15.1. List of abbreviations used in the text.

Category	Abbreviation	Meaning
Trait	LDMC	Leaf dry matter content
	LNC	Leaf nitrogen concentration
	RGR	Relative growth rate
	RGR_{max}	Maximum relative growth rate
	SLA	Specific leaf area
	PH	Plant height
	DFl	Date of flowering
Ecosystem properties	AGB_{max}	Maximum above-ground standing biomass
	ANPP	Above-ground net primary productivity
	NPP	Net primary productivity
	NNI	Nitrogen nutrition index
	PNI	Phosphorus nutrition index
Others	BRH	Biomass ratio hypothesis
	CWM	Community-level weighted mean of trait

Although these are the main environmental factors influencing the structure and functioning of grasslands, the integration of temperature and water availability, both in isolation and in interaction with management practices, should be taken into account in the description of the environment (Fig.15.1), if we are to compare contrasting situations, especially in the context of current climatic changes.

15.3 Trait Response to Management Practices and Community Structure

15.3.1 Species traits

Data from the European programme mentioned above showed that management practices in grasslands have a large impact on trait responses, being responsible for 15% of variations in trait values (to be compared with the 40% due to climate and 21% to soil properties) over the 11 sites studied (Pakeman *et al.*, 2009). Which traits are likely to be affected by these practices?

The stature and morphology of the plants, assessed through a panel of different traits, are an obvious aspect to start with. Specific leaf area (SLA, the ratio of leaf surface to leaf dry mass) and leaf nitrogen concentration (LNC) are of particular importance as they directly relate to resource economy of plants (Wright *et al.*, 2004). Leaf dry matter content (LDMC, the ratio of leaf dry mass to leaf water-saturated fresh mass), although less tightly related to resource economy (Poorter and Garnier, 2007), can be used to assess digestible organic matter (Al Haj Khaled *et al.*, 2006, and see below). Plant height (PH) depicts species' carbon gain in relation to plant ability to compete for light (Westoby *et al.*, 2002) and is part of a suite of life-history traits including reproductive traits (Moles and Leishman, 2008). Date of flowering (DFl) is a key trait for understanding the accumulation of herbage mass, as it determines the date on which the peak biomass occurs (Ansquer *et al.*, 2009b).

Grazing affects a large series of traits: a high grazing pressure generally favours annual over perennial plants, short plants over tall plants, prostrate over erect plants, and species with stoloniferous and rosette architecture over those with a tussock architecture; the timing of defoliation also impacts phenology, including leaf senescence (Duru *et al.*, 1998; Hodgson *et al.*, 2005; Louault *et al.*, 2005; Diaz *et al.*, 2007c). Mowing favours species with stoloniferous and rosette architecture (Pakeman *et al.*, 2009), increases LDMC and decreases LNC, without effect on SLA (Pontes *et al.*, 2007). In abandoned grasslands, tall species with high LDMC and low LNC leaves become dominant (Kahmen and Poschlod, 2004). With increasing nutrient availability, especially nitrogen, both SLA and LNC tend to increase (Lavorel and Garnier, 2002; Lavorel *et al.*, 2007; Ordoñez *et al.*, 2009), while LDMC tends to decrease (Al Haj Khaled *et al.*, 2005; McIntyre, 2008).

15.3.2 The functional structure of communities

The functional structure of communities, as highlighted in Fig. 15.1, can be described through two components, value and range of traits, corresponding to two types of quantitative variables. The first ones allow one to estimate the average value of traits in the communities and can be written as:

$$CFP = \sum_{i=1}^{n} p_i \times trait \qquad (15.1)$$

where *CFP* represents the functional parameter of the community (Violle *et al.*, 2007), p_i and $trait_i$ are, respectively, a weighting factor and the trait value for species *i*, and *n* is the total number of species in the community. When only presence/absence data for species are available, all p_i are fixed at the same value (*1/n*), and the calculated *CFP* are unweighted averages; when species abundances are known, a common procedure is to use the relative abundance of species as p_i values, thereby calculating community weighted mean (CWM) values (Garnier *et al.*, 2004; see also Hodgson *et al.*, 2005 for different weighing coefficients).

The second type of variables aims at describing the functional dissimilarity among the species present in the communities. Although numerous indices have been designed to do this (see Garnier and Navas, 2011), there is an ongoing debate as to the actual form these should take in order to deal with several traits if necessary, take into account species abundance, and measure all facets of functional diversity.

One of the few experiments documenting the change in functional structure of grasslands was performed in temperate mountain grasslands that were mown or grazed and displayed a range of nutrient status (Duru *et al.*, unpublished results). Enhanced nutrient availability induced an increase in CWM_SLA and CWM_PH, a decrease in CWM_LDMC and a decrease in dissimilarity of these traits. The magnitude of response depended on the defoliation regime, with larger responses for pastures than for meadows, despite equal range of nutrient status.

15.4 Functional Structure of Communities and Grassland Properties

Ecosystem properties and services are controlled directly by environmental factors (for a detailed account see Chapin *et al.*, 2002) and by the functional structure of communities (Fig. 15.1). Here, we will focus on studies that have explicitly investigated the controls that plants, as major players in the biotic component, might exert on ecosystem properties (lower part of Fig. 15.2).

15.4.1 Community weighted means and the biomass ratio hypothesis

The 'biomass ratio hypothesis' (BRH: Grime, 1998) states that the effect traits of locally abundant species determine the rate and magnitude of instantaneous ecosystem processes (Diaz and Cabido, 2001; Lavorel and Garnier, 2002; cf. Fig. 15.2). An increasing number of studies shows that this hypothesis holds for major ecosystem properties (Garnier *et al.*, 2004; Diaz *et al.*, 2007a; Violle *et al.*, 2007), including the three major components of agronomic value of grasslands: amount and timing of biomass production and digestibility.

Productivity: amount and timing

Lavorel and Garnier (2002) have proposed a formalization of the BRH as applied to the productivity of ecosystems as:

$$NPP = \frac{\sum_{i=1}^{n} N_i \times Mo_i \times (e^{RGR_i \times (tf - t0)_i} - 1)}{\Delta T} \qquad (15.2)$$

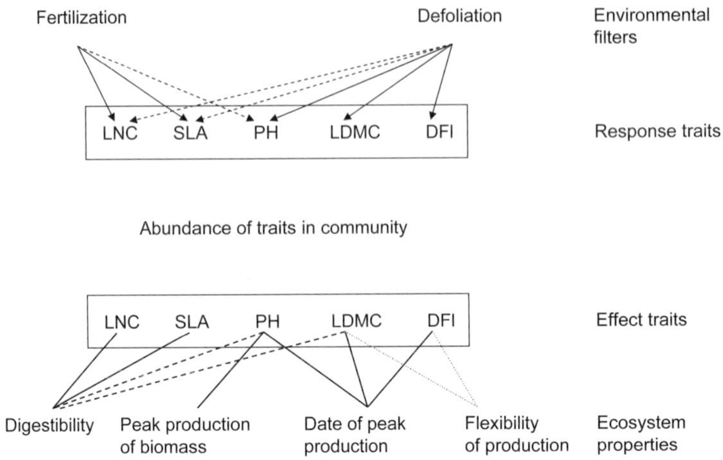

Fig.15.2. Summary of trait responses to environmental filters pertaining to management practices, and effects of traits on ecosystem properties in grasslands. The upper part of the scheme shows the positive (black line) or negative (dashed line) impact of nutrient status and defoliation regime of grasslands on leaf nitrogen concentration (LNC), specific leaf area (SLA), plant height (PH), leaf dry matter content (LDMC) and date of flowering (DFI). The lower part of the scheme presents the positive (black line) or negative (dashed line) impact of community-weighted means of traits on grassland properties; the dotted line indicates a positive impact of the functional divergence of date of flowering and LDMC on flexibility of production.

where n is the total number of species in the community, N_i is the number of individuals of species i per unit ground area, Mo_i is the initial average biomass of individuals of species i, RGR_i and $(tf - t0)_i$ are, respectively, the relative growth rate and period of active growth of species i, and ΔT is the period over which NPP is assessed. Due to the difficulty in accessing the underground compartment, to date, most tests of Equation 2 have been conducted taking only above-ground plant parts into account.

Since realized RGR is particularly difficult to assess in the field, this relationship has been tested using either RGR data measured in the lab (RGR_{max}), or independent measurements of traits (especially leaf traits) known to be related to RGR. Looni *et al.* (2009) report that inter-specific differences in RGR, measured under controlled conditions on dominant species of grazed and ungrazed plots of Uruguayan grasslands, tend to scale up with above-ground NPP (ANPP) measured in the plots (Altesor *et al.*, 2005). ANPP was found to be related to CWM_leaf angle in Australian grasslands (Mokany *et al.*, 2008), to CWM_leaf area in subalpine grasslands in the French Alps (Gross *et al.*, 2008), and to a combination of CWM_LDMC and traits of roots and rhizomes in mesocoms of mountain grasslands of Central France (Klumpp and Soussana, 2009).

In studies conducted where maximum above-ground standing biomass (AGB_{mass}) was used as an estimate of ANPP (see Scurlock *et al.*, 2002 for discussion), AGB_{mass} was found to be positively related to CWM_leaf area (Thompson *et al.*, 2005; Mokany *et al.*, 2008) and CWM_PH (Ansquer *et al.*, 2009b), and negatively related to CWM_SLA and CWM_LDMC (Garnier *et al.*, 2004), whereas no relationship was found with any trait by Díaz *et al.* (2007a).

Although an important component of the agronomic value of grasslands, we are aware of only one study where the timing of biomass production has been related to traits. This study, conducted in the same temperate mountain grasslands mentioned above, showed that the date of peak vegetative biomass production was positively related to CWM_DFl, and negatively with CWM_PH (Ansquer *et al.*, 2009b), indicating that shorter vegetation produces biomass and flowers earlier than taller vegetation.

Digestibility

In monospecific stands, the digestibility of the biomass produced depends on characteristics of leaves and stems, and on proportion thereof (Al Haj Khaled *et al.*, 2006; Pontes *et al.*, 2007; Duru *et al.*, 2008). Traits related to the digestibility of organs include their chemical composition, such as fibre, (hemi)celluloses and lignin concentrations (Al Haj Khaled *et al.*, 2006), all of which are known to be higher in stems than in leaves (Poorter and Bergkotte, 1992). This leads to a lower digestibility in the former and to a strong impact of the leaf/stem ratio on digestibility of the whole plant (Duru *et al.*, 2008). At the organ level, LDMC, SLA and LNC are traits that are easy to measure, and relate best to digestibility (Al Haj Khaled *et al.*, 2006; Pontes *et al.*, 2007).

In multi-specific grasslands, the BRH is likely to apply to digestibility as well, but an actual formalism, comparable with that proposed above for NPP, has not been proposed as yet. A limited number of empirical tests have shown that CWM_LDMC (Ansquer *et al.*, 2009b; P. Cruz *et al.*, unpublished observations) and CMW_PH (Ansquer *et al.*, 2009b) tend to scale negatively with community level digestibility. The latter appears to be a consequence of taller vegetation having a higher stem/leaf ratio (Duru *et al.*, 2008).

15.4.2 Functional dissimilarity among species

So far, only a few studies have examined the impacts of functional dissimilarity of species on ecosystem properties (reviewed in Garnier and Navas, 2011). A synthesis is currently difficult to conduct for at least two reasons. First, the restricted number of studies available limits our potential to generalize safely. A result of interest is the correlation between the functional dissimilarity in LDMC and the variation in biomass production around the peak of herbage mass, a variable of importance for managers as it depicts the temporal flexibility of herbage use (Ansquer *et al.*, 2009b). Second, the outcomes of the analyses strongly depend on the way the different variables are considered in the statistical procedures (see Garnier and Navas, 2011, for more details). A striking example is given by a study conducted in grasslands of Germany (Schumacher and Roscher, 2009): the effects of dissimilarity of traits on AGB_{mass} were positive using a stepwise procedure that did not place any constraint on the selection of variables introduced in the multiple regression, whereas these effects were found to be negative using the selection procedure proposed by Diaz *et al.* (2007a).

15.5 Traits and Grassland Management

A major question for research is to define which traits could be used to direct management of grasslands for production or restoration, taking into account the context of current climatic changes.

15.5.1 Management for production

Research on the response of traits to grassland management led to the identification of LDMC and flowering phenology as easily measured traits accounting for change in herbage digestibility, date of peak production and flexibility of production (Ansquer *et al.*, 2009a) (Fig. 15.2). As the direct assessment of CWM_LDMC remains difficult to use in an applied context, some simplifications have been proposed. The first one was to use a pre-existing functional classification of species into plant functional types, defined on the basis of LDMC measured in monocultures grown under high nitrogen supply (Ansquer *et al.*, 2004; Al Haj Khaled *et al.*, 2006). Such typologies of grasses were successfully used to predict fodder production and digestibility in European (Duru *et al.*, 2008) and South-American grasslands (Cruz *et al.*, 2010). The second simplification was to assess ecosystem processes from traits measured on dominant grasses. This was done because (i) grasses are often dominant species in grasslands, and therefore have a major impact on ecosystem processes according to BRH (Grime, 1998; see above); and (ii) traits of grasses and dicots respond similarly to gradients of fertilization and defoliation, although they differ in average values of traits: grasses show delayed phenology, higher LDMC and lower whole-plant digestibility than dicots (Ansquer *et al.*, 2009a; Duru *et al.*, 2010). A third step in the simplification procedure was to use trait values extracted from databases instead of using traits measured at the sites of study (Duru *et al.*, 2010).

15.5.2 Management for restoration

In the context of a decrease of anthropogenic pressure on grasslands in Europe, substitutes for grazing are tested in order to restore these particular ecosystems. The success of restoration can be evaluated by comparing the spectra of CSR (competitive, stress tolerator and ruderal) strategies (Grime, 1974) or of traits related to the leaf–height–seed strategies (Westoby, 1998) between restored and reference sites. Recent studies showed that there is little overlap between restored and adjacent grassland communities even after periods of time up to 60 years, with legacy of phosphorus fertilization (Fagan *et al.*, 2008). Mulching, burning or mowing induces changes in vegetation across time that highly contrast with the stability of the vegetation recorded under constant grazing (Moog *et al.*, 2005; Romermann *et al.*, 2009).

15.6 Conclusions

In this chapter, we have shown that a trait-based approach could be successfully followed to describe the effect of management practices on the structure of grasslands as well as several aspects of their functioning related to their agronomic value. This functional approach to plant diversity can be applied to other aspects of grassland management and functions, such as soil carbon storage (Klumpp and Soussana, 2009) or prevention of soil erosion (McIntyre, 2008).

Challenges ahead to trait-based research on grasslands cover a number of areas. In terms of plant functioning, whether any easily measured traits can be related to the various aspects of the regeneration niche and to demographic parameters are central issues, since they are likely to be key aspects of responses to disturbances. At the community level, understanding the mechanisms through which species traits, as driven by environmental factors and biotic interactions, determine its functional structure will require theoretical, experimental and modelling approaches. At the ecosystem level, the understanding of how and which traits affect processes, properties and services remains very preliminary. A major issue also pertains to a better identification and quantification of the environmental gradients driving trait responses, taking explicitly into account the identity and strength of the selective pressures acting in these environments (e.g. types and return times of specific disturbances in particular). A general agreement on a minimum set of required environmental variables and standardized protocols to measure them is still lacking.

Finally, a major challenge will be to switch from an evaluation of impacts of management practices to an active driving of grassland dynamics, allowing managers to select for targeted functional structures depending on specific environmental conditions. In particular, the sustainable services that grasslands can deliver, in the new climatic conditions they are going to face (Soussana and Duru, 2007), will depend on changes in their functional properties, which can be assessed through a trait-based approach.

Acknowledgements

This chapter has benefited from the continuous exchanges of people and ideas in the context of the European programme VISTA (Vulnerability of Ecosystem Services to Land Use Change in Traditional Agricultural Landscapes: contract no. EVK2-2001-000356) and within the framework of GDR 2574 'TRAITS' (CNRS, France).

References

Al Haj Khaled, R., Duru, M., Theau, J.P., Plantureux, S. and Cruz, P. (2005) Variation in leaf traits through seasons and N-availability levels and its consequences for ranking grassland species. *Journal of Vegetation Science* 16, 391–398.

Al Haj Khaled, R., Duru, M., Decruyenaere, V., Jouany, C. and Cruz, P. (2006) Using leaf traits to rank native grasses according to their nutritive value. *Rangeland Ecology and Management* 59, 648–654.

Altesor, A., Oesterheld, M., Leoni, E., Lezama, F. and Rodrìguez, C. (2005) Effect of grazing exclosure on community structure and productivity of a Uruguayan grassland. *Plant Ecology* 179, 83–91.

Ansquer, P., Theau, J.P., Cruz, P., Viegas, J., Khaled, R.A.H. and Duru, M. (2004) Caractérisation de la diversité fonctionnelle des prairies à flore complexe: vers la construction d'outils de gestion. *Fourrages* 179, 353–368.

Ansquer, P., Duru, M., Theau, J.P. and Cruz, P. (2009a) Convergence in plant traits between species within grassland communities simplifies their monitoring. *Ecological Indicators* 9, 1020–1029.

Ansquer, P., Duru, M., Theau, J.P. and Cruz, P. (2009b) Functional traits as indicators of fodder provision over a short time scale in species-rich grasslands. *Annals of Botany* 103, 117–126.

Chapin, F.S., III, Matson, P.A. and Mooney, H.A. (2002) *Principles of Terrestrial Ecosystem Ecology*. Springer-Verlag, New York.

Cruz, P. and Soussana, J.-F. (1997) The nitrogen requirement of major agricultural crops. Mixed crops. In: Lemaire, G. (ed.) *Diagnosis of the Nitrogen Status in Crops*. Springer-Verlag, Berlin, pp. 131–144.

Cruz, P., De Quadros, F.L.F., Theau, J.P., Frizzo, A., Jouany, C., Duru, M. and Carvalho, P.C.F. (2010) Leaf traits as functional descriptors of the intensity of continuous grazing in native grasslands in the south of Brazil. *Rangeland Ecology and Management* 63, 350–358.

Diaz, S. and Cabido, M. (2001) Vive la difference: plant functional diversity matters to ecosystem processes. *Trends in Ecology and Evolution* 16, 646–655.

Diaz, S., Lavorel, S., Bello, F.d., Quétier, F., Grigulis, K. and Robson, T.M. (2007a) Incorporating plant functional diversity effects in ecosystem service assessments. *Proceedings of the National Academy of Sciences USA* 104, 20,684–20,689.

Diaz, S., Lavorel, S., Chapin, F.S., Tecco, P.A., Gurvich, D.E. and Grigulis, K. (2007b) Functional Diversity- at the Crossroads between Ecosystem Functioning and Environmental Filters. In: Canadell, J. G., Pataki, D. and Pitelka, L. (eds) *Terrestrial Ecosystems in a Changing World*. Springer-Verlag, Berlin, pp. 81–91.

Diaz, S., Lavorel, S., McIntyre, S., Falczuk, V., Casanoves, F., Milchunas, D.G., Skarpe, C., Rusch, G., Sternberg, M., Noy-Meir, I., Landsberg, J., Zhang, W., Clark, H. and Campbell, B.D. (2007c) Plant trait responses to grazing - a global synthesis. *Global Change Biology* 13, 313–341.

Duru, M. and Ducrocq, H. (1997) A nitrogen and phosphorus herbage nutrient index as a tool for assessing the effect of N and P supply on the dry matter yield of permanent pastures. *Nutrient Cycling in Agroecosystems* 47, 59–69.

Duru, M., Lemaire, G. and Cruz, P. (1997) Grasslands. In: Lemaire, G. (ed.) *Diagnosis of the Nitrogen Status in Crops*. Springer-Verlag, Berlin, pp. 59–72.

Duru, M., Balent, G., Gibon, A., Magda, D., Theau, J.P., Cruz, P. and Jouany, C. (1998) Fonctionnement et dynamique des prairies permanentes. Exemple des Pyrénées Centrales. *Fourrages* 153, 97–113.

Duru, M., Cruz, P., Khaled, R.A.H., Ducourtieux, C. and Theau, J.P. (2008) Relevance of plant functional types based on leaf dry matter content for assessing digestibility of native grass species and species-rich grassland communities in spring. *Agronomy Journal* 100, 1622-1630.

Duru, M., Cruz, P. and Theau, J.P. (2010) A simplified method for characterising agronomic services provided by species-rich grasslands. *Crop and Pasture Science* 61, 420–433.

Fagan, K.C., Pywell, R.F., Bullock, J.M. and Marrs, R.H. (2008) Do restored calcareous grasslands on former arable fields resemble ancient targets? The effect of time, methods and environment on outcomes. *Journal of Applied Ecology* 45, 1293–1303.

Garnier, E. and Navas, M.-L. (2011) A trait-based approach to comparative functional plant ecology: concepts, methods and applications for agroecology. *Agronomy for Sustainable Development*, doi: 10.1007/s13593-011-0036-y

Garnier, E., Cortez, J., Billès, G., Navas, M.-L., Roumet, C., Debussche, M., Laurent, G., Blanchard, A., Aubry, D., Bellmann, A., Neill, C. and Toussaint, J.-P. (2004) Plant functional markers capture ecosystem properties during secondary succession. *Ecology* 85, 2630–2637.

Garnier, E., Lavorel, S., Ansquer, P., Castro, H., Cruz, P., Dolezal, J., Eriksson, O., Fortunel, C., Freitas, H., Golodets, C., Grigulis, K., Jouany, C., Kazakou, E., Kigel, J., Kleyer, M., Lehsten, V., Leps, J., Meier, T., Pakeman, R., Papadimitriou, M., Papanastasis, V.P., Quested, H., Quetier, F., Robson, M., Roumet, C., Rusch, G., Skarpe, C., Sternberg, M., Theau, J.P., Thebault, A., Vile, D. and Zarovali, M.P. (2007) Assessing the effects of land-use change on plant traits, communities and ecosystem functioning in grasslands: A standardized methodology and lessons from an application to 11 European sites. *Annals of Botany* 99, 967–985.

Grime, J.P. (1974) Vegetation classification by reference to strategies. *Nature* 250, 26–31.

Grime, J.P. (1998) Benefits of plant diversity to ecosystems: immediate, filter and founder effects. *Journal of Ecology* 86, 902–910.

Gross, N., Robson, T.M., Lavorel, S., Albert, C., Bagousse-Pinguet, Y.L. and Guillemin, R. (2008) Plant response traits mediate the effects of subalpine grasslands on soil moisture. *New Phytologist* 180, 652–662

Hodgson, J.G., Montserrat-Martí, G., Cerabolini, B., Ceriani, R.M., Maestro-Martí, M., Peco, B., Wilson, P.J., Thompson, K., Grime, J.P., Band, S.R., Bogard, A., Castro-Díez, P., Charles, M., Jones, G., Pérez-Rontomé, M.C., Caccianiga, M., Alard, D., Bakker, J.P., Cornelissen, J.H.C., Dutoit, T., Grootjans, A.P., Guerrero-Campo, J., Gupta, P.L., Hynde, A., Kahmen, S., Poschlod, P., Romo-Díez, A., Rorison, I.H., Rosén, E., Schreiber, K.-F., Tallowin, J., de Torres Espuny, L. and Villar-Salvador, P. (2005) A functional method for classifying European grasslands for use in joint ecological and economic studies. *Basic and Applied Ecology* 6, 119–131.

Jouany, C., Stroia, M.C., Farruggia, A. and Duru, M. (2002) Plant and soil indicators for P management in grassland. *Grassland Science in Europe* 7, 698–699.

Kahmen, S. and Poschlod, P. (2004) Plant functional trait responses to grassland succession over 25 years. *Journal of Vegetation Science* 15, 21–32.

Kleyer, M. (1999) Distribution of plant functional types along gradients of disturbance intensity and resource supply in an agricultural landscape. *Journal of Vegetation Science* 10, 697–708.

Klumpp, K. and Soussana, J.F. (2009) Using functional traits to predict grassland ecosystem change: a mathematical test of the response-and-effect trait approach. *Global Change Biology* 15, 2921–2934.

Lavorel, S. and Garnier, E. (2002) Predicting changes in community composition and ecosystem functioning from plant traits: revisiting the Holy Grail. *Functional Ecology* 16, 545–556.

Lavorel, S., Diaz, S., Cornelissen, H., Garnier, E., Harrison, S.P., McIntyre, S., Pausas, J.G., Perez-Harguindeguy, N. and Urcelay, C. (2007) Plant functional types: are we getting any closer to the Holy Grail? In: Canadell, J. G., Pataki, D. and Pitelka, L. (eds) *Terrestrial Ecosystems in a Changing World.* Springer-Verlag, Berlin, pp. 150–164.

Lemaire, G. and Gastal, F. (1997) N uptake and distribution in plant canopies. In: Lemaire, G. (eds) *Diagnosis of the Nitrogen Status in the Crops.* Springer-Verlag, Berlin, pp. 3–44.

Leoni, E., Altesor, A. and Paruelo, J.M. (2009) Explaining patterns of primary production from individual level traits. *Journal of Vegetation Science* 20, 612–619.

Louault, F., Pillar, V.D., Aufrere, J., Garnier, E. and Soussana, J.F. (2005) Plant traits and functional types in response to reduced disturbance in a semi-natural grassland. *Journal of Vegetation Science* 16, 151–160.

McGill, B.J., Enquist, B.J., Weiher, E. and Westoby, M. (2006) Rebuilding community ecology from functional traits. *Trends in Ecology and Evolution* 21, 178–185.

McIntyre, S. (2008) The role of plant leaf attributes in linking land use to ecosystem function in temperate grassy vegetation. *Agriculture Ecosystems and Environment* 128, 251–258.

Mokany, K., Ash, J. and Roxburgh, S. (2008) Functional identity is more important than diversity in influencing ecosystem processes in a temperate native grassland. *Journal of Ecology* 96, 884–893.

Moles, A.T. and Leishman, M.R. (2008) The seedling as part of a plant's life history strategy. In: Leck, M.A., Parker, V.T. and Simpson, R.L. (eds) *Seedling Ecology and Evolution.* Cambridge University Press, Cambridge, pp. 217–238

Moog, D., Kahmen, S. and Poschlod, P. (2005) Application of CSR- and LHS-strategies for the distinction of differently managed grasslands. *Basic and Applied Ecology* 6, 133–143.

Ordoñez, J.C., van Bodegom, P.M., Witte, J.P.M., Wright, I.J., Reich, P.B. and Aerts, R. (2009) A global study of relationships between leaf traits, climate and soil measures of nutrient fertility. *Global Ecology and Biogeography* 18, 137–149.

Pakeman, R.J., Leps, J., Kleyer, M., Lavorel, S. and Garnier, E. (2009) Relative climatic, edaphic and management controls of plant functional trait signatures. *Journal of Vegetation Science* 20, 148–159.
Pontes, L.D.S., Soussana, J.F., Louault, F., Andueza, D. and Carrère, P. (2007) Leaf traits affect the aboveground productivity and quality of pasture grasses. *Functional Ecology* 21, 844–853.
Poorter, H. and Bergkotte, M. (1992) Chemical composition of 24 wild species differing in relative growth rate. *Plant Cell and Environment* 15, 221–229.
Poorter, H. and Garnier, E. (2007) Ecological significance of inherent variation in relative growth rate and its components. In: Pugnaire, F. I. and Valladares, F. (eds) *Functional Plant Ecology*, 2nd Edn. CRC Press, Boca Raton, pp. 67–100.
Romermann, C., Bernhardt-Romermann, M., Kleyer, M. and Poschlod, P. (2009) Substitutes for grazing in semi-natural grasslands - do mowing or mulching represent valuable alternatives to maintain vegetation structure? *Journal of Vegetation Science* 20, 1086–1098.
Schumacher, J. and Roscher, C. (2009) Differential effects of functional traits on aboveground biomass in semi-natural grasslands. *Oikos* 118, 1659–1668.
Scurlock, J.M.O., Johnson, K. and Olson, R.J. (2002) Estimating net primary productivity from grassland biomass dynamics measurements. *Global Change Biology* 8, 736–753.
Soussana, J.F. and Duru, M. (2007) Grassland science in Europe facing new challenges: biodiversity and global environmental change. *CAB Reviews: Perspectives in Agriculture, Veterinary Science, Nutrition and Natural Resources* 1, 1–11.
Suding, K.N., Lavorel, S., Chapin, F.S., Cornelissen, J.H.C., Diaz, S., Garnier, E., Goldberg, D., Hooper, D.U., Jackson, S.T. and Navas, M.L. (2008) Scaling environmental change through the community-level: a trait-based response-and-effect framework for plants. *Global Change Biology* 14, 1125–1140.
Thompson, K., Askew, A.P., Grime, J.P., Dunnett, N.P. and Willis, A.J. (2005) Biodiversity, ecosystem function and plant traits in mature and immature plant communities. *Functional Ecology* 19, 355–358.
Violle, C., Navas, M.L., Vile, D., Kazakou, E., Fortunel, C., Hummel, I. and Garnier, E. (2007) Let the concept of trait be functional! *Oikos* 116, 882–892.
Westoby, M. (1998) A leaf-height-seed (LHS) plant ecology strategy scheme. *Plant and Soil* 199, 213–227.
Westoby, M., Falster, D.S., Moles, A.T., Vesk, P.A. and Wright, I.J. (2002) Plant ecological strategies: Some leading dimensions of variation between species. *Annual Review of Ecology and Systematics* 33, 125–159.
White, P.S. and Pickett, S.T.A. (1985) Natural disturbance and patch dynamics, an introduction. In: Pickett, S.T.A. and White, P.S. (eds) *The Ecology of Natural Disturbance and Patch Dynamics*. Academic Press, New York, pp. 3–13.
Wright, I.J., Reich, P.B., Westoby, M., Ackerly, D.D., Baruch, Z., Bongers, F., Cavender-Bares, J., Chapin, T., Cornelissen, J.H.C., Diemer, M., Flexas, J., Garnier, E., Groom, P.K., Gulias, J., Hikosaka, K., Lamont, B.B., Lee, T., Lee, W., Lusk, C., Midgley, J.J., Navas, M.L., Niinemets, U., Oleksyn, J., Osada, N., Poorter, H., Poot, P., Prior, L., Pyankov, V.I., Roumet, C., Thomas, S.C., Tjoelker, M.G., Veneklaas, E.J. and Villar, R. (2004) The worldwide leaf economics spectrum. *Nature* 428, 821–827.

16 Exploiting Genetic and Phenotypic Plant Diversity in Grasslands

Mike Humphreys, Athole Marshall, Rosemary Collins and Michael Abberton

16.1 Introduction

The grasslands of the world face a range of challenges from climate change, including the effects of elevated atmospheric carbon dioxide, increasing temperatures, changes in precipitation regime and higher concentrations of ground level ozone. These factors threaten productivity and species composition and quality, with potential impacts not only on livestock production but also on other important aspects that contribute to the multifunctional role of grasslands. Fortunately, grasslands harbour a vast resource of genetic diversity and species with genome plasticity that, once accessed, regulated and utilized in crop improvement programmes, can serve to safeguard the environment and its scarce water and nutrient resources. Grass and legume genotypes may be reconstructed to help combat some of the worst scenarios emanating from climate change, including providing for grassland perpetuity in response to increased exposure to abiotic stress effects, more efficient water and nutrient use, and physiological changes that improve animal nutrition and mitigate against release of greenhouse gases (GHGs).

The major species of grazed temperate pastures are outbreeding and heterozygous and, even within populations, harbour considerable genetic variation. Further evolved genetic change has facilitated the widespread distribution of such grassland species and provides a virtually limitless source of novel genetic variants of potential value in crop improvement programmes. Comparative studies of genetic and phenotypic diversity found in natural grasslands adapted to contrasting climates provides us with insights into the key mechanisms involved in resistance to specific abiotic stress extremes, and reveals components of their genetic control. By selection and transfer from adapted ecotypes of trait-specific alleles, centuries of evolved adaptations may frequently be utilized over a single generation to reproduce a comparable phenotype in a commercial cultivar. Following initial hybridization involving advanced breeders' lines and adapted ecotypes and subsequent genome introgression during backcross breeding programmes, cultivars may achieve greater resilience to stress conditions, to aid in their persistency and production. Having incorporated the relevant adaptive alleles, areas where edaphic and climatic extremes were considered previously too harsh for commercial exploitation may become accessible and suitable for the use of high-quality grassland cultivars.

Although a successful variety released in the EU will satisfy the requirement of distinctiveness, uniformity and stability (as well as value for cultivation and use), it will nevertheless be composed of genetically distinct individuals. Over time, genetic change is likely to occur within pastures, to favour genotypes best adapted to the edaphic and climatic stresses and the relevant land management regimes that prevail. In the case of white clover, stoloniferous spread can be substantial, so that large areas of a pasture consist of only one clonal genotype.

For a perennial grassland species, not only is cultivar performance and stability over years important, but also its ability to coexist within complementary species' mixtures and its competitive capabilities against invasive species. Whilst within-species variation is an important aspect of grassland diversity, another key feature is species ubiquity, which depends largely on climate and species-mixture interaction. This interaction involves both competition, and facilitation. In this regard, an extensive literature has explored the basis of grass/clover competition, with both above- and below-ground factors found to be important. Shading of clover by grass can occur, depending on clover leaf size, management and, crucially, soil nitrogen (N) status. There is thus a strong interaction between interspecific plant competition and soil, particularly for the clover/*Rhizobium* symbiosis. A further element is the interaction of the soil and plant components with the grazing animal, typically sheep or cattle. This is a complex system within which the plant breeder aims to deliver not only production and quality outputs but also ecosystem services. As an example, early spring growth of white clover (aided by effective symbiosis at lower soil temperatures than would usually be expected) facilitates growth of the grass companion (without N fertilizer addition) and brings the possibility of extended grazing periods. Hydraulic lift is another positive species interaction that could be incorporated in species mixtures. The phenomenon, reported initially amongst tree and shrub species, describes how shallow-rooting species can benefit from improved acquisition of water and nutrients because of the release of water and nutrients into the topsoil from deep-rooting neighbours (e.g. Snyder *et al.*, 2008).

The role of enhancing species diversity in agriculture is an important topic at the interface of agriculture and ecology. There is evidence that increasing the number of species in natural grassland systems results in gains in terms of productivity and resilience (e.g. Hooper *et al.*, 2005). Work carried out under the auspices of an EU COST programme further explored this relationship in an agronomic context, through a novel experimental and modelling framework using four species (two legumes and two grasses) at a large number of sites within Europe. The results showed that even a modest increase in species diversity had benefits in terms of increased productivity and resistance to weed invasion (Kirwan *et al.*, 2007). Furthermore, there was clear evidence of 'transgressive over-yielding' in most mixtures (i.e. mixture yields were greater than those of the highest yielding monocultures), and this effect persisted over time (Lüscher *et al.*, 2008). The modelling approach used by Kirwan *et al.* (2007) was based on the proposition that the success with which pairs of species in a mixture contribute to ecosystem services, such as primary productivity and resistance to weed invasion, would depend on their initial relative abundance in the mixture and on their propensity to interact. This approach is being refined to provide more insight into the mechanisms driving diversity effects (Connolly *et al.*, 2010). In practical terms, developing management practices that establish and maintain balanced grass/legume mixtures, to optimize productivity through diversity effects, presents a major challenge for research in grassland agronomy. Difficulties in maintaining well-balanced mixtures and the tendency to lose key species in multispecies swards (Guckert and Hay, 2001) are major reasons for the prevalence of grass monocultures in temperate agriculture.

16.2 Genetic Diversity amongst Grass Species of the *Lolium–Festuca* Complex

The *Poaceae* represent around 11,000 diverse and widely dispersed grass species that have evolved from a common ancestor and, despite their vast heterogeneity, still retain considerable genome macrosynteny. This property assists all those involved in grass trait genetics, firstly by aiding gene location, and secondly by aiding the determination of gene function, the logic being that homoeoalleles are likely to occur at orthologous genome regions and therein perform equivalent duties, irrespective of their species origin. However, despite this, the understanding of the multi-gene controls of complex traits remains generally poor, despite the lauded progress made in high-throughput grass genomics and the sequence completion of entire model grass genomes. Amongst the less-investigated species, the forage grasses from within the *Lolium* and *Festuca* genera comprise species considered to be optimal choices for grassland agriculture and where some progress has been made in trait genetics. In some cases, investigations have utilized the genome synteny between these outbreeding forage grass species and related, more studied grass genomes, such as rice, the *Triticeae* and, more recently, *Brachypodium distachyon*. The latter are considered by many to be a good model for temperate crops because of their taxonomic co-location within the sub-family *Pooideae*.

By combining detailed comparative studies in genomics and phenomics, valuable insights will be achieved into evolution within the *Pooideae*. An understanding will develop of the adaptations that gave rise to increased genome diversity and led to speciation, either through gene duplication or modification, genome hybridization and introgression or polyploidy, and changes in gene expression due to complex epigenetic and pleiotropic genome interactions within the newly assembled genomes. By studying plant phenotype alone, the genome potential in trait expression may not be fully realized. This was demonstrated in research in *Lolium* × *Festuca* spp. hybrids. Androgenesis is used generally to produce a haplotype, and from that inbred pure breeding homologous dihaploids and homogeneous crop varieties. With *Lolium* × *Festuca* amphiploid hybrids, however, the tecnhique has been used to reveal heterogeneity and the extent of 'hidden' and hitherto unexpressed genome variation that may reside within their combined genomes. This is achieved through genome 'dissection', a consequence of gene segregation at meiosis. Rapacz *et al.* (2005) reported how androgenesis induced genome reorganization, so that genes for winter hardiness and snow mould resistance became functional and effective in the dihaploid progeny of parental genotypes that were themselves highly susceptible to both these abiotic and biotic stresses. Using an alternative approach to achieve a similar objective, a complete set of monosomic chromosome addition lines was developed, where each of the seven *Lolium perenne* chromosomes that comprise its genome were replaced by homologous *Festuca pratensis* chromosomes, in a backcross breeding programme (King *et al.*, 2002). During further backcross generations using a *L. perenne* genotype, the extraordinarily high potential for interspecific chromosome pairing and genome exchange was exploited, to 'dissect' each alien *F. pratensis* chromosome and produce a recombination series for each chromosome. This introgression-mapping population is aiding understanding of the genetic control of complex traits, and especially gene heterosis, that appear to underlie much of the optimal expression of adaptive traits in grasses. This unique population provides a procedure to physically map alien *Festuca* genes in an otherwise intact *Lolium* genetic background, and evaluate their impact on plant phenotype. Using these procedures, root growth quantitative trait loci (QTLs) were located on chromosome 3 and heterosis effects between *Lolium* and *Festuca* alleles observed, including gene interactions and root growth responses associated with the onset of drought stress

(Turner *et al.*, 2010; Humphreys *et al.*, 2010). Direct correlations were possible between leaf elongation and root growth responses following the onset of moderate drought.

The widespread occurrence of *Lolium* and *Festuca* species' ecotypes is the result of their genome plasticity and heterogeneity. Where edaphic and climatic stresses require further increase in genome resilience, intergeneric hybridization may ensue. *Festulolium loliaceum* (*F. pratensis* × *Lol. perenne*) exists in either stable diploid or triploid forms (Humphreys and Harper, 2008) in mature water meadows, where the hybrids have increased capabilities to withstand waterlogged soils compared with their parental species. Alternatively, genome instability may follow initial hybridization where, in extreme circumstances, the entire *Lolium* or *Festuca* parental genome is lost due to specific selection pressures and/or a meiotic drive that favours transmission of only one of the parental genomes (e.g Canter *et al.*, 1999). Preliminary studies indicate a preferential retention of the *Festuca* genome due to extreme low temperatures in North America (Casler *et al.*, 2009), whilst in the flooded saline Pampas grasslands of Argentina (Latorre, 2010), retention of the *Lolium* genome would appear to offer an adaptive advantage.

Whilst intergeneric hybridization will clearly broaden the genetic variation available to grass breeders and may prove useful, especially where improved climatic or edaphic adaptations are sought, the extent of acquiring, selecting, and, over generations, gaining new intraspecific genetic variation, either from species ecotypes or even from within breeding populations, should never be discounted. For example, Bonos *et al.* (2004) improved thr root/shoot ratio over two generations in *F. arundinacea* and *Lol. perenne* cultivars, and, at the gene level, Skøt *et al.* (personal communication) found a direct association between an allelic form of the salt-stress root protein gene *Srs-1* on chromosome 3 and tiller re-growth after drought, using three related high sugar *Lolium perenne* cultivars. In both examples, no attempt had been made previously to select genotypes with superior drought resistance traits from within the existing cultivars and the extent of potential variation therein for the trait was unknown.

QTL maps developed from diverse parental genotypes give good indications of the location of genes for adaptive traits and have been developed for both *Lolium* and *Festuca* species (Turner *et al.*, 2008; Alm *et al.*, 2011). Whilst common QTLs for drought resistance were found on chromosomes 1 and 5, a large QTL for resistance to severe drought stress not detected in *L. perenne* was found on chromosome 3 of *F. pratensis* (*QDts3*F), indicating its potential as a source of novel variation for the trait in *Lolium*. Indeed the *QDts3F* QTL has been used as a source of enhanced drought resistance traits for *Lolium*. Humphreys and Pasakinskiene (1996) and Humphreys *et al.* (2005) report transfers of genes for drought resistance from south European ecotypes of *F. arundinacea* and *F. arundinacea* var. *glaucescens*, respectively. Both alien *Festuca* sequences were demonstrated recently to contribute also to increased salinity tolerance (Latorre, 2010). Whilst no QTLs for drought resistance were detected in *Lolium* on chromosome 3, syntenic regions of rice chromosome 1 were a rich source of genes for drought resistance and root-growth traits (Kamoshita *et al.*, 2008).

Festuca species with extreme xerophytic traits have been sourced from the Atlas Mountains of North Africa, as new donors of drought resistance and rooting traits for *Lolium*. Chen *et al.* (1997) described a backcross breeding procedure aimed at the introgression of *F. mairei* genes for drought resistance into perennial ryegrass, and Wang *et al.* (2003), using molecular markers, confirmed the efficacy of the introgression procedures.

For perennial grass species growing in temperate grasslands, especially in northern latitudes, winter hardiness is a necessity. *F. pratensis* is an important contributor of genes for enhanced winter hardiness for *Lolium*. Transfers of genes from *F. pratensis* chromosome 4 onto *Lolium* enhanced freezing tolerance (Humphreys *et al.*, 2006). New evidence of the importance in *Lolium* and *Festuca* species of the adaptive capabilities of Photosystem II (PSII) during cold acclimation in relation to subsequent freezing tolerance was demonstrated. Non-photochemical quenching (NPQ) mechanisms for expulsion of excess light energy were found in *F. pratensis* but were not expressed to the same extent in *Lolium*. A direct relationship between cold-acclimation-induced increases in NPQ and freezing tolerance was reported.

16.3 Enhancing Genetic Diversity within Forage Legumes

Forage legumes have an important role in grassland production systems due to their high-quality forage and capacity to fix atmospheric nitrogen. White clover (*Trifolium repens* L.), can fix up to 250 kg N/ha/year and is often sown in mixtures with perennial ryegrass (*Lol. perenne*). To obtain the benefits of white clover in mixed swards, a clover content of 30% of the total dry matter (DM) is regarded as essential. Dry matter yield, tolerance of biotic and abiotic stress, and ability to grow and compete with companion grasses are therefore important characteristics for white clover varieties. There is considerable intraspecific variation in leaf size and stolon density, which influences the ability of white clover to tolerate grazing or cutting and therefore the management system for which each type should be used. Small leaf types with a dense stolon network are more suited to intensive sheep grazing and large or very large leaf types are more appropriate for cutting systems. Plant breeders have successfully developed varieties of different leaf size that are appropriate for different management systems.

As with the grasses, interspecific hybridization with closely related species can be used to introgress novel traits, to improve agronomic performance or tolerance of biotic and abiotic stress. Drought tolerance is an increasingly important trait in white clover for which there is limited genetic variation. Hybrids have been developed between white clover and Caucasian clover (*Trifolium ambiguum* M.Bieb), with the objective of introgressing the rhizomatous trait into white clover (Marshall *et al.*, 2001) to improve drought resistance and persistence. An alternative approach has been the development of a 'fertile bridge' between *T. ambiguum* and white clover, enabling the transfer of traits between the two species without the use of embryo rescue (Hussain and Williams, 1997). Both approaches have resulted in fertile backcross hybrids that are essentially white-clover-like, but with rhizomes as well as stolons. The rhizomatous trait from Caucasian clover has also been introgressed into alsike clover, *T. hybridum* L. (Paplauskiene *et al.*, 2004). Interspecific hybrids have also been developed between red clover and the rhizomatous zigzag clover (*Trifolium medium*; Isobe *et al.*, 2002), with the objective of introgressing the rhizomatous trait into red clover and increasing persistence.

Good seed yields are important for commercial success but this must be achieved without compromise to agronomic performance. In white clover, hybrids between white clover and the annual, profuse flowering ball clover (*T. nigrescens* Viv.) have been developed, with the objective of introgressing reproductive traits into white clover (Marshall *et al.*, 2002). Transfer of clover cyst nematode resistance from *T. nigrescens* into white clover has also been a target (Hussain *et al.*, 1997).

16.4 Using Genetic Diversity in Grassland Systems for Ecosystem Services

Diffuse pollution of nitrogen and phosphorus from agricultural sources is high on the environmental protection agenda of many governments. Considerable effort is now going into reducing the direct (leaching) and indirect (through animal returns) pollution. Recent insights into mechanisms affecting protein breakdown in the rumen and silo offer new prospects for breeding interventions to reduce environmental impacts. Traits that will reduce the environmental impact are being exploited as a basis of future varieties. The specific areas for improving the perennial clovers in this regard have been reviewed recently (Abberton and Marshall, 2005).

Condensed tannins help to stabilize protein as it passes through the rumen, reducing loss of protein and preventing bloat. Although absent from the leaves of white and red clover, they are present in *Lotus corniculatus* and *Lot. uliginosus* (Marshall *et al.*, 2008). Extensive variation in condensed tannin content of the leaves of *Lotus* has been found. Development of varieties with appropriate levels of condensed tannins, to reduce protein loss without reducing intake or the agronomic yield and persistence in mixed swards is a current target.

Phosphorus pollution from farmland is widely regarded as a growing environmental problem and one which is likely to be addressed by regulation in many parts of the world (e.g. the EU Water Framework Directive). It seems likely that selection for growth at low P will increase in importance, and several approaches may prove viable in addition to direct selection. Phosphorus requirements for good clover performance are relatively high, and research has been undertaken, particularly in New Zealand, to select lines that can yield well at lower levels of P applications (Caradus, 1994). The extent of inorganic phosphorus incorporation into organic compounds shows genetic variation (Caradus *et al.*, 1998) and may be an important selection criterion for future studies. The symbiosis between white clover and arbuscular mycorrhizal fungi (AMF or vesicular arbuscular mycorrhizae, VAM) has been explored, with a view to increasing the efficiency of P uptake (e.g. Crush, 1995), although the role and importance of this association on fertilized agricultural soils is unclear. Similar research is underway in *Lolium* and will include redesigns of root systems better able to scavenge P from soils.

The environmental impact of livestock production has become an area of increasing concern (Steinfeld *et al.*, 2006) and the need to mitigate the effects of climate change is widely seen as urgent (Stern, 2007). The main GHG emissions from grassland-based livestock agriculture are nitrous oxide and methane. Nitrous oxide emission from temperate grasslands is poorly quantified, although Mummey and Smith (2000) reported estimates from USA grasslands of approximately 67 Gg nitrous oxide N/year (based on simulated emissions × area). Gregorich *et al.* (2005) found that emissions of nitrous oxide from soils increased linearly with the amount of mineral nitrogen fertilizer applied, and, because systems containing legumes produce lower annual nitrous oxide emissions, lucerne and other legume crops should be considered differently when deriving national inventories of GHG from agriculture. The two major sources of agricultural methane emissions are enteric fermentation in livestock and livestock manures.

16.5 Reducing Nitrous Oxide Emissions

The rapid breakdown of herbage proteins in the rumen and inefficient incorporation of herbage N by the rumen microbial population are major causes of N loss and gaseous emissions. Scarcity of readily available energy during the time of maximal protein degradation restricts microbial protein synthesis. Ammonia accumulates as a waste product and is absorbed from the rumen and excreted as waste nitrogen in urine. When sheep (MacRae and Ulyatt, 1974) and cattle (Ulyatt *et al.*, 1988) are given fresh forages they can waste 25–40% of forage protein.

Genetic improvement of the forage grasses and legumes that constitute important components of the ruminant diet has the potential to reduce emissions to air. Two possible strategies of increasing the efficiency of conversion of forage N to microbial N have been suggested: (i) increasing the amount of readily available energy accessible during the early part of the fermentation; and (ii) providing a level of protection to the forage proteins, and thereby reducing the rate at which their breakdown products are made available to the colonizing microbial population. One approach is to develop forage species with a better balance between water soluble carbohydrate (WSC) and crude protein (CP), by increasing the WSC content of the grass or the clover component or reducing the protein content of the legume.

The most advanced of these approaches is the development of high WSC ryegrasses, which are already showing considerable commercial success, particularly in the UK, and for which there is evidence that increased production accompanies reduced emissions (Miller *et al.*, 2001). There is also significant variation and opportunities to select within white clover for traits such as lower protein content and higher WSC. Unique non-fixing inbred genotypes of white clover were used to demonstrate that lower leaf protein content results in slower protein degradation in the silo (Kingston-Smith *et al.*, 2006). Following this, it was found that genotypic variation within elite gene pools of white clover is much greater than was previously thought. Interspecific hybrids between white clover and *Trifolium ambiguum* have a CP content 14.2 g/kg DM lower than white clover.

Another approach to enhance ruminant nutrition, it is hypothesized, involves use of *Festuca*-derived adaptations to extreme drought- and heat-stress conditions. Grass forage enters the rumen where it is subjected to anaerobic conditions, a neutral pH of 6.8 and an ambient temperature of 39°C. This induces abiotic stresses to intact ingested cells, such as oxygen depletion and elevated temperature, and, as a result, causes the plant mediated proteolysis that contributes to poor nitrogen-use efficiency in ruminants (Shaw, 2005). There is evidence to support that living cells of *F. arundinacea* var. *glaucescens*, which grows in Mediterranean regions, are affected significantly less than those of *Lolium* by exposure to these rumen stresses.

A further area where genetic approaches can have an impact is in improving the nitrogen-use efficiency (NUE) of crops, to allow lower fertilizer application and hence reduce nitrogenous emissions through the soil–plant–animal–soil cycle. NUE from soil to crop is lower generally, on a whole-farm basis, for grass-based livestock production than with arable crop production, ranging from 10–40% for whole dairy systems compared with 40–80% for arable systems (Neeteson *et al.*, 2004).

Sustainable management of water in landscapes requires balancing demands of agricultural production whilst moderating downstream effects such as flooding, the occurrence of which is expected to increase due to ongoing climate change. Durand *et al.* (2007) demonstrated how *Lol. multiflorum* × *F. arundinacea* var. *glaucescens* hybrids in combination, due to their enhanced rooting capability, are able to extract water from greater soil depths than the *Lolium* parent when the plants are exposed to severe drought stress.

A recent study demonstrated how root growth within a *Lolium/Festuca* hybrid cultivar modifies soil structure to ease soil compaction and reduce runoff of water. Extensive root growth in the interspecific cultivar compared with the parent species, followed by the widespread degeneration of the root system at depth in the soil, resulted in changes to soil structure and the increased water-holding capabilities of the soil (Gregory *et al.*, 2010; Humphreys *et al.*, 2010). The benefits were 50% reduced water runoff compared with *Lol. perenne*, which would also reduce nutrient runoff and, by increased water storage, provide an improved reservoir to combat a future water shortage given a prolonged dry period. Similar effects have been observed under swards of white clover, and combined use of the grass hybrid with clover would probably bring several complementary environmental benefits.

To conclude, new strategies for grass and clover crop development must be holistic, seeking not only to sustain and increase crop yields in a changing climate, but also to incorporate measures for increased crop persistency, improved soil structure, improved uptake and use of scarce water and nutrient resources and systems to mediate against their ecosystem loss. Fortunately the genetic diversity required exists. It will be the challenge for the breeder and the farmer to generate and manage a stable and sustainable grassland system.

References

Abberton, M.T. and Marshall, A.H. (2005) Progress in breeding perennial clovers for temperate agriculture. *Journal of Agricultural Science* 143, 117–135.

Alm, V., Busso C. S., Ergon, Å., Rudi H., Larsen, A., Humphreys, M.W. and Rognli, O.A. (2011) A QTL analyses and comparative genetic mapping of frost tolerance, winter survival and drought tolerance in meadow fescue (*Festuca pratensis* Huds.). *Theoretical and Applied Genetics* 123 (3), 369–382.

Bonos, S.A., Rush, D., Hignight, K. and Meyer, W.A. (2004) Selection for deep root production in tall fescue and perennial ryegrass. *Crop Science* 44, 1770–1775.

Canter, P.H., Pasakinskiene, I., Jones, R.N. and Humphreys, M.W. (1999) Chromosome substitutions and recombination in the amphiploid *Lolium perenne* × *Festuca pratensis* cv Prior (2n=4x=28). *Theoretical and Applied Genetics* 98, 809–814.

Caradus, J.R. (1994) Selection for improved adaptation of white clover to low phosphorus and acid soils. *Euphytica* 77, 243–250

Caradus, J.R., Kennedy, L.D. and Dunn, A. (1998) Genetic variation for the ratio of inorganic to total phosphorus in white clover leaves. *Journal of Plant Nutrition* 21, 2265–2272.

Casler, M.D., van Santen, E., Humphreys, M.W., Yamada, T., Tamura, K., Ellison, N.W., Jackson, R.D., Undersander, D.J., Gildersleeve, R., Stendal, C., Reiter, M. and Opitz, C. (2009). Remnant oak savanna acts as refugium for meadow fescue introduced during nineteenth century human migrations in the USA. In: *5th International Symposium on the Molecular Breeding of Forage and Turf, Sapporo, Japan, 2–6 July, 2007*, pp. 91–101.

Chen, C., Sleeper, D.A., Chao, S., Johal, G.S. and West, C.P. (1997) RFLP Detection of 2n pollen formation by first and second division restitution in perennial ryegrass. *Crop Science* 37, 76–80.

Connolly, J., Bell, T., Brophy, C., Carnus, T., Finn, J., Kirwan, L., Lüscher, A., Sebastià, M.T. and Wiegelt, A. (2010) Multi-species mixtures – new perspectives on models and mechanisms. *Grassland Science in Europe* 15, 782–784

Crush, J.R. (1995) Phosphorus response of white clover (*Trifolium repens* L.) genotypes selected for tolerance of aluminium toxicity. *New Zealand Journal for Agriculture Research* 38, 451–456.

Durand, J.L., Bariac, T., Ghesquière, M., Brion, P., Richard, P., Humphreys, M.W. and Zwierzykowski, Z. (2007) Ranking of the depth of water extraction by individual grass plants using natural 18O isotope abundance. *Environmental and Experimental Botany* 60, 137–144.

Gregorich, E.G., Rochette, P., VandenBygaart, A.J. and Angers, D. (2005) Greenhouse gas contributions of agricultural soils and potential mitigation practices in Eastern Canada. *Soil and Tillage Research* 81, 53–72

Gregory, A.S., Webster, C.P., Watts, C.W., Whalley, W.R., Macleod, C.J.A., Joynes, A., Papadopoulos, A., Haygarth, P.M., Binley, A., Humphreys, M.W., Turner, L.B. and Skøt, L. (2010) Soil management and grass species effects on the hydraulic properties of shrinking soils. *Soil Science Society of America Journal* 74, 753–761.

Guckert, A. and Hay, R.K.M. (2001) The overwintering, spring growth, and yield in mixed species swards of white clover in Europe. *Annals of Botany* 88, 667–668.

Hooper, D.U., Chapin, III F.S., Ewel, J.J., Hector, A., Inchausti, P., Lavorel, S., Lawton, J.H., Lodge, D.M., Loreau, M., Naeem, S., Schmid, B., Setälä, H., Symstad, A.J., Vandermeer, J. and Wardle D.A. (2005) Effects of biodiversity on ecosystem functioning: a consensus of current knowledge. *Ecological Monographs* 75, 3–35.

Humphreys, J., Harper, J.A., Armstead, I.P. and Humphreys, M.W. (2005) Introgression-mapping of genes for drought resistance transferred from *Festuca arundinacea* var. *glaucescens* into *Lolium multiflorum*. *Theoretical and Applied Genetics* 110, 579–587.

Humphreys, M.W. and Harper, J.A. (2008). *Festulolium loliaceum*, an understudied natural UK grass hybrid species that may procide benefits to UK grasslands withstanding the onsets of climate change. *European Crop Wild Relative Newsletter*, January, (6).

Humphreys, M.W. and Pasakinskiene, I. (1996) Chromosome painting to locate genes for drought resistance transferred from *Festuca arundinacea* into *Lolium multiflorum. Heredity* 77, 530–534.

Humphreys, M.W., Gasior, D., Lesniewska-Bocianowska, A., Zwierzykowski, Z. and Rapacz, M. (2006) Androgenesis as a means of dissecting complex genetic and physiological controls: selecting useful gene combinations for breeding freezing tolerant grasses. *Euphytica* 158, 337–345.

Humphreys, M.W., Turner, L.B., O'Donovan, S.A., Macleod, C.J.A., Whalley, W.R., and Haygarth, P.M. (2010) Grass root turn-over for improved soil hydrology to combat flooding. In: Schnyder, H., Isselstein, J., Taube, F., Auerswald, K., Schelberg, J. *et al.* (eds) *General Meeting of the European Grassland Federation, Kiel, Germany, 29 August - 2 September 2010.* European Grassland Federation, ISBN:9783869440217, pp. 764–766.

Hussain, S. W. and Williams, W.M. (1997) Development of a fertile bridge between *Trifolium ambiguum* M. Bieb. and *T. repens* L. *Theoretical and Applied Genetics* 95, 678–690.

Hussain S.W., Williams, W.M., Mercer, C.F. and White, D.W.R. (1997) Transfer of clover cyst nematode resistance from *Trifiolium nigrescens* Viv. to *T. repens* by interspecific hybridisation. *Theroretical and Applied Genetics* 95, 1274–1281.

Isobe, S., Sawai, A., Yamaguchi, H., Gau, M. and Uchiyama, K. (2002) Breeding potential of the backcross progenies of a hybrid between *Trifolium medium* x *T. pratense* to *T. pratense. Canadian Journal of Plant Science* 82, 395–399.

Kamoshita, A., Babu, R.C., Boopathi, N.M. and Fukai, S. (2008) Phenotypic and genotypic analysis of drought-resistance traits for development of rice cultivars adapted to rainfed environments. *Field Crops Research* 109, 1–23.

King, J., Armstead, I.P., Donnnison, I.S., Thomas, H.M., Jones, R.N., Kearsey, M. J., Roberts, L.A., Jones, A. and King, I.P. (2002). Physical and genetic mapping in the grasses *Lolium perenne* and *Festuca pratensis. Genetics* 82, 107–112.

Kingston-Smith, A.H., Bollard, A.L. and Minchin, F.R. (2006) The effect of nitrogen status on the regulation of plant-mediated proteolysis in ingested forage; an assessment using non-nodulating white clover *Annals of Applied Biology* 149, 35–42.

Kirwan, L., Sebastià, M.T., Finn, J.A., Collins, R.P., Porqueddu, C., Helgadóttir, Á., Baadshaug, O.H., Brophy, C., Coran, C., Dalmannsdóttir, S., Delgado, I., Elgersma, A., Fothergill, M., Frankow-Lindberg, B.E., Golinski, P., Grieu, P., Gustavsson, A.M., Höglind, M., Huguenin- Elie, O., Iliadis, C., Jørgensen, M., Kadziuliene, Z., Karyotis, T., Lunnan, T., Malengier, M., Maltoni, S., Meyer, V., Nyfeler, D., Nykänen-Kurki, P., Parente, J., Smit, H.J., Thumm, U. and Connolly, J. (2007). Evenness drives consistent diversity effects in an intensive grassland system across 28 European sites. *Journal of Ecology* 95, 530–539.

Latorre, E.M. (2010) Characterisation of novel *Lolium multiflorum* germplasm for abiotic stress tolerance. PhD Thesis, Aberystwyth University, Wales.

Lüscher, A., Finn, J.A., Connolly, J., Sebastià, M.T., Collins, R.P., Fothergill, M., Porqueddu, C., Brophy, C., Huguenin-Elie, O., Kirwan, L., Nyfeler, D. and Helgadóttir, A. (2008) Benefits of sward diversity for agricultural grasslands. *Biodiversity* 9, 29–32.

MacRae, J.C. and Ulyatt. M.J. (1974) Quantitative digestion of fresh herbage by sheep. *Journal of Agriculture Science (Cambridge)* 82, 309–319.

Marshall, A.H., Rascle, C., Abberton, M.T., Michaelson-Yeates, T.P.T. and Rhodes, I. (2001) Introgression as a route to improved drought tolerance in white clover (*Trifolium repens* L.). *Journal of Agronomy and Crop Science* 187, 11–18.

Marshall, A.H., Michaelson-Yates, T.P.T., Abberton, M.T., Williams, A. and Powell, H.G. (2002) Variation for reproductive and agronomic traits among *T. repens* x *T. nigrescens* third generation backcross hybrids in the field. *Euphytica* 126, 95–201.

Marshall, A.H., Bryant, D.N., Latypova, G.A., Hauck, B., Olyott, P., Morris, P. and Robbins, M.P. (2008) A high throughput method for the quantification of proanthocyanidins in forage crops and its application in assessing variation in condensed tannin content in breeding programmes for *Lotus corniculatus* L. and *L. uliginosus. Journal of Agriculture and Food Chemistry* 56, 974–981.

Miller, L.A., Moorby, J. M., Davies, D.R., Humphreys, M.O., Scollan, N.D., MacRae, J.C. and Theodorou, M.K. (2001) Increased concentration of water-soluble carbohydrate in perennial ryegrass (*Lolium perenne* L.): Milk production from late-lactation dairy cows. *Grass and Forage Science* 56, 383–394.

Mummey, D.L. and Smith, J.L. (2000) Estimation of nitrous oxide emissions from US grasslands. *Environmental Management* 25, 169–175.

Neeteson, J.J., Schroder, J.J., and Jakobsson, C. (2004) Drivers towards sustainability: Why change? In: Hatch,

D.J., Chadwick, D.R., Jarvis, S.C. and Roker, J.A. (eds) *Controlling Nitrogen Flows and Losses.* Wageningen Academic Publishers, The Netherlands, pp. 29–38.

Paplauskiene, V., Dabkeviciene, G and Pasakinskiene, I. (2004) Identification of interspecific hybrids of *Trifolium ambiguum* x *Trifolium hybridum* by inter-SSR fingerprinting. *Grassland Science in Europe* 9, 398–400.

Rapacz, M., Gasior, D., Humphreys, M.W., Zwierzykowski, Z., Plazek, A. and Lesniewska-Bocianowska, A. (2005) Variation for winter hardiness generated by androgenesis from *Festuca pratensis* x *Lolium multiflorum* amphidiploid cultivars with different winter susceptibility. *Euphytica* 142, 65–73.

Shaw, R.K. (2005) Effects of gene transfer from *Festuca* to *Lolium* on plant-mediated proteolysis. PhD Thesis, Aberystwyth University, Wales.

Snyder, K.A., James, J.J., Richards, J.H., and Donovan, L.A. (2008) Does hydraulic lift or nighttime transpiration facilitate nitrogen acquisition? *Plant Soil,* 306, 159–166.

Steinfeld, H., Gerber, P., Wassenaar, T., Castel, V., Rosales, M. and de Haan, C. (2006) *Livestock's Long Shadow. Environmental Issues and Options.* Food and Agriculture Organisation, Rome.

Stern, N (2007) *The Economics of Climate Change. The Stern Review.* Cambridge University Press, UK.

Turner, L.B., Cairns, A.J., Armstead, I.P., Thomas, H., Humphreys, M.W. and Humphreys, M.O. (2008) Does fructan have a functional role in physiological traits? Elucidation by QTL mapping. *New Phytologist* 179, 765–775.

Turner, L.B., Macleod, C.J.A., Watts, C., Whalley, W.R., Binley, A., Papadopoulos, T., O'Donovan, S., Haygarth, P.M., King, J. and Humphreys, M.W. (2010) Dissecting *Festulolium* chromosome 3 to locate rooting and drought resistance traits. In: Huyghe, C. (ed.) *Sustainable Use of Genetic Diversity in Forage and Turf Breeding.* Springer, Dordrecht, The Netherlands, pp. 471–477.

Ulyatt, M.J., Thomson, D., Beever, D.E., Evans, R.T. and Haines, M.J. (1988) The digestion of perennial ryegrass (*Lolium perenne* cv. Melle) and white clover (*Trifolium repens* cv. Blanca) by grazing cattle. *British Journal of Nutrition* 60, 137–149.

Wang, J.P., Bughrara, S.S. and Sleper, D.A. (2003) Genome introgression of *Festuca mairei* into *Lolium perenne* detected by SSR and RAPD markers. *Crop Science* 43, 2154–2161.

17 Soil Microbial Diversity in Grasslands and its Importance for Grassland Functioning and Services

Xavier Le Roux, Sylvie Recous and Eléonore Attard

17.1 Soil: a Hotspot of Biodiversity in Grasslands

A wide range of life forms exist on Earth. For instance, the total numbers of bird and plant species on Earth are around 10,000 and 300,000, respectively. Moreover, we know that some habitats, ecosystems or regions harbour high levels of biodiversity. For instance, Colombia harbours around 1800 bird species whereas 8000 tree species can be found in Brazil! Even at a smaller scale, biodiversity can be huge: a single hectare of forest in French Guyana can harbour 600 trees representing more than 200 tree species, whereas 10 to 50 plant species can be found within one hectare of permanent grassland. More generally, Earth harbours plenty of individuals of many different species: scientists have evaluated that around 200 billion (2×10^{11}) adult or sub-adult birds are present at a given time on planet Earth. Thus, understanding biodiversity dynamics, and the links that exist between living organisms including humans, represents one of the grand challenges of our time.

Astrophysicists could argue that such numbers are not so high and that they have to face a bigger challenge. Indeed, the Universe encompasses 10^{22} stars, including 4×10^{11} in our galaxy! In addition, stars can hardly be seen clearly by the naked eye and cannot be approached to be observed. Astrophysicists thus have to work on objects to which they have no direct access. They develop large and advanced instruments to study them and develop refutable theories to analyse the functioning of the Universe.

One type of ecologist could, however, say that the numbers and difficulties raised by astrophysicists are not so challenging: soil microbial ecologists. Indeed, when you take 10 g of grassland soil in your hand, you handle around 10^{11} bacteria, encompassing 10,000 to 100,000 bacterial 'species', although the concept of bacterial species remains more open and debated than that used for higher organisms (Torsvik *et al.*, 1998; Gans *et al.*, 2005). Actually, when accounting for plants (plant roots), meso- and macro-fauna and microflora, soil is the reservoir of a huge diversity of living forms (Lavelle and Spain, 2002; Bardgett, 2005): around 3×10^8 animals can be found below ground in 1 m^2 of grassland! But knowledge on soil biodiversity is still sparse: in particular, soil microbial diversity is difficult to characterize and study, with more than 90% of bacterial taxa reluctant to be cultivated by humans. Thus, as for astrophysicists, working on grassland soil microbial diversity also implies working on abundant and diverse objects that are hardly accessible, and developing advanced approaches and instruments and new theories to characterize those organisms and understand soil functioning. The big difference is that microbes are living objects, which makes the challenge even greater!

Let's stop the comparisons of numbers and research areas but let's remember this in the context of this book on grassland ecosystem functioning and services: the level of biodiversity is much higher in soil than in any other system we are used to considering when we are thinking about life on Earth. Both the number of species and organism abundance in grassland soil reach levels well above those of above-ground, higher organisms.

This leads to the first question: why is grassland soil biodiversity much higher than the diversity of plants and animals above-ground? A first explanation is that the size of soil organisms is often very small. Whereas an earthworm is large enough to be seen, most grassland soil organisms are too small to be seen with the naked eye: from a few microns for bacteria, 100 times smaller for viruses, to around 1 mm for nematodes. Concurrently, grassland soil is chemically and physically highly heterogeneous at a range of scales: there are soil aggregates of various sizes; the soil volume can be influenced by roots (rhizosphere), the amount or proximity to dead organic matter (detritusphere), or the activity of soil macrofauna such as earthworms, etc.

Grassland soil thus offers a range of diverse habitats to living organisms; in addition, all its porosity and particles create a huge surface to be colonized by life. Thus, this allows the co-existence of large numbers of species, and huge numbers of individuals, able to exploit the different 'microhabitats'. In such a complex ecosystem, the life of a small, weakly mobile organism can strongly differ among microhabitats. A significant proportion of bacteria are 'asleep', like the 'Sleeping Beauty' of the classic fairy tale, because environmental conditions at their exact location are not favourable to their activity: those bacteria are waiting, not for the kiss of their Prince Charming, but that of a root (exudates), that of a soil organism (e.g. earthworm mucus) or an event, such as soil ploughing (e.g. when a temporary grassland is converted into a cropping system), that would bring more oxygen and energy to them and eventually awaken them (Lavelle and Spain, 2002). This is equivalent to a situation where the biomass of a grassland vegetation cover would include a substantial proportion of plant individuals present with no or very weak photosynthetic activity, but those plant individuals would be able to become active, after a few months or years, according to environmental changes. This would probably change much of grassland ecologists' approaches and theories! In addition, the evolutionary capacity of a number of grassland soil organisms is remarkable: with generation times ranging from a few hours to days and a striking ability to exchange genetic material (even between species, through 'horizontal transfer' of DNA); bacterial diversification and bacterial species radiation are fast. Given all those features, one can easily understand why grassland biodiversity is so high in soil. This leads to a second question: how and to what extent does such a huge biodiversity influence grassland soil functioning and the services delivered by grasslands to humans?

17.2 Importance of Soil Biodiversity for Grassland Functioning

Many studies have been performed to understand to what extent and through which mechanisms biodiversity plays a role in the functioning of grassland systems. Most of those studies focused on macroorganisms, often plants (Hooper *et al.*, 2005; Balvarena *et al.*, 2006). The results showed that high levels of biodiversity can enhance grassland ecosystem performance. In addition, those studies demonstrated that biodiversity can influence the stability of grasslands facing disturbances. Indeed, the presence of a diversity of species that can differ in their response to perturbations could decrease the instability of grassland functioning following perturbations. However, fewer studies have focused on grassland soil organisms.

Beyond the methodological difficulties associated with the characterization of grassland soil biodiversity, many authors consider that the level of functional redundancy in soil should be particularly high, given the high level of biological diversity. These authors thus assume that one can disregard soil biodiversity for understanding and predicting grassland ecosystem functioning. This assumption has been tested by different types of approaches (Fig. 17.1).

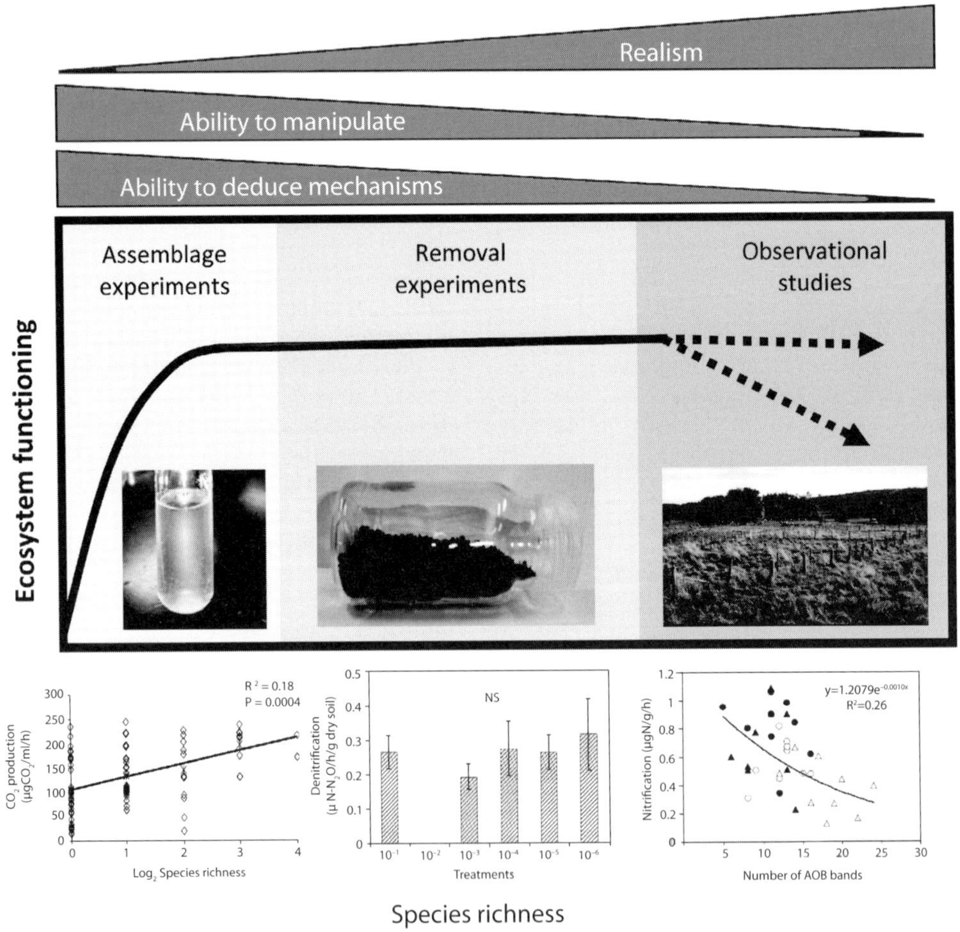

Fig. 17.1. Synthesis of different approaches used to analyse the relationships between the microbial diversity and functioning of grassland soils. (Left) an assemblage experiment (Salles *et al.*, 2009) demonstrates the positive effect of diversity for cultivated denitrifying bacterial taxa derived from bulk and rhizospheric soil; in particular, the authors demonstrated the importance of the functional dimension of diversity and of complementary effects. (Middle) A species removal approach via suspension–dilution applied on a native grassland soil community demonstrates the high level of functional redundancy in native, complex nitrifying and denitrifying communities (Wertz *et al.*, 2006, 2007). (Right) Observatory approaches show a broad spectrum of possible, apparent relationships between the microbial diversity of grassland soil and its functioning; in this case, analysing the co-variations of diversity, abundance and environmental conditions is a prerequisite to assess the actual role of biodiversity in changes in soil functioning (Patra *et al.*, 2005, 2006; Le Roux *et al.*, 2008). This figure has been modified according to an output of the EFS workshop '*Microbial Diversity and Ecosystem Functioning*' held in Lunz, Germany, in 2007.

17.2.1 Species assemblage approaches

Some scientists have used a species assemblage approach to study the importance of microbial biodiversity for the functioning of grassland soil. As performed in ecology on higher organisms, such an approach consists of creating synthetic microbial communities and studying their performance (level of functioning, resistance and resilience to model disturbances) in relation to their diversity. Assemblage experiments are suitable to study biodiversity–functioning relationships in the context of (micro)-habitat colonization from a more or less diverse pool of taxa (Diaz *et al.*, 2003). These experiments are useful in analysing the mechanisms of interaction and complementarity that can exist between taxa and can shape biodiversity–functioning relationships.

Assemblage approaches have been used for litter-decomposing fungi (Setälä *et al.*, 2004), myccorhiza (Van der Heijden *et al.*, 1998), detritivorous macro-fauna (Heemsbergen *et al.*, 2004) and bacteria isolated from grassland soil (Salles *et al.*, 2009), among other studies. Several of these studies reported a positive relationship between the level of soil functioning and the diversity level of the studied communities. They also stress that soil functioning depends mainly on functional diversity (i.e. the diversity of functional characteristics or traits of organisms) rather than on taxa number per se. Salles *et al.* (2009) even demonstrated that a simple index of functional diversity accounting for complementarity between bacterial taxa ('bacterial community niche') predicted threefold better the variations in community functioning than did taxa number.

However, the maximum number of species used in typical assemblage experiments is low, from a dozen to tens, and much lower than the number present in the native, complex microbial communities encountered in grassland soils.

17.2.2 Species removal approaches

Another approach used to study the importance of biodiversity for the functioning of grassland soil and take account of the naturally high level of diversity is species removal. In such an approach, the native soil biodiversity is progressively eroded on purpose and the induced modifications of soil functioning are assessed. A study conducted by Griffiths *et al.* (2000) demonstrated that a function involving a broad range of soil organisms, i.e. organic matter decomposition, was not affected by a significant decrease in the diversity of the soil microbial community. Even more specialized bacterial communities, such as nitrifiers and denitrifiers, proved to be largely non-sensitive to a strong decrease in their diversity (Wertz *et al.*, 2006, 2007). In the latter studies, the authors showed that removing 99% of the bacterial taxa present in grassland soil slurry used to inoculate sterile grassland soil microcosms did not affect the functioning and resistance/resilience capacity of the nitrifying and denitrifying communities. Such results highlight the huge level of functional redundancy that exists within grassland soil bacterial communities.

Do such results imply that soil microbial diversity is so high that one can disregard it for a better understanding of, and to predict, soil function? This would be a misleading conclusion. First, the concept of functional redundancy is context dependent (Loreau, 2004). The huge diversity of the soil microbiota actually represents a large reservoir of genes and functions that can become useful when soil faces variable environmental conditions, and particularly extreme or new environmental conditions. This is clearly illustrated by the response of soil microbial communities to anthropogenic inputs of a pesticide: lindane. At the time of the first input of lindane in agro-ecosystems, no bacteria were able to degrade this new molecule: it is synthesized by humans so lindane had never been a selection pressure before for soil bacteria. After some decades, however, exposed soils were found to harbour bacteria able to degrade lindane. Some studies (Boubakri *et al.*, 2006) strongly suggest that a lindane-degradation gene appeared *de novo* through a reshuffling of genetic material from different soil bacterial taxa: some bacteria might have 'gone shopping' in the huge reservoir of biological and genetic diversity of soil, to create a new gene allowing them to degrade lindane and thus cope with this new stressor. This illustrates well that the huge soil biodiversity is of paramount importance for soil functioning. More generally, humans are not always aware of the key role that soil biodiversity plays in the functioning of grassland soils and the services delivered by grasslands. But a challenge remains for soil microbial ecologists: how do we have to account for soil biodiversity, in particular soil microbial biodiversity, to model and predict changes in grassland soil functioning and services delivery?

17.3 How to Represent Grassland Soil Biodiversity in Models of Grassland Functioning and Service Delivery

Soil biodiversity plays a key role in the services delivered by grassland ecosystems, including recycling of organic detritus; maintenance of soil fertility; discrete and continuous soil ploughing throughout the year and maintenance of a good soil porosity and drainage capacity that can restrict soil erosion and promote crop yield; provision of organisms able to control crop pests or to stimulate plant growth; carbon sequestration and control of greenhouse gas emission; and runoff and sub-surface water purification. However, for ecologists who develop conceptual frameworks and models to understand the ecology of grassland ecosystems better and predict their functioning and services delivery, adequately representing soil biodiversity often remains a challenge. Can they disregard it? Or do they have to deal with the increasing view that more reductionism and increased application of modern molecular approaches of microbial communities are needed to improve the simulation of grassland soil functioning in response to environmental changes? Recent studies have addressed this issue, creating guidelines for grassland ecosystem ecologists and modellers.

In a first study, Attard *et al.* (2011) assessed to what extent can changes in denitrification be predicted by representing the denitrifying community as a black box, i.e. without an adequate representation of the biological characteristics (abundance and composition) of this community. They analysed the effect of changes in land use on denitrifiers for two different agricultural systems: (i) crop/temporary grassland conversion and (ii) cessation/application of tillage. Changes in potential denitrifying activity, that were tightly linked to time-integrated N_2O emissions, were partly due to changes in denitrifier abundance but were not related to changes in the diversity of the denitrifier (nitrite reducer) community. Actually, changes in denitrification were more related to changes in soil environmental conditions than in denitrifier abundance: soil organic carbon explained more than 80% of the variance observed in denitrification at the crop/temporary grassland site, while soil organic carbon, water-filled pore space and nitrate explained more than 90% of the denitrification variance at the till/no-till site (Attard *et al.*, 2011). These results show that an accurate simulation of carbon, oxygen and nitrate availability to denitrifiers is more important, than an accurate simulation of denitrifier abundance and community structure, in adequately understanding and predicting changes in denitrification in response to land-use changes. Avoiding finely characterizing the diversity of the denitrifying community seems reasonable here. The authors noted that their conclusions probably hold for diverse microbial communities with high functional redundancy that realize a process through a facultative activity, as for denitrifiers. They stated that the balance between functional groups (i.e. NO_2^-, NO and N_2O reducers when studying denitrification) could also be more important for soil functioning than the accurate composition of each of these functional groups.

In another study, Attard *et al.* (2010) assessed to what extent understanding the changes in nitrification requires an adequate representation of the biological characteristics (abundance and composition) of the nitrifying community. They analysed the effect of changes in land use on nitrifiers (nitrite-oxidizing bacteria, NOB). Nitrification (nitrite oxidation) was strongly and positively correlated with the abundance of *Nitrobacter*-like NOB and was also strongly related to gross mineralization, a proxy of N availability. In contrast, nitrification was weakly and negatively correlated with the abundance of *Nitrospira*-like NOB. Nitrification was loosely correlated with the genetic structure of the *Nitrobacter*-like NOB community. These results demonstrate that *Nitrobacter*-like NOB are the key functional players within the NOB community in soils with high N availability and high activity level, and that changes in nitrification are due to shifts between *Nitrospira*-like and *Nitrobacter*-like NOB rather than shifts of populations within *Nitrobacter*-like NOB. The results of this second study thus show that the coupling between the activity and abundance/community structure can be strong for less diverse soil microbial communities and obligatory activity (e.g. nitrifiers and nitrification).

This leads us to propose a general framework for guiding grassland or, more generally, ecosystem ecologists and modellers when they try to include some insights into the soil microbial component in their conceptual frameworks or models (Fig. 17.2).

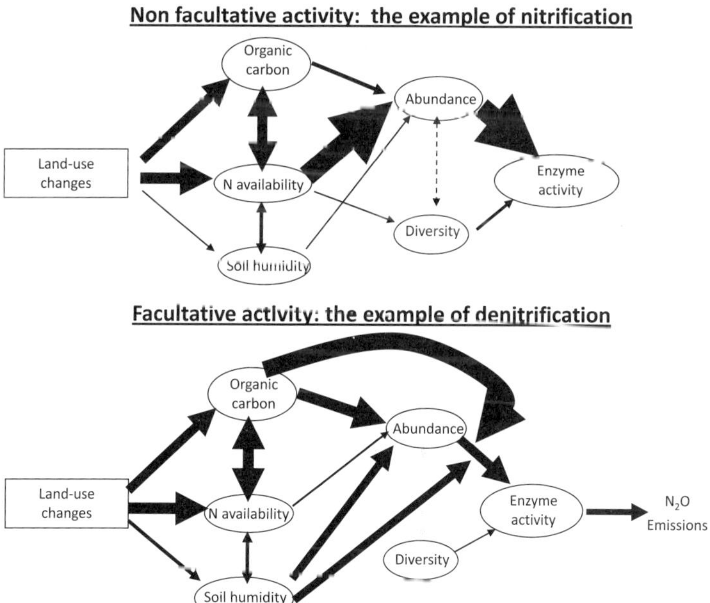

Fig. 17.2. Importance of soil bacterial diversity and abundance along with soil environmental conditions for predicting changes in soil functioning after changes in land use: adapted from the studies of Attard *et al.* (2010, 2011). Field situations involving crop/grassland conversion and cessation/application of tillage were studied. General linear analysis and path analysis were used to synthetize the observed relationships (the larger an arrow, the stronger the link between variables for (top) the nitrifying community and (bottom) the denitrifying community). The activity of the nitrite-oxidizing community is strongly linked to abundance and diversity (the key component of diversity being the balance between *Nitrobacter* and *Nitrospira*) that are themselves strongly linked to N availability. The activity of the denitrifying community is only partly explained by changes in denitrifier abundance and not related to the fine genetic structure of the denitrifying community; the activity is actually mainly explained by changes in organic carbon and, to a lesser extent, oxygen (through water-filled pore space) that influence the specific activity of denitrifiers.

Figure 17.2 demonstrates that there is no general rule for accounting for soil microbial diversity in models of grassland/ecosystem functioning. For some key functions like nitrification, which is a non-facultative activity (nitrifiers nitrify to acquire the energy needed to maintain their activity and to grow), accounting for the abundance of the functional group and key characteristics of their diversity is a major step forward in understanding and predicting changes in soil functioning. For instance, the ratio between *Nitrobacter*- and *Nitrospira*-like bacteria affects soil nitrite oxidizing activity, since *Nitrobacter*-like NOB are r-strategists favoured under high N availability, whereas *Nitrospira*-like NOB are K-strategists favoured under low N availability. For other important functions like denitrification, which is a facultative activity (denitrifiers reduce nitrate or nitrite under conditions of low oxygen availability, but behave as other heterotrophs with sufficient oxygen availability) realized by a broad range of organisms, representing in detail the changes in the abundance and diversity within the microbial functional group is, to some extent, useless. In this case, the activity depends much more on environmental conditions (organic carbon, oxygen and to a lesser extent nitrate availability) than on the microbial community characteristics. A good representation of soil environmental conditions is of paramount importance here.

More generally, the main objective of ecosystem ecologists in this context should be to account for the balance between soil microbial functional groups (e.g. NO_2^-, NO and N_2O reducers when studying denitrification) and/or between main groups differing in terms of their effect/response traits within a given functional group (e.g. *Nitrobacter/Nitrospira* within NOB). Given the wealth of diversity encountered in soil, this broad approach based on the abundance and activity of targeted microbial sub-groups is probably a more efficient way of understanding and predicting soil functioning and service or disservice delivery than the traditional characterization of the diversity/composition/genetic structure of whole functional groups. As Râmakrishna stated, 'knowledge leads to unity, but ignorance to diversity'. We can surely assess our real level of knowledge about the relationships between soil biodiversity and grassland functioning in our ability to provide comprehensive pictures of such functioning and biodiversity, disregarding useless details, and not only in our sole ability to apply advanced methodologies to deepen the characterization of the huge biodiversity harboured by grassland soils. This calls for a significant cooperation between, at least, soil microbial ecologists, grassland ecosystem ecologists and general/theoretical ecologists.

References

Attard, E., Poly, F., Laurent, F., Commeaux, C., Terada, A., Smets, B., Recous, S. and Le Roux, X. (2010) Shifts between *Nitrospira*- and *Nitrobacter*-like nitrite oxidizers underly the response of soil nitrite oxidizing enzyme activity to changes in tillage practices. *Environmental Microbiology* 12, 315–326.

Attard, E., Recous, S., Chabbi, A., De Berranger, C., Guillaumaud, N., Labreuche, J., Philippot, L., Schmid, B. and Le Roux, X. (2011) Soil environmental conditions rather than denitrifier abundance and diversity drive potential denitrification after changes in land-uses. *Global Change Biology* 17, 1975–1989.

Balvanera, P., Pfisterer, A.B., Buchmann, N., He, J.-S., Nakashizuka, T., Raffaelli, D. and Schmid, B. (2006) Quantifying the evidence for biodiversity effects on ecosystem functioning and services. *Ecology Letters* 9, 1146–1156.

Bardgett, R.D. (2005) *The Biology of Soil: a Community and Ecosystem Approach.* Oxford University Press, Oxford.

Boubakri, H., Beuf, M., Simonet, P. and Vogel, T.M. (2006) Development of metagenomic DNA shuffling for the construction of a xenobiotic gene. *GENE* 375, 87–94.

Diaz, S., Symstad, A.J., Chapin III, F.S., Wardle, D.A. and Huenneke, L.F. (2003) Functional diversity revealed by removal experiments. *Trends in Ecology and Evolution* 18, 140–146.

Gans, J., Murray, W. and Dunbar, J. (2005) Computational improvements reveal great bacterial diversity and high metal toxicity in soil. *Science* 309, 1387–1389.

Griffiths, B.S., Ritz, K., Bardgett, R.D., Cook, R., Christensen, S. and Ekelund, F. (2000) Ecosystem response of pasture soil communities to fumigation-induced microbial diversity reductions, an examination of the biodiversity-ecosystem function relationship. *Oikos* 90, 279–294.

Heemsbergen, D.A., Berg, M.P., Verhoef, H.A., Van Hal, J.R., Faber, J.H. and Loreau, M. (2004) Biodiversity effects on soil processes explained by interspecific functional dissimilarity. *Science* 306, 1019–1020.

Hooper, D.U., Ewel, J.J., Hector, A., Inchausti, P., Lavorel, S., Lodge, D., Loreau, M., Naeem, S., Schmid, B., Setälä, H., Symstad, A.J., Vandermeer, J. and Wardle, D.A. (2005) Effects of biodiversity on ecosystem functioning: a consensus of current knowledge and needs for future research. *Ecological Monographs* 75, 3–35.

Lavelle, P. and Spain, A. (2002) *Soil Ecology*. Kluwer Academic Press, Amsterdam.

Le Roux, X., Poly, F., Currey, P., Commeaux, C., Hai, B., Nicol, G.W., Prosser, J.I., Schloter, M., Attard, E. and. Klumpp, K. (2008) Effects of aboveground grazing on coupling among nitrifier activity, abundance and community structure. *ISMF Journal*, 2, 221–232.

Loreau, M. (2004) Does functional redundancy exist? *Oikos* 104, 606–611.

Patra, A.K., Abbadie, L., Clays, A., Degrange, V., Grayston, S., Loiseau, P., Louault, F., Mahmood, S., Nazaret, S., Philippot, L., Poly, F., Prosser, J.I., Richaume, A. and Le Roux, X. (2005) Effect of grazing on microbial functional groups involved in soil N dynamics. *Ecological Monographs* 75, 65–80.

Patra, A.K., Abbadie, L., Clays, A., Degrange, V., Grayston, S., Guillaumaud, N., Loiseau, P., Louault, F., Mahmood, S., Nazaret, S., Philippot, L., Poly, F., Prosser, J.I. and Le Roux, X. (2006) Effects of management regime and plant species on the enzyme activity and genetic structure of N-fixing, denitrifying and nitrifying bacterial communities in grassland soils. *Environmental Microbiology* 8, 1005–1016.

Salles, J.F., Poly, F., Schmid, B. and Le Roux, X. (2009) Community niche predicts the functioning of denitrifying bacterial assemblages. *Ecology* 90, 3324–3332.

Setälä, H. and McLean, M.A. (2004) Decomposition rate of organic substrates in relation to the species diversity of soil saprophytic fungi. *Oecologia* 139, 98–107.

Torsvik, V., Daae, F.L., Sandaa, R.A. and Øvreås, L. (1998) Novel techniques for analysing microbial diversity in natural and perturbed environments. *Journal of Biotechnology* 64, 53–56.

Van der Heijden, M.G.A., Klironomos, J.N., Ursic, M., Moutoglis, P., Streitwolf-Engel, R., Boller, T., *et al.* (1998) Mychorrhizal fungal diversity determines plant biodiversity, ecosysteem variability and productivity. *Nature* 396, 69–72.

Wertz, S., Degrange, V., Prosser, J., Poly, F., Commeaux, C., Freitag, T., Guillaumaud, N. and Le Roux, X. (2006) Maintenance of soil functioning following erosion of microbial diversity. *Environmental Microbiology* 8, 2162–2169.

Wertz, S., Degrange, V., Prosser, J.I., Poly, F., Commeaux, C., Guillaumaud, N. and Le Roux, X. (2007) Decline of soil microbial diversity does not influence the resistance and resilience of nitrifiers and denitrifiers following a model disturbance. *Environmental Microbiology* 9, 2211–2219.

18 Soil Fauna Diversity and Ecosystem Functions in Grasslands

Thibaud Decaëns and Patrick Lavelle

18.1 Introduction

Soil faunal communities comprise representatives of almost all the high order taxonomic groups that have colonized terrestrial ecosystems. They are hugely diversified both at local and global scales and may represent as much as 23% of the total diversity of known living organisms (Decaëns, 2010). Their sizes range across three orders of magnitude, from the smallest invertebrate species of the microfauna (less than 200 μm) and the mesofauna (0.2 to 2 mm) to the largest invertebrates, comprising the macrofauna (> 2 mm), and vertebrates of the megafauna (Swift *et al.*, 1979). They are implicated in most soil functions that are essential for a wide diversity of ecosystem goods and services of benefit to human populations (Lavelle *et al.*, 2006). Establishing the linkage between soil functioning and biodiversity is thus a critical scientific challenge. The scales across which soil services are provided, as well as the nature of the benefits they provide, may vary to a large extent. They are mainly controlled by a suite of abiotic factors operating at large scales (climate, mineral substrate) and biotic interactions achieved at smaller scales (vegetation and soil biota) (Fig. 18.1; Lavelle

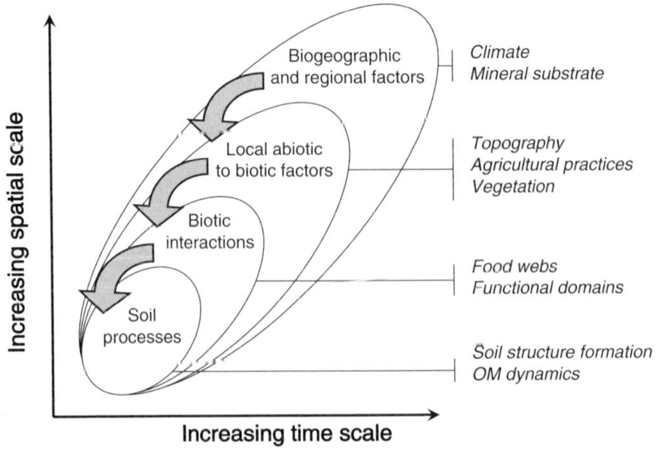

Fig. 18.1. Scale domains involved in soil functioning controls (adapted from Lavelle *et al.*, 1993), including organic matter (OM) dynamics.

et al., 1993). The functional traits of soil organisms determine their trophic position or the nature of their physical activities, and aggregative effects of biochemical and biophysical processes achieved by individual species influence soil processes and services at higher scales of organization (Fig. 18.2).

Grasslands, including natural and man-made herbaceous ecosystems, are mainly characterized by their high turnover of shoot and root biomass, and by heavy herbivore loads that strongly influence both vegetation production and composition (Bardgett and Cook, 1998). Grassland soils tend to have small amounts of surface litter and large amounts of organic nitrogen and carbon. These features combine to produce a soil environment that sustains abundant and diverse faunal communities that are usually numerically dominated by small invertebrates, while ecosystem engineers, i.e. species that physically modify their surrounding environment (described by Jones *et al.*, 1994), often comprise the main part of the biomass. Earthworms probably represent the most important group from both quantitative and functional points of view (Decaëns *et al.*, 1994, 1997). They may significantly influence soil functioning through bioturbation and the production of a range of pores and aggregates that strongly impact soil structure formation, organic matter dynamics and life conditions for other soil biota (Lavelle *et al.*, 2006).

The management of soil biological activities in managed grasslands has been recently identified as a promising research topic for the enhancement of system sustainability (Yeates *et al.*, 1997). In this context, understanding the relationships between soil biodiversity and functions is of particular concern (Bardget and Cook, 1998; Barrios, 2007; Decaëns *et al.*, 2008). In this chapter, we focus on a few questions that have been and are still important for the understanding of biodiversity–function relationships in grassland ecosystems: How many animal species comprise the communities of grassland soils? How do grassland management practices impact soil biodiversity? What are the relationships between species biological traits, soil functions and the delivery of ecosystem services? Does soil animal biodiversity matter?

18.2 The Diversity of Soil Animals in Grasslands

Grassland soils have very rich animal communities (Bardgett and Cook, 1998),

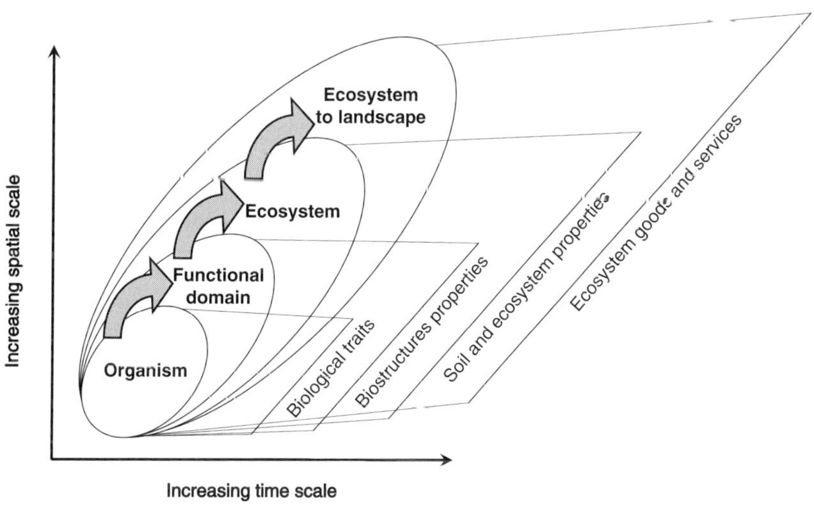

Fig. 18.2. Scale domains involved in the provision of earthworm-mediated ecosystem services in soils.

although probably not as diverse as comparable wooded ecosystems. For instance, on the one hand, Lavelle and Pashanasi (1989) and Decaëns *et al.* (1994) found a higher number of large macrofaunal taxonomic groups in the tropical rainforest than in derived grasslands. On the other hand, Decaëns *et al.* (1998) found a higher taxonomic richness in temperate chalk grasslands than in the older forested stages of a secondary succession.

As mentioned earlier, earthworms may represent the main group of the macrofauna, with fresh biomass ranging from 5 to 150 g/m^2, representing 50–90% of macroinvertebrate total biomass. Their numerical contribution may be lower, ranging from 30 to 800 individuals per m^2, representing from <10% in natural grasslands to >80% in sowed pastures (Decaëns *et al.*, 1994, 1997). Earthworm species assemblages rarely comprise more than eight to ten species, with little if any influence of the size of the regional species pool, suggesting that high levels of competition limit local diversity to a low species number (Lavelle and Spain, 2001; Wardle, 2002; Decaëns *et al.*, 2008).

Microarthropods (Collembola, Acari and Protura) are usually one of the most abundant groups of non-aquatic soil animals, with individual densities in grasslands ranging from 50,000 to 300,000 per m^2 (Bardgett and Griffiths, 1997; Bardgett and Cook, 1998), often predominantly composed of Collembola and Acari (Seastedt, 1984). In temperate grasslands, Collembola and Acari biomasses are often reported to be similar, whereas mites usually dominate in the tropics (Petersen and Luxton, 1982; Bardgett and Griffiths, 1997; Bargett and Cook, 1998). In terms of species diversity, Siepel and Van de Bund (1988) reported that there were up to 108 species of microarthropods in 500 cm^2 soil of unmanaged temperate grasslands.

Nematodes are very abundant in grassland soils, with individual densities ranging from 1 million to 10 million per m^2, depending on grassland productivity and pedoclimatic conditions (Coulson and Whittaker, 1978; Bardgett *et al.*, 1997; Yeates *et al.*, 1997). Local species richness in grasslands often ranges from 75 to more than 150 species (Hodda and Wanless, 1994; Yeates *et al.*, 1997). In temperate agricultural grassland, nematode populations are dominated by plant parasites and micropredators, with omnivores, fungal feeders and predators being less represented (Bardgett and Cook, 1998).

Despite these published quantitative data, we are probably far from reaching a relevant estimation of animal biodiversity in grassland soils, whatever the scale considered (Decaëns, 2010). As for soil fauna in general, grassland soil invertebrates have only weakly engaged the attention of taxonomists compared with other groups like higher plants and vertebrates. Taxonomic deficit for soil biota is on average 76%, and usually above 90% for organisms smaller than 100 mm (André *et al.*, 1994; Decaëns, 2010). On the other hand, even in groups that have been intensively studied, such as lumbricid earthworms, using molecular identification tools (DNA barcodes) reveals an unsuspected number of cryptic species that are often impossible to separate on a morphological basis (King *et al.*, 2008; Rougerie *et al.*, 2009).

18.3 Impact of Grassland Management on Soil Animal Communities

A number of studies have illustrated how grassland management practices (e.g. stocking rate management, fertilizer inputs, etc.) impact soil invertebrate communities (Decaëns *et al.*, 1994; Bardgett and Cook, 1998; Lavelle and Spain, 2001). These practices probably operate through modifying both the levels of stress (resource availability, temperature, pH, etc.) and disturbances (fires, grazing, cultivation, etc.) that will act on the soil system (Wardle, 2002). Along gradients of stress or perturbation, species richness theoretically follows a humpback pattern of variation. In strongly stressed or disturbed ecosystems, diversity is reduced because only a few species survive under such constraints, whereas in more productive or less disturbed conditions, species richness may decrease as a result of competitive exclusion or a decreased spatial heterogeneity of resource distribution. While this peaked pattern has long been presented as ubiquitous in nature, recent studies have revealed that positive or negative monotonic and unimodal responses, as well as non-significant relationships, may all exist (Mackey and Currie, 2000; Mittelbach *et al.*, 2001).

Currently available results for soil biota are rather inconsistent, suggesting that many soil taxa are not strongly regulated by competition, and that increasing resource availability is not likely to result in competitive exclusion (Bardgett *et al.*, 2005b; Wardle, 2006). As an example, chilopod or woodlouse diversity (Paoletti, 1988) seems to increase monotonically along a gradient of resource availability. On the other hand, Decaëns (2010) reviewed the available literature and concluded that humpback patterns have also been described for some taxa. Variations in soil biodiversity along disturbance gradients have also been extensively illustrated in the scientific literature (Wardle, 2002; Bardgett *et al.*, 2005b). In grassland ecosystems, livestock grazing pressure is known to affect soil invertebrate communities significantly. Decaëns *et al.* (1994) described how cattle introduction in native tropical savannahs enhances earthworm populations, to the detriment of other groups such as termites. Reasonable levels of stocking, however, usually allow abundant and diverse invertebrate communities to be maintained, even in the more productive systems, as demonstrated for nematodes (Freckman *et al.*, 1979; Bardgett *et al.*, 1997), mites and springtails (Bardgett *et al.*, 1993), and soil macrofauna (Decaëns *et al.*, 1994, 1998). This may be due to an appropriate balance between an increase in trophic resource availability and quality (animal faeces, litter and dead roots, microbial communities) and a slight increase in disturbances (animal trampling and changes in soil microclimate). While grazing may greatly enhance system carrying capacity for soil invertebrates, disturbances maintain dominant species at a sufficiently low population level, weakening competitive exclusion mechanisms.

In cases where this equilibrium is not respected any more, the story can be completely different. Overgrazing may, for example, result in the disappearance of whole groups of invertebrates that are not adapted to resist increasing disturbance levels. In grazed tropical savannahs, overgrazing strongly alters the structure and decreases the taxonomic richness of macrofaunal communities (Decaëns *et al.*, 1994). Similar responses have been reported for soil microarthropod communities (Bardgett and Cook, 1998). On the other hand, grazing cessation in highly productive pastures may allow aggressive competitors to dominate species assemblages and deplete taxonomic richness through competitive exclusion (Decaëns, 2010). This has been described in Amazonian pastures sowed after clearance of the original rainforest, where the earthworm *Pontoscolex corethrurus* builds into large and monospecific populations, preventing colonization by other ground invertebrates and leading to severe physical degradation of superficial soil layers (Chauvel *et al.*, 1999).

18.4 Responses of Soil Biodiversity to Successional Gradients

Grassland afforestation classically occurs after cessation of grazing activities. The early process of vegetation succession, described as the aggradation phase, corresponds to an increase of soil micro-organism biomass and diversity and an increasing relative dominance of fungi (Bardgett *et al.*, 2005a). Soil animal diversity may present different responses to these changes, making any generalization difficult. For instance, Scheu and Schulz (1996) and Decaëns *et al.* (1998) observed an increase of macro-invertebrate biomass and taxonomic richness during grassland afforestation, while Scheu (1992) and Decaëns *et al.* (1997) found no change for earthworm assemblage in the same successional contexts. These discrepancies underline different responses to top-down and bottom-up controls and to abiotic factors during the chronosequence (Doblas-Miranda *et al.*, 2008).

A number of agricultural or sylvicultural systems of production use management cycles that drive the system toward succession-like dynamics. In these anthroposystems, soil communities may present similar temporal patterns to those described for true successions. In both temperate and tropical systems, the diversity of soil fauna usually increases during the first years after vegetation sowing and then stabilizes (Chauvat *et al.*, 2007; Decaëns and Jiménez, 2002), suggesting a positive response to increased resource availability until niche saturation is obtained.

18.5 Soil Ecosystem Engineers and Ecosystem Functioning

Soil functioning is controlled by a suite of factors ranging from large-scale abiotic constraints to small-scale biotic interactions (Fig. 18.1; Lavelle *et al.*, 1993). Soil fauna influence soil functions at a very local scale, through its engineering activities or its trophic interactions in soil food webs (Brussaard, 1998; Lavelle *et al.*, 2006). These functions are local drivers of important ecosystem services provided at scales that are orders of magnitude above those at which biological controls are achieved (Fig. 18.2; Andrén *et al.*, 1999; Lavelle *et al.*, 2006). As mentioned earlier, soil ecosystem engineers, and especially earthworms, often dominate faunal communities in grassland soils, and their impact on soil functioning is thus likely to be critical.

18.5.1 Water supply

The contribution of soil invertebrates to water infiltration, storage and detoxification is likely to be very important (Lavelle *et al.*, 2006). Edaphic invertebrates affect soil surface roughness and porosity, and may decrease water surface runoff while enhancing infiltration (Blanchart *et al.*, 2004). These impacts may vary greatly between species or functional groups, and the diversity of pore shapes and sizes is probably important in allowing soils to store water at a wide range of potentials (Lavelle *et al.*, 2006). As an example, small-body-sized earthworms may reduce soil bulk density and the percentage of macro-aggregates, inducing a decrease in water retention capacity and an increase in water infiltration. Conversely, larger species may induce an increase in bulk density and macro-aggregation, resulting in higher water retention capacity and lower infiltration rates (Blanchart *et al.*, 2004).

18.5.2 Organic matter and nutrient cycling

Soils are essential in sustaining the service of organic matter decomposition, as they are responsible for the decay of 60–90% of terrestrial primary production (Lavelle *et al.*, 2006). Soil fauna influences organic matter and nutrient cycling through litter comminution and the stimulation of microbial activities (Swift *et al.*, 1979; Lavelle and Spain, 2001). Ecosystem engineers also create biogenic structures that are short-term incubators for microbial activities and long-term microsites for carbon and nutrient sequestration (Lavelle *et al.*, 2006). They affect nutrient cycling in a contrasting way, according to the spatio–temporal scale, contributing significantly to nutrient cycling and to the prevention of leaching from impaired ecosystems towards low-lying aquifers, streams and oceans. This effect, however, has never been quantified in any ecosystem or large landscape unit. Some studies on dissolved organic carbon flow in terrestrial ecosystems have emphasized the importance of micro-organism activation through the soil profile (e.g. Michelsen *et al.*, 2004), but the participation of large invertebrates in this process remains little studied (Jiménez and Lal, 2006).

18.5.3 Primary production

The mechanistic basis by which primary productivity is regulated by soil fauna has been well documented (Wardle, 2002; Wardle *et al.*, 2004; Lavelle *et al.*, 2006). Parasites and rhizophagous species, for example, can directly remove carbon and nutrients from plant tissues and reduce root uptake capacity, producing a negative feedback on plant growth (Bever *et al.*, 1997). In contrast, saprophagous and/or engineer species can significantly enhance plant productivity. This has been described in many experiments in the presence of Protista (Bonkowski, 2004), nematodes and Enchytraeidae (Setälä and Huhta, 1991), Collembola (Curl and Harper, 1990; Harris and Boerner, 1990), combinations of these organisms (Setälä and Huhta, 1991), termites or ants (Okello-Oloya and Spain, 1986), and earthworms (Scheu, 2003; Brown *et al.*, 2004; Laossi *et al.*, 2010). These effects are attributed to five main mechanisms: (i) the enhancement of soil organic matter mineralization and nutrient cycling; (ii) the production of plant growth regulators via the stimulation of microbial activity; (iii) the control of pests and parasites; (iv) the stimulation of symbionts; and (v) modifications of soil porosity and aggregation, which induce changes in water and oxygen availability for plant roots. Recently, earthworms were shown to play a significant role in adjusting plant phenotypes to current environmental conditions in soils (Blouin *et al.*, 2005), probably as a result of millions of years of co-evolution. Research is only starting to unravel the mechanisms behind such interactions.

18.5.4 Soil formation and erosion controls

The implication of soil faunal activities in the long-term process of soil formation has not been much considered to date (Lavelle *et al.*, 2006), probably because of the time lag that exists between the microsites where invertebrates directly regulate soil processes and the scale at which pedogenesis operates. It is, however, likely that small-scale faunal activity, accumulated over time periods of 1000 or more years (i.e. the time taken to create 1 m of soil in temperate areas), has large-scale consequences on soil formation (Lavelle *et al.* 2006). It is, for example, acknowledged that earthworms and termites play a significant role in the creation of the soil structures of 'vermisols' and oxisols, respectively. Soil animals also have recognized effects in soil creeping (Nooren *et al.*, 1995; Nye, 1954) and burial of stones (Darwin, 1881). Examples of rapid changes in soil profiles following invasion of faunal communities by exotic earthworms has also been described in Amazonian pastures (Chauvel *et al.*, 1999).

The soil fauna, through its impact on soil structure, is also likely to affect soil erodibility and erosion. The abundant literature dedicated to this topic highlights, however, the complex and probably strongly species-dependent mechanisms of faunal impacts on soil erosion. Earthworms, for example, may improve resistance to erosion by enhancing soil infiltration and aggregate stability, but can also increase erosion when they deposit fragile casts at the soil surface (Blanchart *et al.*, 2004). In the tropics, the effects of 'compacting' and 'decompacting' species may differ substantially and vary drastically according to soil type and organic matter contents. The impact of other groups of soil invertebrates on soil erodibility remains largely unknown.

18.5.5 Carbon storage and climate regulation

Carbon sequestration, the process whereby atmospheric CO_2 is transferred into a long-lived C pool, is a critical issue because of its potential role in off-setting fossil fuel emissions and the resulting climate changes (Jiménez and Lal, 2006). The amount of C stored in soils is estimated to be 2344 Pg, that is, the third largest global C pool after the lithosphere and deep oceans, and also the third largest in terms of accumulation rate (1.4 Pg/year), after oceanic net uptake (1.7) and atmospheric uptake (3.2) (Jiménez and Lal, 2006). Carbon fixed in plant biomass by photosynthesis is incorporated into the soil through the decomposition of decaying plant tissues, under the action of different soil organisms. Soil invertebrate activities can influence this pathway through the long-term accumulation of small-scale effects and/or limitations imposed on decomposition that increase the residence time of organic compounds in soil (Lavelle *et al.*, 2006). First, soil aggregates produced by soil engineers physically protect some fractions of the soil organic matter, thus increasing the residence time of C in soil and preventing its rapid release in the form of greenhouse gases (Beare *et al.*, 1994, Guggenberger *et al.*, 1996). Large food-web processes may also control humification, thus modulating the amounts of relatively labile C transformed into forms that are much more resistant to further decomposition, especially in nutrient-poor ecosystems (Wardle, 2002). The direction and strength of this control, however, is likely to vary greatly according to nutrient and water availability in the soil (Wardle, 2002; Filser, 2002).

18.6 Does Diversity Matter?

Although a large body of literature has accumulated on the importance of diversity–functioning relationships in above-ground plant and animal communities, the question of how much soil biodiversity matters for the functioning of grasslands or other ecosystems has seldom been addressed in the scientific literature. The hypotheses provided for above-ground communities, which are equally applicable below ground (Wardle, 2002), are that more diverse ecosystems should be more resistant and resilient (McNaughton, 1994; Pimm, 1984). While there is growing empirical evidence to support this hypothesis when considering the productivity of plant communities (Tilman, 1996; Naeem *et al.*, 1995), the outstandingly high species richness in soil communities compared with their functional diversity suggests a high level of functional redundancy for many soil species (Wardle, 2002; Barrios, 2007). Despite the recognized impact of soil fauna on soil functioning (Lavelle *et al.*, 2006), the relationships between soil animal diversity and the rates of below-ground processes remains obscure (Hedde *et al.*, 2010).

Several authors have highlighted the problems that soil ecologists have to face when addressing the complex question of the diversity–functioning relationships. The historical existence of two separate approaches in the study of soil biology (community versus functional) is probably one such problem, although it cannot be considered as a specific constraint of soil ecology (Barrios, 2007). An obvious limitation is the poor taxonomical knowledge that has been reached so far for most soil taxa, and the very high species diversity that characterizes most soil communities. This has impaired the most tentative descriptions of soil biodiversity in any holistic way (Decaëns, 2010; Rougerie *et al.*, 2009), and this may explain the limited progress that has been reached in the understanding of soil diversity–function relationships. Finally, the poor knowledge of the functional ecology of most soil invertebrates, and the existing difficulty in conducting species removal experiments under field conditions, have also impaired the development of experimental and conceptual studies that would have provided significant insight (Barrios, 2007). Microcosm and modelling approaches have been proposed as an interesting option, but the incongruent conclusions of some studies based on food-web approaches have pointed out the limits of these approaches in addressing the full complexity of soil community interactions and the multifunctional nature of many soil organisms (Barrios, 2007).

One recent and promising research avenue for the understanding of soil diversity–function relationships is the use of trait-based approaches, which have provided more reliable insights than the simple use of species richness as a measure of diversity (Diaz and Cabido, 2001). Many works in plant ecology have focused on physiological and morphological traits, such as specific leaf area and seed size or number (Díaz *et al.*, 2004). Recently, some studies have successfully explained soil macrofauna biodiversity effects on soil processes by the level of functional dissimilarity or of morphological trait dispersion within species assemblages (Heemsbergen *et al.*, 2004; Hedde *et al.*, 2010). Experimental and field studies on earthworm impacts on soil structure have also pointed out the importance of maintaining a minimum level of functional diversity for the sustainability of grassland ecosystems. The replacement of diverse native communities by monospecific populations of the endogeic *P. corethrurus* has, for instance, been recognized as having lasting effects on the porosity of pasture soils in Manaus (Chauvel *et al.*, 1999; Blanchart *et al.*, 2004). Similar examples of negative effects of functionally simplified earthworms assemblages on soil structure have been reported in other studies (see e.g. Rose and Wood, 1980). As suggested by Hedde *et al.* (2010), these results confirmed that soil macro-invertebrate assemblages are more likely to be driven by complementarity or facilitation rather than by competition.

Further studies are now needed to integrate new and promising methodological approaches that link identity and function of soil organisms, experimental approaches under field conditions, and conceptual approaches that look at soil communities and processes at different scales (Barrios, 2007). The use of genomic approaches (DNA barcoding and environmental barcoding), by allowing more rapid and efficient characterization of soil animal diversity at different spatial scales, will, for example, facilitate and improve studies on soil diversity–function relationships (Decaëns *et al.*, 2008; Rougerie *et al.*, 2009). Near infrared reflectance spectrometry (NIRS), which allows the attribution of biogenic structures to specific soil organisms as well as a rapid and reliable characterization of soil properties across large areas, may also represent a relevant approach to improve our understanding of how soil invertebrates affect both soil structure formation and organic matter dynamics (Shepherd and Walsh, 2002; Hedde *et al.*, 2005; Jouquet *et al.*, 2010). Finally, all these approaches, together with others not mentioned here, will have to be used in an integrative approach if we want to address efficiently the question of how soil invertebrate diversity matters for the functioning of soil in grasslands as well as in other ecosystems.

References

André, H.M., Noti, M.-I. and Lebrun, P. (1994) The soil fauna: the other last biotic frontier. *Biodiversity and Conservation* 3, 45–56.

Andrén, O., Brussaard, L. and Clarholm, M. (1999) Soil organisms influence on ecosystem-level processes – bypassing the ecological hierarchy? *Applied Soil Ecology* 11, 177–188.

Bardgett, R.D. and Cook, R. (1998) Functional aspects of soil animal diversity in agricultural grasslands. *Applied Soil Ecology* 10, 263–276.

Bardgett, R.D. and Griffiths, B. (1997) Ecology and biology of soil protozoa, nematodes and microarthropods. In: van Elsas, J.D., Wellington, E. and Trevors J.T. (eds) *Modern Soil Microbiology*. Marcell Dekker, New York, pp. 129–163.

Bardgett, R.D., Frankland, J.C. and Whittaker, J.B. (1993). The effects of agricultural practices on the soil biota of some upland grasslands. *Agriculture, Ecosystems and Environment* 45, 25–45.

Bardgett, R.D., Leemans, D.K., Cook, R. and Hobbs, P.J. (1997) Seasonality of soil biota of grazed and ungrazed hill grasslands. *Soil Biology and Biochemistry* 29, 1285–1294.

Bardgett, R.D., Bowman, W.D., Kaufmann R. and Schmidt, S.K. (2005a) A temporal approach to linking aboveground and belowground ecology. *Trends in Ecology and Evolution* 20, 634–641.

Bardgett, R.D., Yeates, G.W. and Anderson, J.M. (2005b) Patterns and determinants of soil biological diversity. In: Bardgett R.D., Usher M.B. and Hopkins D.W. (eds) *Biological Diversity and Function in Soils.* Cambridge University Press, Cambridge, pp. 100–118.

Barrios, E. (2007) Soil biota, ecosystem services and land productivity. *Ecological Economics* 64, 269–285.

Beare, M.H., Hendrix, P.F. and Coleman, D.C. (1994) Water-stable aggregates and organic matter fractions in conventional- and no-tillage soils. *Soil Science Society of America Journal* 58, 777–786.

Bever, J.D., Westover, K.M. and Antonovics, J. (1997) Incorporating the soil community into plant population dynamics: the utility of the feedback approach. *Journal of Ecology* 85, 561–573.

Blanchart, E., Albrecht, A., Brown, G., Decaens, D., Duboisset, A., Lavelle, P., Mariani, L. and Roose, E. (2004) Effects of tropical endogeic earthworms on soil erosion. *Applied Soil Ecology* 104, 303–315.

Blouin, M., Zuily-Fodil, Y., Pham-Thi, A.-T., Laffray, D., Reversat,G., Pando, A., Tondoh, J. and Lavelle, P. (2005) Belowground organism activities affect plant aboveground phenotype, inducing plant tolerance to parasites. *Ecology Letters* 8, 202–208.

Bonkowski, M. (2004) Protozoa and plant growth: the microbial loop in soil revisited. *New Phytologist* 162, 617–631.

Brown, G.G., Edwards, C.A. and Brussaard, L. (2004) How earthworms effect plant growth: burrowing into the mechanisms. In: Edwards, C.A. (ed) *Earthworm Ecology*. CRC Press, Boca Raton, Florida, pp. 13–49.

Brussaard, L. (1998) Soil fauna, guilds, functional groups and ecosystem processes. *Applied Soil Ecology* 9, 123–135.

Chauvat, M., Wolters, V. and Dauber, J. (2007) Response of collembolan communities to land-use change and grassland succession. *Ecography* 30, 183–192.

Chauvel, A., Grimaldi, M., Barros, E., Blanchart, E., Desjardins, T., Sarrazin, M. and Lavelle, P. (1999) Pasture damage by an Amazonian earthworm. *Nature*, 398, 32–33.

Coulson, J.C. and Whittaker, J.B. (1978) Ecology of moorland animals. In: Heal O.W. and Perkins D.F. (eds) *Production Ecology of British Moors and Montane Grasslands*. Springer, London, pp. 52–93.

Curl, E.A. and Harper, J.D. (1990) Fauna-microflora interactions. In: Lynch J.M. (ed.) *The Rhizosphere.* John Wiley and Sons, Chichester, pp. 369–388.

Darwin, C. (1881) *The Formation of Vegetable Mould through the Action of Worms with Observations of their Habits.* Murray, London.

Decaëns, T. (2010) Macroecological patterns in soil communities. *Global Ecology and Biogeography* 19, 287–302.

Decaëns, T. and Jiménez, J.J. (2002) Earthworm communities under an agricultural intensification gradient in Colombia. *Plant and Soil* 240, 133–143.

Decaëns, T., Lavelle, P., Jiménez, J.J., Escobar, G. and Rippstein, G. (1994) Impact of land management on soil macrofauna in the Oriental Llanos of Colombia. *European Journal of Soil Biology* 30, 157–168.

Decaëns, T., Dutoit, T. and Alard, D. (1997) Earthworm community characteristics during afforestation of abandoned chalk grasslands (Upper Normandy, France). *European Journal of Soil Biology* 33, 1–11.

Decaëns, T., Dutoit, T., Alard, D. and Lavelle, P. (1998) Factors influencing soil macrofaunal communities in post-pastoral successions of western France. *Applied Soil Ecology* 9, 361–367.

Decaëns, T., Margerie, P., Aubert, M., Hedde, M. and Bureau, F. (2008) Assembly rules within earthworm communities in north-western France—a regional analysis. *Applied Soil Ecology*, 39, 321–335.

Diaz, S. and Cabido, M. (2001). Vive la difference: plant functional diversity matters to ecosystem processes. *Trends in Ecology and Evolution* 16, 646–655.

Díaz, S., Hodgson, J.G., Thompson, K., Cabido, M., Cornelissen, J.H.C., Jalili, A., Montserrat-Martí, G., Grime, J.P., Zarrinkamar, F., Asri, Y., Band, S.R., Basconcelo, S., Castro-Díez, P., Funes, G., Hamzehee, B., Khoshnevi, M., Pérez-Harguindeguy, N., Pérez-Rontomé, M.C., Shirvany, F.A., Vendramini, F., Yazdani, S., Abbas-Azimi, R., Bogaard, A., Boustani, S., Charles, M., Dehghan, M., de Torres-Espuny, L., Falczuk, V., Guerrero-Campo, J., Hynd, A., Jones, G., Kowsary, E., Kazemi-Saeed, F., Maestro-Martínez, M., Romo-Díez, A., Shaw, S., Siavash, B., Villar-Salvador, P. and Zak, M.R. (2004) The plant traits that drive ecosystems: evidence from three continents. *Journal of Vegetation Science* 15, 295–304.

Doblas-Miranda, E., Wardle, D.A., Peltzer, D.A. and Yeates, G.W. (2008) Changes in the community structure and diversity of soil invertebrates across the Franz Josef Glacier chronosequence. *Soil Biology and Biochemistry* 40, 1069.

Filser, J. (2002) The role of Collembola in carbon and nitrogen cycling in soil. *Pedobiologia* 46, 234–245.

Freckman, D.W., Duncan, D.A. and Larson, J. (1979) Nematode density and biomass in an annual grassland ecosystem. *Journal of Rangeland Management* 32, 418–422.

Guggenberger, G., Thomas, R.J. and Zech, W. (1996) Soil organic matter within earthworm casts of an anecic–endogeic tropical pasture community, Colombia. *Applied Soil Ecology* 3, 263–274.

Harris, K.K. and Boerner, R.E.J. (1990) Effects of belowground grazing by Collembola on growth, mycorrhizal infection, and P uptake of *Geranium robertianum*. *Plant Soil* 129, 203–210.

Hedde, M., Lavelle, P., Joffre, R., Jiménez, J.J. and Decaens, T. (2005) Specific functional signature in soil macro-invertebrate biostructures. *Functional Ecology* 19, 785–793.

Hedde, M., Bureau, F., Chauvat, M. and Decaëns, T. (2010) Patterns and mechanisms responsible for the relationship between the diversity of litter macro-invertebrates and leaf degradation. *Basic and Applied Ecology* 11, 35–44

Heemsbergen, D.A., Berg, M.P., Loreau, M., van Hal, J.R., Faber, J.H. and Verhoef, H.A. (2004) Biodiversity effects on soil processes explained by interspecific functional dissimilarity. *Science* 306, 1019–1020.

Hodda, M. and Wanless, F.R. (1994) Nematodes from an English chalk grassland: species distribution. *Nematologica* 40, 116–132.

Jiménez, J.J. and Lal, R. (2006) Mechanisms of C sequestration in soils of Latin America. *Critical Review in Plant Sciences* 25, 337–365.

Jones, C.G., Lawton, J.H. and Shachak, M. (1994) Organisms as ecosystem engineers. *Oikos* 69, 373–386.

Jouquet, P., Henry-des-Tureaux, T., Mathieu, J., Doan Thu, T., Tran Duc, T. and Orange, D. (2010) Utilization of near infrared reflectance spectroscopy (NIRS) to quantify the impact of earthworms on soil and carbon erosion in steep slope ecosystem: a study case in northern Vietnam. *Catena* 81, 113–116.

King, R.A., Tibble, A.L. and Symondson, W.O.C. (2008) Opening a can of worms: unprecedented sympatric cryptic diversity within British lumbricid earthworms. *Molecular Ecology* 17, 4684–4698.

Laossi, K.R., Decaëns, T., Jouquet, P. and Barot, S. (2010) Can we predict how earthworm effects on plant growth vary with soil properties? *Applied and Environmental Soil Science*, article ID 784342, doi:10.1155/2010/784342.

Lavelle, P. and Pashanasi, B. (1989) Soil macrofauna and land management in Peruvian Amazonia (Yurimaguas, Loreto). *Pedobiologia* 33, 283–291.

Lavelle, P. and Spain, A.V. (2001) *Soil Ecology*. Kluwer Academic Publishers, Dordrecht, The Netherlands.

Lavelle, P., Blanchart, E., Martin, A., Martin, S., Spain, A.V., Barois, I., Toutain, F. and Schaefer, R. (1993) A hierarchical model for decomposition in terrestrial ecosystems: application to soils of the humid tropics. *Biotropica* 25, 130–150

Lavelle, P., Decaëns, T., Aubert, M., Barot, S., Blouin, M., Bureau, F., Margerie, P., Mora, P. and Rossi, J. (2006) Soil invertebrates and ecosystem services. *European Journal of Soil Biology* 42, S3–S15.

Mackey, R.L. and Currie, D.J. (2000) A re-examination of the expected effects of disturbance on diversity. *Oikos* 88, 483–493.

McNaughton, S.J. (1994) Biodiversity and function of grazing ecosystems. In: Schulze E.D. and Mooney H.A. (eds) *Biodiversity and Ecosystem Function*. Springer, London, pp. 361–383.

Michelsen, A., Andersson, M., Jensen, M., Kjøller, A. and Gashew, M. (2004) Carbon stocks, soil respiration and microbial biomass in fire-prone tropical grassland, woodland and forest ecosystems. *Soil Biology and Biochemistry* 36, 1707– 1717.

Mittelbach, G.G., Steiner, C.F., Scheiner, S.M., Gross, K.L., Reynolds, H.L., Waide, R.B., Willig, M.R., Dodson, S.I. and Gough, L. (2001) What is the observed relationship between species richness and productivity? *Ecology* 82, 2381–2396.

Naeem, S., Thompson, L.J., Lawler, S., Lawton, J.H. and Woodfin, R.M. (1995) Empirical evidence that declining species diversity may alter the performance of terrestrial ecosystems. *Philosophical Transactions of the Royal Society of London B* 347, 249–262.

Nooren, C.A.M., Vanbreemen, N., Stoorvogel, J.J. and Jongmans, A.G. (1995) The role of earthworms in the formation of sandy surface soils in a tropical forest in Ivory-Coast. *Geoderma* 65, 135–148.

Nye, P.H. (1954) Some soil-forming processes in the humid tropics. I. A field study of a catena in the West African forest. *Journal of Soil Science* 5, 7–21.

Okello-Oloya, T. and Spain, A.V. (1986) Comparative growth of two pasture plants from northeastern Australia on the mound materials of grass and litter-feeding termites (Isoptera: Termitidae) and on their associated surface soils. *Revue d'Ecologie et de Biologie du Sol* 23, 381–392.

Paoletti, M. (1988) Soil invertebrates in cultivated and uncultivated soil in northeastern Italy. *Redia* 71, 501–563.

Petersen, H. and Luxton, M. (1982) A comparative analysis of soil fauna populations and their role in decomposition processes. *Oikos* 39, 287–388.

Pimm, S.L. (1984) The complexity and stability of ecosystems. *Nature* 307, 321–326.

Rose, C.J. and Wood, A.W. (1980) Some environmental factors affecting earthworm populations and sweet potato production in the Tari Bassin, Papua New Guinea Highland. *Papua New Guinea Agriculture Journal* 31, 1–10.

Rougerie, R., Decaëns, T., Deharveng, L., Porco, D., James, S.W., Chang, C.-H., Richard, B. and Hebert, P.D.N. (2009) DNA barcodes for soil animal taxonomy: transcending the final frontier. *Pesquisa Agropecuaria Brasileira* 44, 789–801.
Scheu, S. (1992) Changes in the lumbricid coenosis during secondary succession from a wheat field to a beechwood on limestone. *Soil Biology and Biochemistry* 24, 1641–1646.
Scheu S. (2003) Effects of earthworms on plant growth: patterns and perspectives. *Pedobiologia* 47, 846–856.
Scheu, S. and Schulz, E. (1996) Secondary succession, soil formation and development of a diverse community of oribatids and saprophagous soil macro-invertebrates. *Biodiversity and Conservation* 5, 235–250
Seastedt, T.R. (1984) The role of microarthropods in decomposition and mineralization processes. *Annual Review of Entomology* 29, 25–46.
Setälä, H. and Huhta, V. (1991) Soil fauna increase *Betula pendula* growth: laboratory experiments with coniferous forest floor. *Ecology* 72, 665–671.
Shepherd, K.D. and Walsh, M.G. (2002) Development of reflectance spectral libraries for characterization of soil properties. *Soil Science Society of America Journal* 66, 988–998.
Siepel, H. and Van de Bund, C.F. (1988) The influence of management practices on the microarthropod community of grassland. *Pedobiologia* 31, 339–354.
Swift, M.J., Heal, O.W. and Anderson, J.M. (1979) *Decomposition in Terrestrial Ecosystems.* Blackwell Scientific, Oxford.
Tilman, D. (1996) Biodiversity: population versus ecosystem stability. *Ecology* 77, 97–107.
Wardle, D.A. (2002) *Communities and Ecosystems – Linking the Aboveground and Belowground Components.* Princeton University Press, Princeton, NJ.
Wardle, D.A. (2006) The influence of biotic interactions on soil biodiversity. *Ecology Letters* 9, 870–886.
Wardle, D.A., Bardgett, R.D., Klironomos, J.N., Setälä, H., Van der Putten, W.H. and Wall, D.H. (2004) Ecological linkages between aboveground and belowground biota. *Science* 304, 1629–1633.
Yeates, G.W., Bardgett, R.D., Cook, R., Hobbs, P.J., Bowling, P.J. and Potter, J.F. (1997) Faunal and microbial diversity in three Welsh grassland soils under conventional and organic management regimes. *Journal of Applied Ecology* 34, 453–471.

Part IV:

Grasslands as Components of Sustainable Farming Systems

Editor: Gilles Lemaire

Grasslands are anthropogenic ecosystems managed for meat, dairy or wool production that have to be included within coherent farming systems. But nowadays, these farming systems have not only to be managed for optimizing economic returns in term of animal outputs, but they have also to be evaluated through their capacity to provide environmental and ecological services to society. So Part IV shows how the different effects and impacts of grassland areas on biogeochemical cycles, environmental fluxes and biodiversity, as analysed in Parts II and III, could be optimized through the use and management of grassland areas within sustainable farming systems. These aspects are analysed in two conditions: those of North-west Europe (Chapter 19) and New Zealand (Chapter 20). Special attention is paid to grazing versus cutting management and to the role of grassland areas in efficiently recycling animal wastes within intensive livestock systems. The extensive use of grasslands within rangeland systems is analysed within Chapter 21, paying special attention to the problem of the conservation of natural resources and the production of different social and economic services not directly linked to grassland productivity itself. Nevertheless, the interest of grassland area is not restricted to livestock production, and the role of grasslands through spatial and temporal interactions with arable cropping systems has to be investigated. These general aspects of the importance of grasslands within mixed farming systems are developed in Chapter 22 for different regions of North America and in Chapter 23 for the subtropical regions of South America. These two series of examples show that both environmental and economical added value could be obtained by integrating grasslands and arable crops within the same agrosystem, instead of managing them separately.

19 The Role of Grasslands in Intensive Animal Production in North-west Europe: Conditions for a More Sustainable Farming System

Jean-Louis Peyraud

19.1 Introduction

High milk prices have encouraged dairying systems using high inputs of chemical fertilizers, concentrate feeds and mechanized methods for silage production, at the expense of grazing, in many European countries (Bourgeois, 2002). This tendency is largely reinforced by the convenience of managing dairy herds indoors as opposed to the difficulties involved in organizing the feeding programme of grazing dairy herds in time and space and from an unstable feed resource. For 40 years, the selection of dairy cows has been almost exclusively oriented towards genetic potential for milk production, which now cannot be adequately reached by grazing alone. This has largely contributed to reductions in the use of grazing, especially in countries in the north of Europe. Increased herd size and the increasing use of automated milking systems may be other reasons for the decrease of grazing, at least for countries in the north of Europe where pressures on land use and stocking rates are high.

The context has gradually changed since the early 1990s, primarily with the emergence of environmental regulation, concerning water quality in the first step (the Nitrate Directive, the Water Framework Directive). Agri-environmental measures have become gradually more restrictive and will define new priorities. Apart from nitrate in runoff and ground waters, ruminant production systems are also considered to be responsible for the emission of large quantities of greenhouse gases (FAO, 2006). More recently, declines or instability in the price of milk and the increasing price of non-renewable energy and mineral fertilizers offer new opportunities for dairying systems based on grasslands and grazing. Indeed, grazing is not only a relatively cheap source of feed for ruminants but grasslands are increasingly recognized for their contributions to the conservation of biodiversity, regulation of physical and chemical fluxes in ecosystems, the mitigation of pollution, carbon sequestration, which might partly counteract methane emissions from ruminants, and the protection of soils (See Introduction, this volume). Additionally, grassland-based systems promote a clean, animal-welfare-friendly image for ruminant production, and open landscapes with grazing ruminants are highly appreciated by the public. The objective of this chapter is to review existing knowledge for developing productive, efficient and environmentally friendly dairying systems based on grazing and grassland utilization and using levels of inputs that are as low as possible.

19.2 Grazing Improves Sustainability of Dairying Systems

19.2.1 Grassland-based dairy systems combine economic and environmental performances

The comparisons made at the world level show that dairying systems maximizing grassland utilization appear to be highly competitive, as shown by Dillon *et al.* (2008) and Fig. 19.1. There is a large variation, both between and within countries, in the grass-growth season and expected yield, but in all cases grazed forage must be at the maximum to reduce costs. For similar climatic conditions, grazing is still more economically attractive than indoor-feeding systems. In Ireland, early grazing will generate an increased profitability of €2.70 per cow and day for each extra day at grazing, through higher animal performance and lower feed costs (Kennedy *et al.*, 2005). In several regions of France, the variable cost of milk production is generally much lower in dairy systems based on grazing than on intensive systems

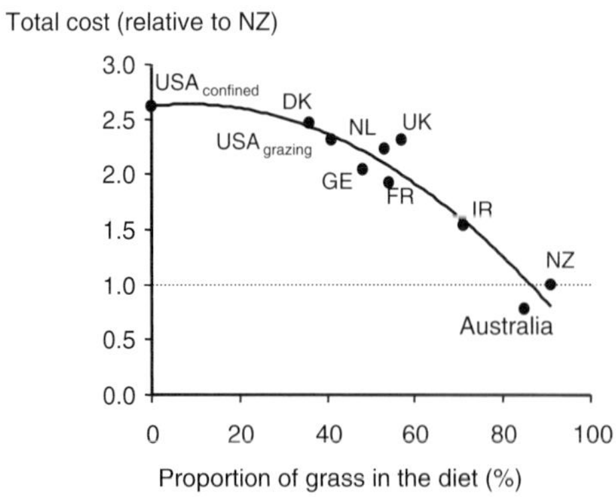

Fig. 19.1. Total cost (expressed relative to New Zealand costs) of milk production according to the proportion of grass in the annual diets of the cows (adapted from Dillon *et al.*, 2008). DK, Denmark; NL, The Netherlands; GE, Germany; FR, France; UK, United Kingdom; IR, Ireland.

(0.11 versus 0.17 €/l; Peyraud *et al.*, 2010), whereas the selling price of milk hardly varies between systems. Similarly in the Netherlands, the more grass the cows eat at pasture, the larger the farmers' income (Van den Pol-van Dasselaar *et al.*, 2010).

Basset-Mens *et al.* (2005) have shown that global warming potential per kg milk is higher in conventional intensive dairying systems encountered in Sweden and in southern Germany, which rely mainly on conserved forages and large amounts of concentrates, than in organic dairy Swedish systems and extensive dairying systems in Germany (1.2 versus 1.0 kg CO_2 eq/kg milk). New Zealand systems, which rely essentially on permanent white clover–grass pastures grazed all year round, produce much lower values (0. 7 kg CO_2 eq/kg milk). It is also noticeable that total consumption of non-renewable energy is reduced in grassland-based systems. Le Gall *et al.* (2009) have calculated that energy consumption to produce 1 kg of milk decreases from 5.0 MJ for intensive dairy farms in Netherlands to 4.0 for French farms using maize silage and fertilized grass pastures, 3.1 for Irish systems based on fertilized ryegrass pastures, and 1.4 for New Zealand farms. Increasing the proportion of grassland in arable land linearly decreases the utilization of pesticides, as shown in the European project Greendairy (Raison *et al.*, 2008).

The relative importance of the multiple functions provided to society by grasslands varies, depending on regional contexts, grassland type and type of management. In particular, grasslands contribute to preserve various components of diversity, although the practices of management largely influence botanical diversity and that of insects and other small fauna (Dumont and Tallowin, Chapter 14, this volume; Decourtye and Bouquet, 2010), with, in general, a reduction of diversity with intensification. The role of grazing in the regulation of N flow is more questionable. Summarizing data from New Zealand, France and Denmark, Ledgard *et al.* (2009) have shown that leached nitrate N remains below 50 kg/ha/year so long as total mineral N inputs are lower than 200 kg/ha/year.

However, in Europe, many dairy production systems are, to a large extent, based on ley–arable rotations that are characterized by three phases: pasture, ploughing out and subsequent arable cropping. These systems present more important risks of nitrate leaching than the systems based on permanent grassland, as shown in the European project Greendairy (Raison *et al.*, 2008), and they require great technical control to reduce the nitrate losses to the minimum.

19.2.2 There is room for manoeuvre to increase the efficiency of grazing

Dry-matter intake and milk yield of grazing dairy cows are limited compared with those of cows on conserved forage-based diets (Kolver and Muller, 1998), and grassland-based systems prevent high-genetic-merit (HGM) cows from fully expressing their milk potential. However, several trials have shown that relatively high milk production (i.e. 7400 kg per lactation) is achievable with HGM cows in grassland-based systems with only 350 kg of concentrate (Kennedy *et al.*, 2002; Horan *et al.*, 2005), at least under areas well suited for giving a high herbage production rate over a prolonged grazing season. Herbage allowance is one of the primary factors influencing herbage intake (Peyraud *et al.*, 2004) and high herbage allowance is required to achieve maximum intake and milk yield per cow. The implication is that grazing systems designed to maximize individual animal performances are inefficient in utilization per hectare. To solve this dilemma (per cow versus per ha), sward structures and grassland management allowing the maintenance of a high intake together with a low residual sward height must be determined.

Grazing also suffers from difficulties of management. The feed resource is not constant during the season and there is large inter-annual variability. Animal performances may therefore fluctuate and this is not well accepted by farmers. The use of on/off grazing (combination of restricted access time to pasture and indoor feeding) should be investigated as a way of increasing grassland utilization and reducing variability in animal performance, even for systems that have a very high stocking rate and short grass-growing season.

19.3 Multi-species Swards with Forage Legumes for Improving the Proportion of Milk Produced from Grazed Forages

Difficulties in maintaining well-balanced grass–legume mixtures contribute to the prevalence in temperate grasslands of grass monocultures, with associated significant inputs of synthetic fertilizers. Yet legumes can make an important contribution to sustainable dairy production systems in the future.

19.3.1 Forage legumes can sustain high animal performance

At grazing, herbage intake is higher on grass–clover mixtures than on pure grass swards. In the studies conducted in Rennes (Ribeiro-Filho *et al.*, 2003), mixed pastures steadily increased dry matter (DM) intake and milk yield (on average by 1.5 kg/day). It is well established that voluntary intake is greater for legumes than for grasses of similar digestibility (INRA, 1989). Higher intake on mixed pastures is mediated through a higher rate of intake compared with that on pure perennial ryegrass (PRG) pastures, indicating that grass–clover mixtures are more favourable for prehension than pure grasses (INRA, 1989). One of the most decisive advantages of white clover is that the rate of decline of nutritional quality throughout the plant-ageing process is far less than for grasses (Peyraud, 1993). Ribeiro-Filho *et al.* (2003) showed that herbage intake declines by 2.0 kg DM/day on PRG pastures, compared with 0.8 kg on mixed pastures, between 20 and 35 days of regrowth. This makes mixed pastures easier to manage than pure grass pastures. Increasing the content of white clover in pasture has increased milk yield by 1–3 kg/cow/day in several short-term trials conducted at similar herbage allowance (Philips and James, 1998; Ribeiro-Filho *et al.*, 2003). The difference increases with clover content and reaches a maximum when white clover content averages 50–60%.

19.3.2 Multi-species swards can sustain high production per unit area

Because grass–clover mixtures are managed with very low N fertilization, the biomass per hectare is often lower than that of highly fertilized grass pastures at a similar stage of regrowth. Therefore, to allow the same amount of herbage, the stocking rate should be reduced on grass–clover mixtures; the milk yield per hectare does not increase. Another alternative is to extend the regrowth period (Pflimlin, 1993). The main advantage of grass–clover mixtures is their high flexibility in use, allowing long intervals between two successive grazings, compensating for lower productivity without penalizing the performance of the cows. Using grass–clover mixtures also offers the possibility of extending the herbage growth season. Characteristics within-season growth patterns favour the grasses in spring, during reproductive growth, and the legumes in summer when temperatures are high. Thus the production of mixed pastures is generally shifted about the summer and, by combining in a grazing system some pure stands and grass–clover stands, the period of pasture growth during the year can be extended.

A pan-European experiment was carried out at 28 sites in 17 countries across Europe. At each site, the two most important forage grasses and the two most important forage legumes were tested and the management of the swards followed local recommendations for best agricultural practice. The results showed strong benefits of grass–clover mixtures containing four species as compared with these species sown in monoculture (Lüscher *et al.*, 2008), for all sites. For the mid-European and north European sites, all the mixtures yielded more than the best monoculture. This occurred even though mixtures were sown with widely varying species proportions (from 10 to 70% for each species). This result persisted over the 3 years of the experiment.

Therefore multi-species swards are very promising for the future of dairy systems, the more so as they will contribute to the sustainability of dairy systems. Forage legumes contribute to reduce total greenhouse gas emissions and global warming potential per kg milk. In New Zealand, total greenhouse gas emissions per kg milk were 10% lower on farmlets receiving 0 kg/ha mineral N than on a farmlet receiving 207 kg/ha mineral N (Ledgard *et al.*, 2009). In Sweden, Cederberg and Mattson (2000) observed lower greenhouse gas emissions in organic systems using grass–clover swards than in conventional systems. Forage legumes have potential for reducing the consumption of non-renewable energy because 55 MJ are required to produce, transport and spread 1 kg of mineral N while legumes only require solar energy to fix N_2 from the air. Energy efficiency, calculated as herbage net energy (NE) produced per MJ of non-renewable energy consumed, is three times higher for white clover–grass pastures than for fertilized grasses pastures (17.7 kJ NE versus 5.7 kJ NE/MJ; Besnard *et al.*, 2006). Finally, legumes have potential for reducing nitrate leaching to some extent (Ledgard *et al.*, Chapter 9, and Lüscher *et al.*, Chapter 11, this volume).

19.4 Grassland and Herd Management for Improving the Proportion of Milk Produced from Grazed Forages

19.4.1 Increasing stocking rate to produce more milk per hectare

In most countries of Europe, pressure on land use is high and maximizing milk yield per unit area is more than just a challenge to maximize profitability per hectare. In experimental herds, annual production of 15,000 kg/ha milk from grass, with less than 500 kg of concentrate, has been reported (Horan *et al.*, 2005). A recent meta-analysis including 131 comparisons of stocking rate concluded that an increase in stocking rate of one cow/ha resulted in an increase in milk yield of 1650 kg/ha (i.e. 20%) and milk solids of 113 kg/ha while milk yield per cow was marginally decreased by 1.3 kg/day and milk protein content by 0.5 g/kg (McCarthy *et al.*, 2010). The principle of increasing stocking rate (or decreasing post-grazing sward height) in order to increase milk production per hectare has strong applications in dairy farms.

The effects of feeding supplement on cow performance were reviewed by Peyraud and Delaby (2001). Efficient response of 1 kg of milk to 1 kg concentrate is now currently reached when the amount of concentrate per cow does not exceed 6 kg/day. Moreover, the efficiency of supplementation at grazing appears to be closely related to energy balance of the cows, and it increases when pasture intake is restricted through increased stocking rate, with economic returns depending on the concentrate to milk price ratio. Therefore, feeding concentrate can be a very efficient tool to maintain a high stocking rate and thus good sward management, which allows the control of post-grazing sward height while achieving high milk yield per cow and per hectare, with high economic returns. For a given stocking rate, the choice of the grazing system is not critical.

19.4.2 Extending the grazing season to consume more grass while increasing cow performance

There is considerable opportunity to extend the grazing season, thereby reducing costs associated with indoor feeding systems. Given the high feeding value of grazed grass relative to conserved forages, there is interest in extending the grazing season as much as possible. In many situations, moderate grass growth occurs in early spring while herbage growth in late autumn is almost entirely lost through senescence during winter, if not grazed. Experiments have been conducted from mid-February to mid-April in Ireland (Kennedy *et al.*, 2005), in Northern Ireland (Sayers and Mayne, 2001) with grass silage fed indoors, and in Brittany (O'Donovan *et al.*, 2004) with maize silage fed indoors. These experiments have all shown that access to pasture for a few hours per day increased milk yield by 1 to 3 kg and reduced silage intake by 4 to 6 kg per day, thus reducing the amount of conserved forage to be harvested and distributed. In the study of Kennedy *et al.* (2005), it is noticeable that early turnout of spring-calving cows to pasture in the early postpartum period resulted in a slight improvement of milk yield and milk protein content when compared with those of cows housed indoors and fed with a total mixed ration (44% grass silage). The grazing season may also be extended in autumn, as shown by Mayne and Laidlaw (1995). The effective use of late autumn grass as part of the diet of dairy cows was further confirmed under Brittany conditions (Chenais and Le Roux, 1996). Cows with access to grazing during daylight (6 h/day) produced 1 kg/day more milk and consumed 5.1 kg/day less maize silage than cows that were fully housed. Obviously, during these transition periods, daily grazing time and stocking rate should be adjusted according to the climatic conditions and soil types, to avoid poaching and to limit the risk of nitrate leaching. Future investigations are required to quantify the risk of nitrate leaching during late-season grazing.

Besides the positive effect on cow performance during the period of part-time grazing, early spring grazing has a positive effect on herbage quality in subsequent grazing rotations. O'Donovan *et al.* (2004) compared grassland utilization and milk production on swards that were previously grazed in March (6 h/day) with swards not grazed before mid-April. Early grazed swards contained a relatively high proportion of green leaf and a relatively low proportion of stems and senescent material, and were relatively more digestible. Moreover, grass growth during the subsequent rotations was not affected by early grazing, although pre-grazing sward height was lower and almost all grass produced was grazed, thus making the system more efficient (Table 19.1).

The grazing season could also be extended by intentionally deferring grazing from the period of production to a period of grass shortage. In area with dry summers, grazing can be deferred to a certain extent by changing the length of the grazing rotation. Paddocks are closed in spring for up to 60 or 80 days before being grazed in July/August. This requires mixed swards to maintain the sward quality for feeding dairy cows. The net energy value of ryegrass–white clover mixed sward averaged 1.3 MJ NE/kg DM after 70 days re-growth (Delaby and Pecatte, 2003), which is equivalent to the values observed for good quality grass silage. Because the sward height of deferred paddocks is high, it is recommended to strip graze these paddocks.

19.4.3 Part-time grazing to maintain grassland utilization for large herds

Grazing becomes more complicated with increasing herd size. It is not the size of the herd by itself which is the most limiting factor because large herds (up to 400 cows or more) graze in England and in New Zealand. The difficulties arise from the increased area that is needed around the milking parlour for grazing. This problem occurs in many regions or countries in Europe, especially in the most intensive dairying areas. In these situations, where stocking rates are high, the time at pasture can be restricted to a few hours per day to maintain grazed forage in the diets. Part-time grazing combined with restricted indoor feeding should be considered as an interesting alternative, to reduce the need for conserved forages, which are always expensive to produce, and to keep as much as possible of the fresh forage for producing milk. Allowing cows to graze for some hours per day can also improve the welfare of the animals (Sairanen *et al.*, 2006).

When time at pasture is restricted, cows increase the proportion of time spent grazing and the pasture intake rate, to compensate. Large increases in rate of intake (more than 25%) were reported when access time decreased from 8–9 h to 4–5 h/day (Kristensen *et al.*, 2007; Pérez-Ramirez *et al.*, 2009). However, these adaptations are insufficient to compensate fully for the large reduction of time at pasture. A recent experiment (Delaby *et al.*, 2010) has examined the response curves of milk to increasing levels of maize silage fed as a supplement to cows with access to pasture for either 4 h (between morning milking and midday) or 8 h a day (between morning and evening milking) (Fig. 19.2). When the access time was restricted to 4 h, 15 kg of maize silage were required to achieve high animal performance. When the access time was 8 h, the response

Table 19.1. Effect of the date at turnout on herbage production and animal performance in spring grazing (from April to end of June) (Adapted from O'Donovan *et al.*, 2004).

	Date of first grazing	
	Early February	Late March
Sward height, 30 March (cm)	5.5	10.0
Herbage DM growth (kg/ha/day)	66	66
Pre-grazing sward height (cm)[a]	11.8	15.2
Milk (kg/day)	23.2	22.1
Herbage lost (kg/day)[b]	15.6	15.9
Herbage utilization (% of growth)	100	85

[a] When cows enter a new paddock. [b] Calculated as (herbage pre-grazing – herbage post-grazing) × sward bulk density.

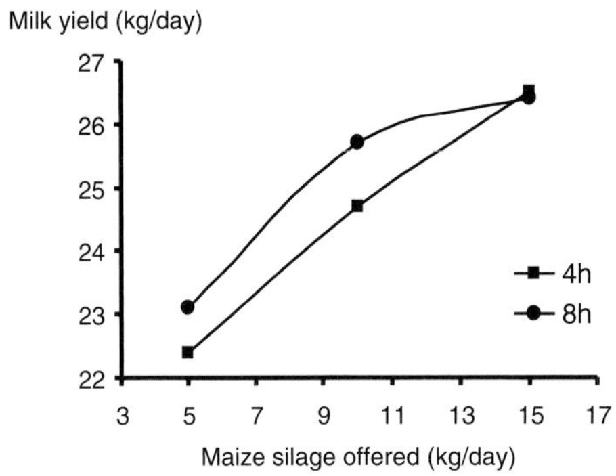

Fig. 19.2. Effect of the supply of maize silage on performance of dairy cows with a restricted access to pasture for grazing (adapted from Delaby *et al.*, 2010).

reached a plateau with 10 kg of maize silage. Feeding only 5 kg of maize silage did not maintain animal performances. These results clearly show that the amount of supplementary forage must be adjusted to the access time. The level of supplementation needs to be chosen in order to maximize milk yield per cow or to maximize pasture utilization.

19.4.4 Appropriate cows for successful temperate pasture-based systems

Several trials have compared cow genotypes in grassland-based systems, both in Ireland (Dillon *et al.*, 2003; Horan *et al.*, 2005) and in France (Delaby *et al.*, 2010), under compacted spring-calving seasonal systems. These experiments have all shown that HGM cows for milk production always produce more milk, even on low-input systems, but such cows were also characterized by degraded reproductive performances and relatively high rates of culling. The undesirable side effects of HGM for milk, on reproduction and survival, do not seem to be counteracted by adjustment in concentrate supplementation (Kennedy *et al.*, 2002; Delaby *et al.*, 2010). Therefore HGM cows do not fit seasonal pasture-based systems. Infertility problems are less acute when compact calving is not required. This is the case when high-quality forages are available all round the year, for example, when maize silage can be used in periods of low or zero grass growth. Although pasture-based systems prevent HGM cows from fully expressing their milk potential, this does not remove the need for having animals of high milk potential, especially in the context of fluctuating prices that require the level of production to be adjusted to the current situation. In periods of high milk prices, it could be efficient to produce more milk by increasing the supply of concentrate, and a good genetic potential for milk production then guarantees an efficient milk response. The objective would be to have animals expressing their ability to produce milk primarily by a long persistency of lactation and with a moderate peak of lactation. This will limit the energy deficit and associated pathological troubles at the beginning of lactation.

It is also advisable to reconsider the use of dual-purpose dairy breeds that have better milk composition and beef merit than Holstein cows. These breeds make it possible to produce 6000–7000 kg of milk per lactation, mainly from grassland with very small amounts of concentrates. Pregnancy rates were higher with Normande cows than with Holstein cows (Dillon *et al.*, 2003; Delaby *et al.*, 2010), thus ensuring only limited culling for infertility problems and replacement costs, in seasonal compact calving systems. Moreover, dual purpose breeds ensure a greater stability because of the double source of income (milk and meat).

19.5 Conclusions

In the future, dairy systems should be more environmentally sound, economically viable and productive. Because grazed pasture is the cheapest feed for dairy cows, and pastures have multifunctional roles, pasture-based dairy systems provide a good basis for developing more sustainable dairy production systems in the long term. This chapter shows that there is considerable scope to improve the performances of dairy systems based on pasture utilization by grazing. Forage legumes in multi-species swards will undoubtedly constitute one of the pillars for the development of future dairy systems with high environmental and economic performances. There is also considerable scope to improve animal and pasture performances with pasture-based systems, given recent developments in our understanding of management factors influencing grass intake and grass growth. Selection of functional traits for more robust cows will constitute another pillar in the development of more sustainable dairy systems in the future.

References

Basset-Mens, C., Ledgard, S. and Carran, A. (2005) First life cycle assessment of milk production from new Zealand dairy farm systems. www.anzsee.org/anzsee2005papers/Basset-Mens_LCA_NZ_milk_production.pdf

Besnard, A., Montarges-Lellahi, A. and Hardy, A. (2006) Systèmes de culture et nutrition azotée. Effet sur les émissions de GES et le bilan énergétique. *Fourrages* 187, 311–320.

Bourgeois, L. (2002) Common agricultural policy and grasslands: the case study of France. *Grassland Science in Europe* 7, 5–15.

Cederberg, C. and Mattsson, B. (2000) Life cycle assessment of milk production – a comparison of conventional and organic farming. *Journal of Cleaner Production* 8, 49–60.

Chenais, F. and Le Roux, M. (1996) Réduction de la part de maïs dans les systèmes d'alimentation des vaches laitières. Résultats expérimentaux obtenus en Bretagne. In: *Document EDE-Chambre d'Agriculture de Bretagne*. Chambre d'Agricuture, Rennes, pp. 12–15.

Decourtye, A. and Bouquet, C. (2010) Managing the grass cover in a manner favourable to the bees and to the small lowland fauna. *Fourrages* 202, 117–124.

Delaby, L. and Peccatte, J.R. (2003) Valeur alimentaire des prairies d'association ray-grass angles / trèfle blanc utilisées entre 6 et 12 semaines d'âge de repousse. *Rencontres autour des Recherches sur les Ruminants*, 10, 389.

Delaby, L., Horan, B., O'Donovan, M., Gallard, Y. and Peyraud, J.L. (2010) Are high genetic merit dairy cows compatible with low input grazing systems? *Grassland Science in Europe* 15, 928–930.

Dillon, P., Snijders, S., Buckley, F., Harris, B., O'Connor, P. and Mee, J.F. (2003) A comparison of different dairy cow breeds on a seasonal grass-grazed system of milk production. 2. Reproduction and survival. *Livestock Production Science* 83, 35–42.

Dillon, P., Hennessy, T., Shalloo, L., Thorne, F. and Horan, B. (2008) Future outlook for the Irish dairy industry: a study of international competitiveness, influence of international trade reform and requirement for change. *International Journal of Dairy Technology* 61, 16–29.

FAO (2006) *Livestock's Long Shadow, Environmental Issues and Options.* United Nations Food and Agriculture Organisation, Rome.

Horan, B., Dillon, P., Faverdin, P., Delaby, L., Buckley, F. and Rath, M. (2005) The interaction of strain of Holstein–Friesian cow and pasture based feed system for milk production, bodyweight and body condition score. *Journal of Dairy Science* 88, 1231–1243.

INRA (1989) *Ruminant Nutrition: Recommended Allowances and Feed Tables.* Institut National de la Recherche Agronomique, Paris, pp. 33–48.

Kennedy, E., O'Donovan, M., Murphy, J.P., Delaby, L. and O'Mara, F. (2005) Effects of grass pasture and concentrate-based feeding systems for spring calving dairy cows in early spring on performances during lactation. *Grass and Forage Science* 60, 310–318.

Kennedy, J., Dillon, P., Faverdin, P., Delaby, L., Buckley, F. and Rath, M. (2002) The influence of cow genetic merit for milk production on response to level of concentrate supplementation in a grass-based system. *Animal Science* 75, 433–445.

Kolver, E.S. and Muller, L.D. (1998) Performance and nutrient intake of high producing Holstein cows consuming pasture or a total mixed ration. *Journal of Dairy Science* 81, 1403–1411.

Kristensen, T., Oudshoorn, F., Munksgaard, L. and Soegaard, K. (2007) Effect of time at pasture combined with restricted indoor feeding on production and behaviour in dairy cows. *Animal* 1, 439–448.

Le Gall, A., Bèguin, E., Dollé, J.B., Manneville, V. and Pflimlin, A. (2009) Nouveaux compromise techniques pour concilier efficacit économique et environnementale en élevage herbivore. *Fourrages* 198, 131–151.

Ledgard, S., Schils, R., Eriksen, J. and Luo, J. (2009) Environmental impacts of grazed clover/grass pastures. *Irish Journal of Agricultural Research* 91, 91–107.

Lüscher, A., Finn, J.A., Connolly, J., Sebastià, M.T., Collins, R., Fothergil,l M., Porqueddu, C., Brophy, C., Huguenin-Elie, O., Kirwan, L., Nyfeler, D. and Helgadottir, A. (2008) Benefits of sward diversity for agricultural grasslands. *Biodiversity* 9, 29–32.

McCarthy, B., Delaby, L., Pierce, K.M., Journot, F. and Horan, B. (2010) Meta-analysis of the impact of stocking rate on the productivity of pasture-based milk production system. *Journal of Dairy Science* 90, 1493–1505.

Mayne, C.S. and Laidlaw, A.S. (1995) Extending the grazing season, a research review. In: *Extending the Grazing Season, Discussion Meeting, Proceedings of the British Grassland Society*, Reaseheath College, Nantwich, pp. 31–35.

O'Donovan, M., Delaby, L. and Peyraud, J.L. (2004) Effect of time of initial grazing date and subsequent stocking rate on pasture production and dairy cow performance. *Animal Research* 53, 489–502.

Pérez-Ramirez, E., Delagarde, R. and Peyraud, J.L. (2009) Restricting daily time at pasture at low and high herbage allowance: effects on herbage intake and behavioural adaptation of lactating dairy cows. *Journal of Dairy Science* 92, 3331–3340.

Peyraud, J.L. (1993) Comparaison de la digestion du trèfle blanc et des graminées prairiales chez la vache laitière. *Fourrages* 135, 465–473.

Peyraud, J.L. and Delaby, L. (2001) Ideal concentrate feeds for grazing dairy cows – Response to concentrates in interaction with grazing management and grass quality. In: Garnsworthy, P.G. and Wiseman, J. (eds) *Recent Advances in Animal Nutrition*, University of Nottingham, Nottingham, pp. 203–220.

Peyraud, J.L., Mosquera-Losada, R. and Delaby, L. (2004) Challenges and tools to develop efficient dairy systems based on grazing: how to meet animal performances and grazing management. *Grassland Science in Europe* 9, 373–384.

Peyraud, J.L., Dupraz, P., Samson, E., Le Gall, A. and Delaby, L. (2010) Producing milk at grazing to reconcile economic and environmental performances. *Rencontres autour des Recherches sur les Ruminants* 17, 17–24.

Pflimlin, A. (1993) Conduite et utilisation des associations graminées-trèfle blanc. *Fourrages* 135, 407–428.

Philips, C.J.C. and James, N.L. (1998) The effects of including white clover in perennial ryegrass pastures and the height of mixed pastures on the milk production, pasture selection and ingestive behaviour of dairy cows. *Animal Science* 67, 195–202.

Raison, C., Chambault, H., Le Gall, A. and Pflimlin, A. (2008) Impact du système fourrager sur la qualité des eaux. Enseignements issus du projet Green Dairy. *Fourrages* 193, 3–18.

Ribeiro-Filho, H.M.N., Delagarde, R. and Peyraud, J.L. (2003) Inclusion of white clover in strip-grazed perennial ryegrass pastures: herbage intake and milk yield of dairy cows at different ages of pasture regrowth. *Animal Science* 77, 499–510.

Sairanen, A., Khalili, H., Virkajärvi, P. and Hakosalo, J. (2006) Comparison of part time grazing and indoor silage feeding on milk production. *Agricultural and Food Science* 15, 280–292.

Sayers, H.J. and Mayne, C.S. (2001) Effect of early turnout to grass in spring on dairy cows performance. *Grass and Forage Science* 56, 259–267.

Van den Pol-van Dasselaar, A., De Haan, M. and Philipsen, A.P. (2010) Simulation of the effect of grass intake on the farmer's income. *Grassland Science in Europe* 15, 100–102.

20 Management of Grazing Systems in New Zealand and Environmental Impacts

M. Greg Lambert and V.O. Snow

20.1 Introduction

New Zealand pastoral systems predominantly involve year-round grazing of grass–clover pastures. Sheep (38 M), dairy cattle (5.3 M), beef cattle (4.4 M) and deer (1.4 M) are grazed on about 11 M ha of grassland (MWNZES, 2009). Approximately one third is flat and rolling, and one third is low-altitude hill country, both developed mainly from indigenous forest and farmed relatively intensively. The remainder is at high altitude and farmed less intensively. Dairying is more intensive than sheep and beef farming and is generally confined to the flat and rolling country.

Exports of agricultural commodity products have been a dominant component of the economy during the 20th and 21st centuries. Dairy farming has expanded in recent years, while sheep and beef farming has declined, being replaced by dairying on easier topography and production forestry or indigenous vegetation on hill country. In 1984, a free-market economic model including liberalization of import regulations and removal of agricultural product price support was adopted. Implementation in 1991 of the Resource Management Act (RMA) provided a national framework for addressing environmental issues; and in 1998 New Zealand signed the Kyoto Protocol.

Environmental issues include nutrient, sediment and microbial pathogen emissions to water; agricultural greenhouse gas (GHG) emissions; soil erosion; decreased soil quality and indigenous biodiversity; and effects on visual landscape attributes. This chapter will focus on nitrogen (N) and phosphorus (P) emissions to water, GHG emissions, and landslide erosion in intensive farming systems.

20.2 Environmental Impacts

20.2.1 Nitrogen and phosphorus emissions

Nitrogen emissions to water occur mainly as nitrate (NO_3) produced by nitrification of soil ammonium and leached in soil drainage. Nitrous oxide (N_2O) is an intermediary in denitrification of NO_3 and emissions are greatest under conditions of high NO_3 supply and low soil aeration (De Klein and Eckard, 2008). Nitrate leaching and N_2O emissions occur mainly from animal excreta, and particularly from urine patches which contain loadings of 400–900 kg N/ha (Haynes and Williams, 1993). The amount of excretal N deposited on pastures is strongly related to the amount of herbage N ingested by animals, most of which is excreted in urine.

Biological N fixation by clover is the major N input to pastoral systems. Relatively little fertilizer N is applied in hill country; however, the input cost/product price relationships in dairying have led to significant use (114 kg N/ha per annum on average, Basset-Mens *et al.*, 2009), at similar magnitude to N inputs from fixation (Ledgard *et al.*, 1998, 2000; Chapter 9, this volume). Further N inputs occur in feed introduced from off-farm sources. Effects of fertilizer N on emissions are mainly indirect, through stimulation of pasture production, increased animal intake, and so increased excretal N.

Soil, dung and fertilizer are the main sources of P losses from grazed pastures. These occur mainly in overland water flow, although sub-surface drainage systems can act as conduits for nutrient transport to waterways. Grazing animals accelerate P losses through dung deposition, and soil disturbance by hooves can increase surface erosion and so P losses in runoff (Monaghan, 2008). Sediment from hill-slope soil erosion can also be a significant source of P in waterways draining hill country catchments.

20.2.2 Methane emissions

Ruminant CH_4 emissions arise mainly from fermentation of feed in the rumen, and total emissions per animal (or unit area) are determined by the amount of dry matter eaten (De Klein *et al.*, 2008).

Legume diets, and those containing condensed tannins and high proportions of grain, produce lower CH_4 emissions than temperate grasses, and tropical grasses have increased emissions; however, these are relatively minor components of pasture-based diets and inclusions of moderate (15–30%) amounts of legume or grain in a grass diet appear not to decrease CH_4 emissions (De Klein *et al.*, 2008).

20.2.3 Landslide erosion

The New Zealand (NZ) landscape is geologically young, and replacement of indigenous forest with shallow-rooted permanent pasture on erosion-susceptible hill country has led to accelerated landslide erosion in significant areas (Dymond *et al.*, 2006). Consequences include lowered pasture production (Lambert *et al.*, 1984) and sedimentation of streams and rivers. Topography, geology and climate are major determinants of erosion susceptibility; 14% of NZ hill country has high and 62% has medium susceptibility to erosion (Basher *et al.*, 2008).

20.3 Current Status of Environmental Impacts and Trends

The Parliamentary Commissioner for the Environment has reported that agricultural intensification is placing natural capital under intense pressure, and waterways and lakes are becoming nutrient enriched and degraded (PCE, 2004). There is increasing public concern regarding the impact of agriculture and its ongoing intensification on water quality and GHG emissions (Macfie, 2010) and a 'dirty dairying' campaign instigated in 2003 by a non-governmental organization led to agreement by stakeholders to work together to develop practical solutions to managing water quality (MAF, 2010a).

Increased nutrient loadings reduce limitations to growth of undesirable plant life and algae in waterways, and lead to eutrophication of lakes. From 1989 to 2007, N and P loadings increased in rivers with catchments dominated by pastoral land use (MfE, 2010a). Agricultural GHGs constitute nearly half of NZ's emissions and increased by 9.3% during 1990 to 2008 (MfE, 2010b). Between 1997 and 2002 the area of high erosion susceptibility land being farmed decreased by 3%, mainly due to conversion to plantation forestry or reversion to indigenous vegetation (MfE, 2010c).

20.4 Management of Environmental Impacts

20.4.1 Reduction in total emissions

Because the amount of feed eaten and the associated stocking rates and fertilizer inputs are the major drivers of N, P and CH_4 emissions, the most potent mitigation approach is to reduce per animal intake or stocking rate. These are usually negatively correlated, so reduction in one will be less effective unless the other is constrained, e.g. by reducing inputs or reducing level of feed utilization. This approach has major financial implications for both individual farmers and the national economy.

Other approaches that are regarded as best management practice and that have less financial impact are also less potent. These include restricting animal access to waterways and their banks; retention/creation of wetlands and riparian margins to trap nutrients in overland flow; and application of fertilizers more precisely according to pasture requirements and with attention to application timing, to reduce potential for direct losses. Best practice recommendations can, however, involve trade-offs, e.g. fertilizer N application should, in theory, be avoided in cold or very wet conditions; unfortunately N-boosted pasture is most valuable in the slow growth, cool and wet periods of the year.

Nitrogen emissions can be reduced by feeding low N concentration supplements, e.g. maize silage with pasture to balance carbohydrate to protein ratio and increase utilization of dietary N. Capture of excreta when animals are off-pasture (e.g. on feed pads or in the farm dairy) and return to the pasture through efficient effluent systems can reduce N and P losses; also, as a side benefit, the nutrients in evenly spread effluent can generate greater pasture production than when concentrated in excretal patches.

Ryegrasses with a high concentration of water-soluble carbohydrate can improve the carbohydrate to protein ratio of ingested feed, and tannin-containing forages can increase partitioning of surplus N into dung rather than urine, making N less immediately available for leaching and denitrification. However, neither option is yet fully developed as a feasible management practice, and will only have an impact where successful pasture and crop establishment and persistence can be achieved.

Nitrification inhibitors applied to pasture, and so to urine patches, reduce N emissions and increase N use efficiency. This practice has been adopted by a proportion of farmers, although best practice and effectiveness for the full range of commercial farming situations and over long periods of time have yet to be fully elucidated (Kelliher *et al.*, 2008; Cichota *et al.*, 2010). If N application rates are not reduced when inhibitors are used it is likely that, over time, N pools will equilibrate and emissions will trend towards previous levels. Also, the increased response efficiency will lead to increased pasture growth, pasture intake and total GHG emissions (Eckard *et al.*, 2010).

Off-pasture grazing during winter can reduce treading damage and associated reductions in soil aeration, and so decrease N_2O emissions. Sub-surface drainage of wet soils can reduce denitrification through improved aeration but may increase N and P losses in drainage water. Some dairying expansion in recent years has been into regions less suited to low-cost year-round grazing systems, including areas with colder winter temperatures and/or poorly drained soils, necessitating different wintering systems. These systems can increase emissions through increased treading damage, and high nutrient loadings from excreta during intensive utilization of pasture and forage crops.

There are currently no practical ways of significantly reducing total CH_4 emissions from grazing systems other than by reducing animal intake.

20.4.2 Reduction in intensity of greenhouse gas emissions

Total nutrient and GHG emissions per unit area are the major drivers of environmental impacts on water and air. However, intensity of GHG emissions (emissions per unit production) may be the basis for assessing GHG contribution of grazing animals in the proposed emissions trading scheme (MAF, 2010b). Use of such metrics would introduce a range of feasible GHG mitigation options, as gains in production efficiency will also reduce intensity of emissions. This is possible through adoption of practices and systems that increase the proportion of feed ingested that is used for production rather than maintenance. Dairy cows bred for higher efficiency, managing pastures for higher quality, increased longevity of breeding animals (Beukes *et al.*, 2010), increased lambing percentage in sheep, and utilizing surplus calves from the dairy industry for beef production rather than maintaining beef cows, will all increase production efficiency and reduce emissions intensity. Where animal production is maintained with less feed input, potential also exists to reduce total emissions.

20.5 Management of Landslide Erosion

The major options for managing landslide erosion on hill slopes where indigenous forest has been replaced with pasture are to space-plant with trees and continue grazing the resultant two-tier system; or, alternatively, to retire the land from grazing and blanket-plant with production forest or let the land revert to indigenous vegetation.

For most effective erosion control, space-planted trees need to be at a sufficient density for their network of roots to stabilize the hill slope (Douglas *et al.*, 2008), at which density pasture production is significantly reduced (Wall *et al.*, 2006) leading to reduced animal production. This disadvantage needs to be offset against the impacts of increased erosion where protective measures are not implemented. Blanket planting is much more effective at slope stabilization than space-planting, but has significant establishment and tending costs, revenue from livestock ceases, and revenue from timber to offset the cost of conversion to forestry is not generated until far into the future. Credits for carbon sequestration may, however, significantly enhance the attractiveness of retirement from grazing.

20.6 System Implications

Environmental issues are often considered in isolation from each other and from economic issues, but they are usually interconnected, e.g. planting trees on eroding hill country is a mitigation option for landslide erosion but trees also sequester carbon that can be offset against GHG emissions. Also, drivers of nutrient and CH_4 emissions are interconnected and so mitigation methods aimed at water-quality improvement will influence GHG emissions. Implementation of mitigation practices can have significant economic implications, so integrated solutions that mitigate environmental impacts whilst maintaining business viability must be developed.

To achieve high levels of animal production from grassland farming, a large amount of high-quality feed should be grown per unit area, with high levels of grazing utilization (McMeekan, 1956). Farm system configuration influences feed eaten, and is normally optimized for economic performance, taking into account input costs and product prices, and the farm's climatic, edaphic and topographical characteristics. System optimization through adoption of best management practices (which may include reduction in inputs) may have little negative impact on, or even increase, financial performance; however, for already well-optimized systems, significant reductions in feed eaten will usually reduce farm profitability.

Increased costs of production to comply with environmental standards will make products less competitive in export markets compared with alternatives from other countries that observe lower standards of environmental performance or where such costs are not imposed directly on farmers. Also, introducing more stringent environmental standards will, in essence, transfer public responsibility for environmental issues to private land managers and, in so doing, increase business risk.

An emissions trading scheme based on intensity of GHG emissions, rather than total emissions, will probably not act to reduce total emissions, but rather will encourage increased efficiency of production. In contrast, regulations to reduce impacts on water quality are likely to focus on reducing total nutrient emissions, so it is possible (because of the strong interrelationships between GHG and nutrient emissions) that changed management practices to restrict emissions to water will also restrict total GHG emissions.

Biological N fixation is commonly regarded as more environmentally friendly than fertilizer N application, in part because of the use of non-renewable hydrocarbons in manufacture. However, biologically fixed N has the same environmental impact as fertilizer N used at the same supply rate. Thus, N fixation has historically been the main driver of N emissions to water and air because the use of fertilizer N in pastoral systems is relatively recent. In addition, P fertilizers, which are applied to grass–clover pastures primarily to stimulate clover growth and N fixation, are a major driver of P emissions and furthermore are derived from phosphate rock, which is a non-renewable resource.

Reliance on biological N fixation indirectly places a limit on emissions because soil N supply is the primary nutrient limitation to pasture production, and clover growth (and so N fixation) is limited by competition with grasses. As clovers fix more N and it becomes available for uptake, grasses become more competitive and this constrains clover growth. Use of fertilizer N circumvents this biological feedback and potential pasture production is much greater.

Dairying has expanded in some cases into areas with greater summer soil moisture deficits, and irrigation is increasingly being used to overcome this limitation. This increases feed production and so consumption by grazing animals, with the associated potential to increase emissions. Increased irrigation availability will lead to optimization of other limitations to pasture growth, so will provide an incentive for increased fertilizer N use, which improves water use efficiency (Neal *et al.*, 2010) but also increases the potential for nutrient emissions.

The spatial scale of assessment of environmental impacts influences perceptions of their severity. Currently, land-use planning occurs mainly at the scale of individual businesses and there is no process for integrated planning across these entities. For example, a very intensive beef system coupled with an extensive lamb-breeding system may yield acceptable total emissions at a farm scale, but similarly a mosaic of sheep and beef farming, dairy farming, forestry and other land uses may yield acceptable emissions and erosion control at a catchment scale. Grazing replacement dairy stock and non-lactating cows in another catchment means feed on the dairy platform can be allocated mainly to lactating cows. This has strong economic advantages and can reduce environmental impacts in intensive dairying regions; it also transfers some impacts to other 'receiving' catchments that may differ in their response to emissions.

The temporal scale for impact assessment also influences perceptions of mitigation practice effectiveness, e.g. the value of trees in sequestering carbon is mainly confined to one production cycle, typically 30 years as currently practised for pine plantations (MacLaren, 1996), so using this approach to offset carbon emissions from pastoral farming will have a time limitation. Also, the effects of intensification on water quality may lag years behind changes in land management (Hadfield *et al.*, 2007; Stewart *et al.*, 2007), which may give a false sense of security when assessing the severity of environmental impacts. Burggraaf and Snow (2010) showed, in a modelling study, that the influence of plant tannins in mediating N cycling might change over time, with lag effects on N leaching.

20.7 Balancing Economic and Environmental Imperatives

20.7.1 Intensification

Jay (2007) suggests the focus of the dairy industry on expansion of production and economic efficiency aligns with an ideology of industrial agriculture or 'productivism'.

Barnett and Pauling (2005) postulate that the economic reforms of the 1980s have contributed to the expansion of the dairy industry and consequent environmental impacts, and that, while the RMA is potentially capable of controlling these effects, its implementation has not been effective in this regard. The removal of substantial price support for sheep farming may have reduced the competitiveness of that industry, contributing to the displacement of sheep farming by relatively more intensive dairy farming in the intervening years. However, given farming had been intensifying since the 1920s (MacLeod and Moller, 2006), the overall drivers for intensification existed long before the economic reforms of the 1980s.

Farm businesses are complex and risky, yet have not historically shown high financial returns compared with low-risk investments such as bank deposits. Return on investment averaged only 4.0% for sheep and beef and 5.8% for dairy farms during the 11 years 1999–2000 to 2009–2010 (MAF, 2010c). This suggests that, rather than farmers adopting a productivist approach, they had been intensifying their systems to maintain business viability in response to economic drivers. Rhodes *et al.* (2000) found that despite a gloomy financial background, farmers had a continued stated commitment to environmental management and their personal values frequently over-rode financial considerations.

20.7.2 Farmer decision making and innovation

Farmer decision making at the individual farm scale is the major determinant of land-management practices. Changing farmer awareness, market requirements, societal pressure and regulation are significant influences on these processes. New Zealand is emerging from a lightly regulated operating environment, to one with increasing use of policy to influence land management, as is currently more the case in Europe. However, development of regulations that diminish farmers' property rights is subject to intense scrutiny, and promulgation of unacceptable regulations runs the risk of social disobedience, serious impacts on individual farm financial viability, and reduction in national economic performance.

Adoption of new practices by farmers is influenced by many factors, e.g. their advantages compared with existing practices, ease of testing and implementation, commercial value, fit to the current farm system, and ability to solve a problem that is acknowledged (Bewsell and Dake, 2008). In what may be a precedent-setting approach, individual farm N-leaching caps have been imposed in the catchment of the iconic Lake Taupo. The consultation process used demonstrated the need for a clear definition of the issues and the underpinning theory; for an ability to demonstrate practical and economic implications of mitigations at the farm scale; for willingness of stakeholders to engage, discuss different points of view and learn; and for provision of time for interactive consultation, assistance packages to be developed, and for practices to be implemented (Clark and Lambert, 2002). Awareness of an issue and solutions does not necessarily mean it is economically rational, and slow rates of adoption are a rational reaction to changes that are economically undesirable (Bewsell and Dake, 2008). Hence adoption is more likely where there are on-farm or financial benefits and, failing this, regulation may be required to achieve desired change (Bewsell *et al.*, 2007). Rhodes *et al.* (2000) found that issues of long-term environmental sustainability were of less priority compared with immediate financial needs; however, financial profitability alone was no guarantee of a commitment to invest in those practices that support sustainability. Progress towards sustainability was most likely where problems were visible and so issues such as water and soil quality generally received less attention.

20.8 Looking Forward

Farms are managed within an uncertain and complex environment with often conflicting economic, environmental and social pressures, and the resolution of these will necessitate a systems approach. MacLeod and Moller (2006) question whether NZ farming is ecologically sustainable because of accelerating agricultural intensification. Their question needs to be extended to include economic sustainability. It would appear that the nutrient and GHG emission standards currently sought by some stakeholders will be difficult to meet if current intensification trends, including conversion from less intensive meat production systems to more intensive dairying, persist. Increased GHG emissions from intensification on the lowland could be offset by accelerated tree planting on erosion-susceptible and less productive hill country, which will reduce soil loss from, and sedimentation of, waterways. However, this will not mitigate the direct impacts of intensive agriculture on water quality further down the catchment.

Assessment of environmental impacts will be complicated where effects lag behind changes in land management. Creating awareness of issues, consulting with stakeholders, development of guidelines and regulations, and implementation all take time. If action is taken only after issues are clearly apparent, then opportunity to manage environmental impacts to acceptable levels is much diminished.

If extensification of agriculture is required to achieve environmental targets, then individual farm profitability is likely to be substantially reduced. If a 'clean-green' environment is in the public interest, an argument can be made for use of tax-payer funded incentives to facilitate on-farm change to meet off-farm targets. Apart from ideological resistance to providing 'subsidies' to agriculture, this approach does have another significant impediment, i.e. should the end result be extensification of farming systems, agricultural production would be reduced, as would export income, leading to a lowered national tax take and, in turn, less ability of tax payers to incentivize farmers.

The desire to reduce environmental emissions in the face of ongoing intensification may lead to implementation of systems that include a significant component of 'off-pasture grazing' and hybrid systems that combine grazing with confinement (Kleinman and Soder, 2008) may become more common. Advantages include ability to optimize diets for cows, greater production efficiency, less energy used in grazing, less damage to soils, opportunity for efficient management of excreta (including use as fertilizer), potential to grow highly productive forages that are intolerant of grazing, and provision of shade and shelter. One objection to animal confinement revolves around perceptions of diminished animal welfare; however, properly designed housing systems can alleviate welfare issues (Von Keyserlingk *et al.*, 2009). It is interesting to note that European agriculture has reduced the level of inputs in response to demand for reduced environmental impacts, and a shift to more outdoor grazing is seen as desirable in some countries.

It will be important to understand fully all the economic, social and environmental implications of differing systems. Some analyses of elements of such systems have been conducted (e.g. De Klein *et al.*, 2000; Basset-Mens *et al.*, 2009; De Klein and Eckard, 2008; White *et al.*, 2010). A study of GHG emissions from international dairy systems (FAO, 2010) suggests that grassland systems have, on average, 50% larger GHG emissions intensity than 'mixed' farming systems. However, the values calculated for grassland systems were considerably higher than New Zealand calculations by Basset-Mens *et al.* (2009) because of the use of different emission factors. In almost all the above cases, results for the full range of relevant outputs (nutrient and GHG emissions, physical production and financial performance) were not presented. It is important that whole farm systems analyses, including economic and environmental outputs, along with life-cycle assessments, are conducted to ascertain the system implications of mitigating environmental impacts (Eckard *et al.*, 2010). Additionally it is important that the assumptions made using these methodologies are clear and, where possible, are harmonized across studies.

The reliance of the NZ economy on sales of commodity products has a strong influence on the farming systems implemented. It could be argued that, if consumers paid more for the products exported, this would enable less intensive systems to be financially viable. This philosophy is undermined by competitors and trading partners who have lower environmental standards and/or who subsidize domestic production of agricultural commodities. If NZ commodities could be substituted with value-added products, this problem might be lessened. Molecular technologies offer exciting possibilities here, but consumer resistance may slow adoption of products developed using these approaches.

The ultimate solution may lay in an integrated land-use approach at the catchment level, customized to suit the particular socio-economic and physical characteristics and to fit the environmental guidelines for that catchment. Ultimately, communities and markets will determine land use through their influence on the formal and informal rules that shape the operating environment for pastoral farming, and the price they are prepared to pay for specified products.

References

Barnett, J. and Pauling, J. (2005) The environmental effects of New Zealand's free-market reforms. *Environment, Development and Sustainability* 7, 271–289.

Basher, L.R., Botha, N., Douglas, G.B., Lynn, I., Marden, M., McIvor, I.R. and Smith, W. (2008) *Hill Country Erosion: a Review of Knowledge on Erosion Processes, Mitigation Options, Social Learning and their Long-term Effectiveness in the Management of Hill Country Erosion.* Contract Report LC0708/081. Landcare Research, Lincoln, New Zealand.

Basset-Mens, C., Ledgard, S. and Boyes, M. (2009) Eco-efficiency of intensification scenarios for milk production in New Zealand. *Ecological Economics* 68, 1615–1625.

Beukes, P.C., Gregorini, P., Romera, A.J., Levy, G. and Waghorn, G.C. (2010) Improving production efficiency as a strategy to mitigate greenhouse gas emission on pastoral dairy farms. *Agriculture, Ecosystems and Environment* 136, 358–365.

Bewsell, D. and Dake, C. (2008) Socio-economic issues in pasture-based farming. In: McDowell, R.W. (ed.), *Environmental Impacts of Pasture-based Farming.* CABI, Wallingford, UK, pp. 98–121.

Bewsell, D., Monaghan, R.M. and Kaine, G. (2007) Adoption of stream fencing among dairy farmers in four New Zealand catchments. *Environmental Management* 40, 201–209.

Burggraaf, V.T. and Snow, V.O. (2010) Effects of condensed tannins on nitrogen dynamics in grazed temperate agricultural systems. In: *Tannins: Types, Foods Containing, and Nutrition.* Nova Science Publishers Inc, New York, pp. 317–326.

Cichota, R., Vogeler, I., Snow, V.O. and Shepherd, M. (2010) Modelling the effect of DCD on N losses from grazed pastures. *Proceedings of the New Zealand Grassland Association Conference* 72, 43–47.

Clark, D.A. and Lambert, M.G. (2002) Implementation of environmental regulations on-farm. *Proceedings of the New Zealand Animal Production Society* 62, 219–224.

De Klein, C., Paton, J. and Ledgard, S. (2000) An evaluation of a strategic de-stocking regime for dairying to improve nitrogen efficiency and reduce nitrate leaching from dairy farms in nitrate-sensitive areas. *Proceedings of the New Zealand Grasslands Association* 62, 105–110.

De Klein, C.A.M. and Eckard, R.J. (2008) Targeted technologies for nitrous oxide abatement from animal agriculture. *Australian Journal of Experimental Agriculture* 48, 14–20.

De Klein, C.A.M., Pinares-Patino, C. and Waghorn, G.C. (2008) Greenhouse gas emissions. In: McDowell, R.W. (ed.) *Environmental Impacts of Pasture-based Farming.* CABI, Wallingford, UK, pp. 1-32.

Douglas, G., McIvor, I., Manderson, A., Todd, M., Braaksma, S. and Gray, R. (2008) *Effectiveness of Space-planted Trees for Controlling Soil Slippage on Pastoral Hill Country.* AgResearch, Hamilton, New Zealand.

Dymond, J.R., Ausseil, A., Shepherd, J.D. and Buettner, L. (2006) Validation of a region-wide model of landslide susceptibility in the Manawatu-Wanganui region of New Zealand. *Geomorphology* 74, 70–79.

Eckard, R.J., Grainger, C. and de Klein, C.A.M. (2010) Options for the abatement of methane and nitrous oxide from ruminant production: a review. *Livestock Science* 130, 47–56.

FAO (2010) *Greenhouse Gas Emissions from the Dairy Sector. A Life Cycle Assessment.* Food and Agriculture Organisation of the United Nations, Rome.

Hadfield, J.C., Morgenstem, U. and Piper, J.J. (2007) Delayed impacts of land-use via groundwater on Lake Taupo, New Zealand. Water Resource Management. *WIT Transactions on Ecology and the Environment* 103, 293–303.

Haynes, R.J. and Williams, P.H. (1993) Nutrient cycling and soil fertility. *Advances in Agronomy* 49, 119–199.

Jay, M. (2007) The political economy of a productivist agriculture: New Zealand dairy discourses. *Food Policy* 32, 266–279.

Kelliher, F.M., Clough, T.J., Clark, H., Rys, G. and Sedcole, J.R. (2008) The temperature dependence of dicyandiamide (DCD) degradation in soils: a data synthesis. *Soil Biology and Biochemistry* 40, 1878–1882.

Kleinman, P.J.A. and Soder, K. (2008) The impact of hybrid dairy systems on air, soil and water quality: focus on nitrogen and phosphorus cycling. In: McDowell, R.W. (ed.) *Environmental Impacts of Pasture-based Farming*, CABI, Wallingford, UK, pp. 249–276.

Lambert, M.G., Trustrum, N.A. and Costall, D.A. (1984) Effect of soil slip erosion on seasonally dry Wairarapa hill pastures. *New Zealand Journal of Agricultural Research* 27, 57–64.

Ledgard, S.F., Crush, J.R. and Penno, J.W. (1998) Environmental impacts of different nitrogen inputs on dairy farms and implications for the Resource Management Act of New Zealand. *Environmental Pollution* 102, 515–519.

Ledgard, S.F., de Klein, C.A.M., Crush, J.R. and Thorrold, B.S. (2000) Dairy farming, nitrogen losses and nitrate sensitive areas. *Proceedings of the New Zealand Animal Production Society* 60, 256–260.

Macfie, R. (2010) The price of milk. In: *New Zealand Listener*, August 14-20, 2010, pp. 16–23.

MacLaren, J.P. (1996) New Zealand's planted forests as carbon sinks. *Commonwealth Forestry Review* 75, 100–103.

MacLeod, C.J. and Moller, H. (2006) Intensification and diversification of New Zealand agriculture since 1960: an evaluation of current indicators of land use change. *Agriculture, Ecosystems and Environment* 115, 201–218.

MAF (2010a) *Dairying and Clean Streams Accord.* Ministry of Agriculture and Forestry. http://www.maf.govt.nz/mafnet/rural-nz/sustainable-resource-use/resource-management/dairy-clean-stream/ (accessed 15 November 2010).

MAF (2010b). *A Guide to Agriculture in the Emissions Trading Scheme.* Ministry of Agriculture and Forestry. http://www.maf.govt.nz/climatechange/agriculture/agriculture-in-nzets-guide/page-03.htm#P92_11457 (accessed 15 November 2010).

MAF (2010c) *Farm Monitoring Reports.* Ministry of Agriculture and Forestry. http://www.maf.govt.nz/mafnet/rural-nz/statistics-and-forecasts/farm-monitoring/. (accessed 15 November 2010).

McMeekan, C.P. (1956) Grazing management and animal production. In: *Proceedings of the VII International Grassland Congress.* Palmerston North, New Zealand, pp. 146–156.

MfE (2010a) *Analysis of National River Water Quality Data for the Period 1998–2007.* Ministry for the Environment. http://www.mfe.govt.nz/publications/water/analysis-of-national-river-water-quality data-1998-2007/index.html (accessed 15 November 2010).

MfE (2010b) *New Zealand's Greenhouse Gas Inventory 1990–2008.* Ministry for the Environment. http://www.mfe.govt.nz/publications/climate/greenhouse-gas-inventory-2010-snapshot/index.html (accessed 15 November 2010).

MfE (2010c) *Soil Intactness of Erosion Prone Land.* Ministry for the Environment. http://www.mfe.govt.nz/environmental-reporting/land/erosion-risk/ (accessed 15 November 2010).

Monaghan, R.M. (2008) The environmental impacts of non-irrigated, pasture-based dairy farming. In: McDowell, R.W. (ed.) *Environmental Impacts of Pasture-based Farming.* CABI, Wallingford, UK, pp. 209–231.

MWNZES (2009). *Compendium of New Zealand Farm Facts.* Meat & Wool New Zealand Ltd Economic Service, Wellington, New Zealand.

Neal, J.S., Fulkerson, W.J. and Sutton, B.G. (2010) Differences in water use efficiency among perennial forages used by the dairy industry under optimum and deficit irrigation. *Irrigation Science* 29, 213–232.

PCE (2004) *Growing for Good: Intensive Farming, Sustainability and New Zealand's Environment.* Parliamentary Commissioner for the Environment, Wellington, New Zealand.

Rhodes, T., Willis, B. and Smith, W. (2000) *A Study of Issues Affecting North Island Hill Country Farmers.* Ministry of Agriculture and Forestry. http://www.maf.govt.nz/mafnet/rural-nz/profitability-and-economics/performance/impediments-to-optimum-performance/impopt-01.htm#P96_1894 (accessed 15 November 2010).

Stewart, M.K., Mehlhorn, J. and Elliott, S. (2007) Hydrometric and natural tracer (oxygen-18, silica, tritium and sulphur hexafluoride) evidence for a dominant groundwater contribution to Pukemanga Stream, New Zealand. *Hydrological Processes* 21, 3340–3356.

Von Keyserlingk, M.A.G., Rushen, J., de Passile, A.M. and Weary, D.M. (2009) The welfare of dairy cattle – Key concepts and the role of science. *Journal of Dairy Science* 92, 4101–4111.

Wall, A.J., Kemp, P.D. and MacKay, A.D. (2006) Predicting pasture production under poplars using canopy closure images *Proceedings of the New Zealand Grassland Association* 68, 325–330.

White, T.A., Snow, V.O. and King, W.McG. (2010) Intensification of New Zealand beef farming systems. *Agricultural Systems* 103, 21–35.

21 Rangeland Management for Sustainable Conservation of Natural Resources

Craig D. Morris

21.1 Introduction

Rangelands comprise the single most extensive land type in all inhabited continents, covering approximately 50–70% of the Earth's land surface (Holechek *et al.*, 1989). Rangeland is 'land on which the indigenous vegetation (climax or semi-natural) is predominantly grasses, grass-like plants, forbs, or shrubs which are grazed or have the potential to be grazed, and is used as a natural ecosystem for the production of grazing livestock and wildlife' (Allen *et al.*, 2011). Included are natural grasslands, savannahs, shrublands, many semi-deserts, steppes, tundras, alpine communities and marshes. By virtue of their wide environmental range, rangelands are compositionally and structurally diverse (West, 1993). Many rangelands are in arid to semi-arid regions where rainfall and forage production is low, patchy and unreliable (Neely *et al.*, 2009).

Forage is the defining product of rangelands, and livestock products (food and fibre) ostensibly their chief economic output, but, because of their diverse flora and fauna, rangelands are multifunctional, providing many non-forage goods and services to numerous and varied users living on or around rangelands and in urban areas (Havstad *et al.*, 2007). People harvest and utilize a remarkable variety of plant, wildlife and abiotic resources (e.g. sand and clay for construction and crafts) for home consumption, local exchange or for sale in towns and cities or to industries (Scoones *et al.*, 1992). The total value of such non-forage natural resources (NRs) can exceed the direct value of livestock commodities to rangeland inhabitants (Shackleton *et al.*, 2001).

Rangelands are not fully natural systems and have been, and continue to be, modified and shaped by the impact of grazing animals and human agricultural activities. Components and processes of grazed ecosystems are closely interlinked (Tainton *et al.*, 1996) and the excessive utilization of one rangeland resource can have numerous positive or negative multiplier effects on other extractable and non-consumptive goods that rangelands provide. The effect of management for improved livestock production on other rangeland resources has, however, received little research attention. In this chapter, the nature and value to users of various non-forage NRs are briefly discussed, followed by a consideration of the direct and indirect impact of rangeland management practices (grazing and fire) on some key NRs, namely thatch grass, medicinal and food plants, and fuelwood. Finally, some important considerations

 Grassland Productivity and Ecosystem Services (eds G. Lemaire, J. Hodgson and A. Chabbi)

for the sustainable management of multi-user, multifunctional rangelands are identified.

21.2 Rangeland Resources and Users

Cattle and small-stock ranching is practised on private or leased public land in the Americas, Europe, Australia and in certain countries in Africa (Holechek *et al.*, 1989) but communal usage of rangelands is the predominant form of land tenure in most developing countries (Neely *et al.*, 2009). The following outline of the utilization of non-forage NRs in communal rangelands is based mainly on the extensive ecological and socio-economic research on natural resource use in the heterogeneous rangelands of southern Africa, where usage patterns are broadly similar to those on rangelands elsewhere in Africa and on other continents (Scoones *et al.*, 1992).

Forage and other NRs are not evenly distributed over space and time in rangelands, and local classifications of resources usually reflect this heterogeneity. An example of such a local resource classification from the mountain rangelands of Lesotho, which have been targeted for improved rangeland management by national and international programmes because of perceived overgrazing of the subalpine and alpine communal grazing areas by transhumant herds (Quinlan and Morris, 1994), is depicted in Fig. 21.1. The classification of the abiotic and biotic resources indicates a complex set of individual and collective access and usage rights to NRs available in various places at different times, controlled by different tribal authorities. Land defined as 'rangeland' by government authorities is just one category ('grassland administered by district chiefs') of grazing land in the classification (Fig. 21.1). Management schemes aimed at improving the quality

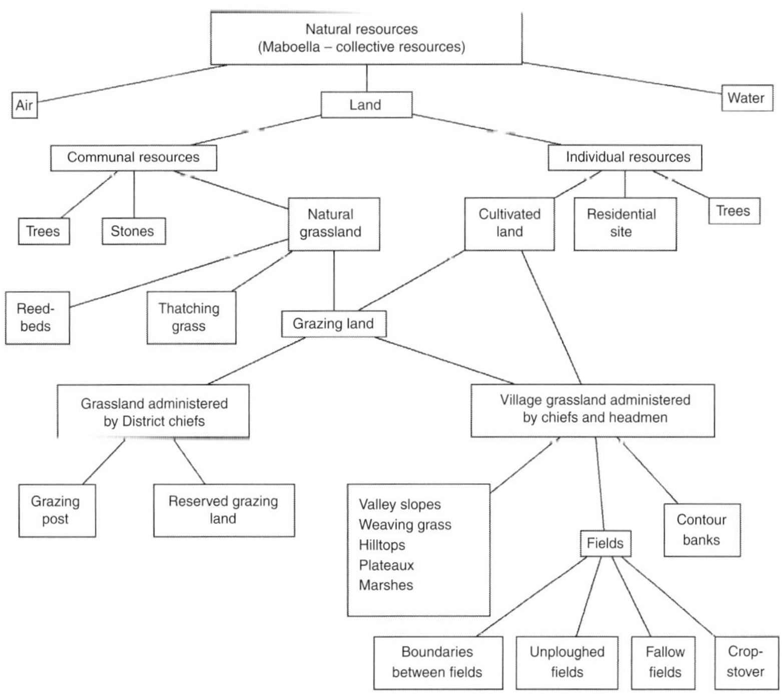

Fig. 21.1. Sesotho classification of local natural resources in the mountains of Lesotho (redrawn with kind permission from the Grassland Society of Southern Africa; from Quinlan, 1995).

and productivity of grazing and other resources would therefore need to encompass the whole resource production landscape rather than focusing on one 'rangeland' area judged to be of primary importance to one group of users (Quinlan and Morris, 1994).

Natural rangeland resources are indispensable to many rural households: they comprise the 'invisible capital' that underpins rural livelihoods (Cousins, 1999). Wild plants (especially leafy vegetables) and wildlife, ranging from large game to insects and their larvae (Butynski and Von Richter, 1974), are consumed regularly by most rural people (Scoones *et al.*, 1992), providing critical supplementary and emergency nutrition (e.g. Harris and Mohammed, 2003). Cash sales of raw or processed NRs, mostly through informal markets, can form a significant proportion of the income of poor and vulnerable rural households (Cavendish, 2000), matching or even exceeding income from wages and other arable and livestock farming activities (Shackleton *et al.*, 2001). Attempts to quantify the value of NRs to the rural economy do, however, often ignore the important aesthetic, spiritual and ecological (especially water-related supportive services) value of resource endowments to local communities (Campbell *et al.*, 1997; Kepe, 2008).

21.3 Effects of Rangeland Management on Key Natural Resources

Thatching grass, medicinal and wild food plants, and fuelwood are three of the most important plant-derived resources harvested from rangelands by rural people for domestic or commercial use. The impact of livestock and other rangeland management practices, particularly burning, on the availability, productivity and conservation of these key non-forage NRs has seldom been directly examined but has to be inferred from studies on the effects of grazing and burning on the plant populations that supply these commodities, as summarized below.

21.3.1 Thatch grass

Fibrous leaf and stem material from a number of graminoid species, including reeds, rushes and grasses, is used to roof most rural dwellings as well as many peri-urban and some urban houses in developing countries (Wells, 1995). It is a low-cost material with desirable thermal properties that is readily available from mid-to-late successional grasslands on fallowed or abandoned cropping land (Smits *et al.*, 1999) and in protected areas (Mengistu *et al.*, 2005). Tall, robust perennial species that produce numerous long, hollow, quick-curing and smooth inflorescence stems are preferred for thatching (Hall, 1988).

Repeated grazing by livestock in the growing season and frequent burning are inimical to the development of thatch grass swards (Shackleton, 1989), but once allowed to develop, such swards can be sustained by regular harvesting in late winter, after seed has dropped and leaves and stems have dried (Wells, 1995). Grazing and/or burning prior to harvesting will diminish available thatch but grazing for a limited period after the harvest appears not to reduce the long-term productivity of thatch grass communities (Sola, 2005). Biennial rather than annual harvesting may maximize yield of thatch grasses (Shackleton, 1989) but their productivity will probably decline and their populations eventually die out if they are not intermittently defoliated (Fynn *et al.*, 2005).

Production of grass thatch is largely unproblematic on private lands and in protected nature reserves or grazing enclosures (Mengistu *et al.*, 2005), where grazing and fire management can be judiciously applied. But in communal rangeland areas, where exploitation of rangeland resources may not be tightly controlled by local customs and rules, thatch grass can be a contested and dwindling resource (Kepe, 2005). Community-supported strategies, derived through a participatory approach and based on indigenous knowledge of the production ecology of local grassland (e.g. Sola, 2005), are thus required to ensure equitable access to, and sustained utilization of, thatch resources.

21.3.2 Medicinal and wild food plants

Many different plants are harvested from rangelands for traditional medicinal, ritualistic or nutritional use at home or for trade in large urban markets (Williams *et al.*, 2000). Harvesting profiles are taxonomically diverse, typically comprising whole plants or plant parts collected from several tens of wild food species (Harris and Mohammed, 2003) and up to several hundred medicinal species (Mander, 1998). The miscellany of exploited species and life forms precludes any general recommendations for the sustainable exploitation of food and medicinal plants, and there is little direct empirical evidence upon which management prescriptions could be based (but see Sher *et al.*, 2010; Zaloumis and Bond, 2010). Human and livestock disturbance can favour populations of some useful leafy vegetables (Dovie *et al.*, 2007) but an overall reduction in the plant diversity of rangelands, especially in the abundance of herbaceous forbs, which typically comprise most of the diversity of grassland (Morris, 2004), will probably reduce the availability of many medicinal and wild food plants.

Generally, herbivory has a larger effect on the plant species composition than the diversity of rangelands (some species may decline but be replaced by others) but grazing has been reported to markedly deplete or enhance diversity (for examples, see Olff and Ritchie, 1998; Bakker *et al.*, 2006). Responses appear contingent on a number of biological and environmental factors, including habitat productivity (i.e. soil fertility), evolutionary grazing history (Milchunas *et al.*, 1988) and the number, type and mix of grazing animals deployed. In the main, species richness is decreased by herbivory in nutrient-poor habitats but enhanced in nutrient-rich habitats (Bakker *et al.*, 2006). Grazing generally reduces diversity in dry, infertile grasslands with a short history of grazing (e.g. Landsberg *et al.*, 2003) but can increase the diversity of fertile grassland that has co-evolved with herbivores (e.g. Anderson *et al.*, 2007). The effect on diversity of large grazing ungulates is equivocal (Bakker *et al.*, 2006) but small-mouthed animals such as sheep can deplete grasslands of herbaceous forbs (O'Connor *et al.*, 2011).

From the modest empirical evidence available, it appears that the (grazing) system employed to distribute animals over the grazing area over time has less influence on plant diversity than the grazing pressure (stocking rate) applied (Hickman *et al.*, 2004; Jacobo *et al.*, 2006). Continuous grazing (stocking) is judged more benign than heavily stocked rotational systems that concentrate grazing and trampling on a small area (O'Connor and Kuyler, 2009), but this assertion remains largely untested.

Burning does not always predictably alter the species (Uys *et al.*, 2004) or the functional (Loiola *et al.*, 2010) diversity of grassland but frequent burning can increase grass at the expense of forbs (e.g. Collins *et al.*, 1998). In mesic grassland, burning interacts strongly with growing-season cutting, with certain combinations of disturbances maintaining, while others markedly reducing, species richness (Fynn *et al.*, 2004). Fire also usually interacts with grazing (Noy-Meir, 1995), which can offset the negative effects of frequent burning by reducing the suppression of herbaceous forbs by dominant grasses (Hartnett *et al.*, 1996).

21.3.3 Fuelwood

Woody plants are key multifunctional resource assets in savannahs and woodlands, providing a large variety of useful materials and products, e.g. timber, roots, bark, fruit (for pulp and juice), leaves, etc. Fuelwood is the most important woodland resource for rural communities (Dovie *et al.*, 2004), used by a large majority of rural households (> 90% in many places) in developing countries, for cooking and heating (Cline-Cole *et al.*, 1990), even where electricity is available (Madubansi and Shackleton, 2006). The demand for firewood from peri-urban and urban areas is also high and growing (Tabuti *et al.*, 2003).

A scarcity of fuelwood resulting from excessive harvesting for household use and for charcoal enterprises (Luoga *et al.*, 2000) has been reported, especially from densely populated rural areas (e.g. Kirkland *et al.*, 2007). Generally, however, the predicted 'fuelwood crisis' has not materialized in most areas (Hiemstra-van der Horst and Hovorka, 2009) because, among other things, savannah trees have co-evolved with fire and herbivory – many species coppice vigorously after harvesting (Shackleton, 2001) – and other factors such as land-use change, climate, fire and animal impacts are likely to be the overriding drivers of woody plant dynamics in savannah and woodland areas (Scholes and Archer, 1997).

A countervailing dynamic to deforestation is the widespread and apparently relentless thickening or encroachment of woody vegetation that has occurred in many grasslands and savannahs over the last century or so (for examples, see Van Auken, 2009; Wigley *et al.*, 2010). Bush encroachment reduces grass production and the grazing capacity of rangeland (Richter *et al.*, 2001), alters hydrological and nutrient cycles (Asner *et al.*, 2004), and reduces plant diversity (Knapp *et al.*, 2008). Some of the main encroaching species may, however, be particularly useful. For example, *Dichrostachys cinerea*, a medium-sized tree that has increased in abundance in a number of savannahs (e.g. Roques *et al.*, 2001), is a multi-purpose medicinal and 'magical' tree (used by shamans) that provides good fuelwood, charcoal and fencing poles (Coates Palgrave, 2003).

The cause of bush encroachment is contested (House *et al.*, 2003) but models and experimental evidence indicate that complex interactions between environment (rainfall, soils) and disturbance (fire, grazing, browsing) directly or indirectly influence woody plant demographics and the resultant dynamic balance between grasses and trees (Sankaran *et al.*, 2004). There is also mounting evidence that the increase in atmospheric CO_2 (which should favour C_3 trees over C_4 grasses) may be accelerating bush encroachment (Wigley *et al.*, 2010), thereby possibly sustaining future fuelwood stocks.

21.4 Management Considerations

Research has revealed the importance of managing animal numbers but has not produced consensus as to what grazing system should be used to sustain long-term productivity (Briske *et al.*, 2008), nor have any convincing examples emerged suggesting that the imposition of a grazing system on communal rangelands is necessary or worthwhile. A rangeland capable of delivering an assortment of natural resources and ecosystem services would need to be complex in structure and composition, diverse in its flora and fauna, and functionally intact, but no single rangeland management system can be prescribed to optimize and sustain the output of multiple NRs across a range of biotic, environmental and socio-economic settings. Nonetheless, the following are some general considerations that can contribute towards better management of multi-user, multi-functional rangelands.

21.4.1 Maintain landscape heterogeneity

Rangelands are inherently highly variable in their environment (soils, topography), climate and vegetation structure, composition and diversity. Forage and other NRs are similarly not uniformly available (see Fig. 21.1) or productive across the landscape. Patchy landscapes are likely to offer a greater variety of natural resources and be more biodiverse than more uniform rangelands (Fischer *et al.*, 2006) and large-scale heterogeneity can provide a buffer against temporal variation in resource productivity and quality, thereby widening options for resource users in times of scarcity (Hobbs *et al.*, 2008). There are, therefore, compelling reasons to seek to maintain landscape heterogeneity and the miscellaneous disturbances that generate and maintain the natural patch mosaic dynamic of rangelands (Meyer *et al.*, 2009).

Rotational stocking systems designed to create a relatively homogeneous palatable and productive sward and ensure even utilization of available forage by livestock can homogenize vegetation and reduce the diversity of wildlife habitats (Fuhlendorf and Engle, 2001). Frequent burning applied relentlessly in the same season also reduces vegetation and functional heterogeneity at all scales (Collins and Smith, 2006). More flexible management strategies are therefore required in which the frequency and intensity of grazing and burning can be varied by, for example, using patch mosaic burning (Fuhlendorf and Engle, 2001) or by tactically moving livestock among paddocks to match varying forage regrowth rates (Venter and Drewes, 1969).

21.4.2 Manage resource patches

Certain NRs occur in patches and flourish only in particular environments or under invariable disturbance regimes. Thatch grass stands, riverine and wetland habits and natural woodlots or bush clumps are some examples of 'resource islands' (Van Auken, 2009) that would need to be managed differently to the surrounding vegetation matrix. Targeted disturbances applied at appropriate set times and intervals (e.g. burning and grazing thatch grass infrequently at the end of the growing season) may be required to maintain these resources. Sensitive or rare plants or animals may also need to be protected from excessive human disturbances (West, 1993).

Conflict over access to resource patches in communal rangelands is not uncommon. Kepe (2005), for example, reports that women have resorted to burning firebreaks around thatch grass patches in a coastal grassland in South Africa, to prevent untimely burning by livestock owners (predominantly men) who desire to burn the grassland annually before winter, for hunting and to stimulate forage regrowth. However, communal access does not necessarily obviate against protection and management of valuable NRs. For example, individual prominent trees in the landscape, which provide cash-earning fruit or are appreciated by rural inhabitants for their aesthetic, amenity and spiritual value (Wilson, 1989), are sometimes owned by groups or individuals and actively managed to increase their productivity (Chivaura-Mususa *et al.*, 2000).

21.4.3 Employ restorative rests

Frequent and chronic exploitation of forage and other NRs can eventually diminish the productivity and abundance of resource stocks, even if plants are afforded periodic rests between harvests (e.g. the period of absence in a rotational stocking system) to allow regrowth and harvestable material to accumulate (Tainton, 1971). Consequently, longer rests, encompassing at least a full growing season, may occasionally be necessary to restore fully the vigour of plants, allow necessary population processes (such as flowering and seedling establishment) to be completed, and to build up natural capital that will sustain the delivery of goods and services (Clewell and Aronson, 2006). Periodic long restorative rests may be particularly important for dryland vegetation with a low restorative potential (Müller *et al.*, 2007). Resource stocks accumulated during rests can provide a crucial reserve for seasons or years of scarcity (Hughes, 1990).

Snyman (1998) noted that '... the way in which rangeland is rested is usually more important than the way in which it is utilized' (p. 646), thus season- or year-long resting should be an integral part of NR management (NRM). The duration of rest required will depend on a number of local factors, including the sensitivity and resilience of the particular resource, its disturbance history and condition and growing conditions during the rest. Resting during wet periods (Müller *et al.*, 2007) and also after droughts (Retzer, 2006) appears most effective for restoring resource productivity. Large-scale migration by pastoralists was an effective means to distribute rests across the landscape but is increasingly being curtailed, resulting in degradation of settled areas (Tsegaye *et al.*, 2010). Multi-year rests coupled with active rehabilitation may be required to restore severely degraded areas (e.g. Visser *et al.*, 2007).

21.5 Conclusions

Managing heterogeneous rangelands for a single product or purpose, such as livestock production or nature conservation, under a single management authority is not straightforward. Ensuring sustained delivery of many different kinds of plant- and animal-derived goods and services to multifarious (and often competing) users on communal rangeland, while maintaining the functional integrity of the ecosystem, presents an even more formidable challenge. The principles of, and lessons learned from, application of community-based natural resource management (CBNRM) in diverse contexts (Fabricius *et al.*, 2004) would be applicable to initiatives aimed at enhancing the production of multiple resources on multi-user rangelands. An in-depth understanding of how resources are used and contribute to rural livelihoods is required before developing CBNRM schemes, which should be carefully designed so not to favour one group of users (e.g. livestock owners) over others. Greatest progress will probably be made by building on local initiatives and innovations and by fully involving users in the planning and management of their own rangeland resources.

References

Allen, V.G., Batello, C., Berretta, E.J., Hodgson, J., Kothmann, M., McIvor, J., Milne, J., Morris, C.D., Peeters, A., Sanderson, M. and Xianglin, L. (2011) An international terminology for grazing lands and grazing animals. *Grass and Forage Science*, 66, 2–28.

Anderson, T.M., Ritchie, M.E. and McNaughton, S.J. (2007) Rainfall and soils modify plant community response to grazing in Serengeti National Park. *Ecology* 88, 1191–1201.

Asner, G.P., Elmore, A.J., Olander, L. P., Martin, R.E. and Harris, A.T. (2004) Grazing systems, ecosystem responses, and global change. *Annual Review of Environment and Resources* 29, 261–299.

Bakker, E.S., Ritchie, M.E., Olff, H., Milchunas, D.G. and Knops, J.M.H. (2006) Herbivore impact on grassland plant diversity depends on habitat productivity and herbivore size. *Ecology Letters* 9, 780–788.

Briske, D.D., Derner, J.D., Brown, J.R., Fuhlendorf, S.D., Teague, W.R., Havstad, K.M., Gillen, R.L., Ash, A.J. and Willms, W.D. (2008) Rotational grazing on rangelands: reconciliation of perception and experimental evidence. *Rangeland Ecology and Management* 61, 3–17.

Butynski, T.M. and Von Richter, T. (1974) In Botswana most of the meat is wild. *Unasylva* 26, 24–29.

Campbell, B.M., Luckert, M. and Scoones, I. (1997) Local-level valuation of savanna resources: a case study from Zimbabwe. *Economic Botany* 51, 59–77.

Cavendish, W. (2000) Empirical regularities in the poverty-environment relationship of rural households: evidence from Zimbabwe. *World Development* 28, 1979–2003.

Chivaura-Mususa, C., Campbell, B. and Kenyon, W. (2000) The value of mature trees in arable fields in the smallholder sector, Zimbabwe. *Ecological Economics* 33, 395–400.

Clewell, A. F. and Aronson, J. (2006) Motivations for the restoration of ecosystems. *Conservation Biology* 20, 420-428.

Cline-Cole, R.A., Main, H.A.C. and Nichol, J.E. (1990) On fuelwood consumption, population dynamics and deforestation in Africa. *World Development* 18, 513–527.

Coates Palgrave, K. (2003) *Trees of Southern Africa*. New Holland Publishers, Cape Town.

Collins, S.L. and Smith, M.D. (2006) Scale-dependent interaction of fire and grazing on community heterogeneity in tallgrass prairie. *Ecology* 87, 2058–2067.

Collins, S.L., Knapp, A.K., Briggs, J.M., Blair, J.M. and Steinauer, E.M. (1998) Modulation of diversity by grazing and mowing in native tallgrass prairie. *Science* 280, 745.

Cousins, B. (1999) Invisible capital: the contribution of communal rangelands to rural livelihoods in South Africa. *Development Southern Africa* 16, 299-318.

Dovie, D.B.K., Witkowski, E.T.F. and Shackleton, C.M. (2004) The fuelwood crisis in southern Africa - relating fuelwood use to livelihoods in a rural village. *GeoJournal* 60, 123–133.

Dovie, D.B.K., Shackleton, C.M. and Witkowski, E.T.F. (2007) Conceptualizing the human use of wild edible herbs for conservation in South African communal areas. *Journal of Environmental Management* 84, 146–156.

Fabricius, C., Koch, E., Turner, S. and Magome, H. (eds) (2004) *Rights, Resources and Rural Development: Community-Based Natural Resource Management in Southern Africa.* Earthscan, London.

Fischer, J., Lindenmayer, D.B. and Manning, A.D. (2006) Biodiversity, ecosystem function, and resilience: ten guiding principles for commodity production landscapes. *Frontiers in Ecology and the Environment* 4, 80–86.

Fuhlendorf, S.D. and Engle, D.M. (2001) Restoring heterogeneity on rangelands: ecosystem management based on evolutionary grazing patterns. *BioScience* 51, 625–632.

Fynn, R.W.S., Morris, C.D. and Edwards, T.J. (2004) Effect of burning and mowing on grass and forb diversity in a long-term grassland experiment. *Applied Vegetation Science* 7, 1–10.

Fynn, R.W.S., Morris, C.D. and Edwards, T.J. (2005) Long-term compositional responses of a South African mesic grassland to burning and mowing. *Applied Vegetation Science* 8, 5–12.

Hall, N. (1988) *Thatching: a Handbook.* Intermediate Technology, London.

Harris, F.M.A. and Mohammed, S. (2003) Relying on nature: wild foods in Northern Nigeria. *Ambio* 32, 24–29.

Hartnett, D.C., Hickman, K.R. and Fischer-Walter, L.E. (1996) Effects of bison grazing, fire and topography on floristic diversity in tallgrass prairie. *Journal of Range Management* 49, 413–420.

Havstad, K.M., Peters, D.P.C., Skaggs, R., Brown, J., Bestelmeyer, B., Fredrickson, E., Herrick, J. and Wright, J. (2007) Ecological services to and from rangelands of the United States. *Ecological Economics* 64, 261–268.

Hickman, K.R., Hartnett, D.C., Cochran, R.C. and Clenton, F. O. (2004) Grazing management effects on plant species diversity in tallgrass prairie. *Journal of Range Management* 57, 58–65.

Hiemstra-van der Horst, G. and Hovorka, A. J. (2009) Fuelwood: the "other" renewable energy source for Africa? *Biomass and Bioenergy* 33, 1605–1616.

Hobbs, N.T., Galvin, K.A., Stokes, C.J., Lackett, J.M., Ash, A.J., Boone, R.B., Reid, R.S. and Thornton, P.K. (2008) Fragmentation of rangelands: implications for humans, animals, and landscapes. *Global Environmental Change* 18, 776–785.

Holechek, J.L., Pieper, R.D. and Herbel, C.H. (1989) *Range Management: Principles and Practices.* Prentice-Hall, Englewood Cliffs, NJ.

House, J.I., Archer, S., Breshears, D.D., Scholes, R.J. and NCEAS Tree-Grass Interaction Participants (2003) Conundrums in mixed woody–herbaceous plant systems. *Journal of Biogeography* 30, 1763–1777.

Hughes, L.E. (1990) Twenty years of rest-rotation grazing on the Arizona Strip: an observation. *Rangelands* 12, 173–176.

Jacobo, E.J., Rodríguez, A.M., Bartoloni, N. and Deregibus, V.A. (2006) Rotational grazing effects on rangeland vegetation at a farm scale. *Rangeland Ecology and Management* 59, 249–257.

Kepe, T. (2005) Grasslands ablaze: vegetation burning by rural people in Pondoland, South Africa. *South African Geographical Journal* 87, 10–17.

Kepe, T. (2008) Beyond the numbers: understanding the value of vegetation to rural livelihoods in Africa. *Geoforum* 39, 958–968.

Kirkland, T., Hunter, L.M. and Twine, W. (2007) "The bush is no more": insights on institutional change and natural resource availability in rural South Africa. *Society and Natural Resources* 20, 337–350.

Knapp, A.K., McCarron, J.K., Silletti, A.M., Hoch, G.A., Heisler, J.C., Lett, M.S., Blair, J.M., Briggs, J.M. and Smith, M.D. (2008) Ecological consequences of the replacement of native grassland by *Juniperus virginiana* and other woody plants. In: Van Auken, O. W. (ed.) *Western North American Juniperus Communities: a Dynamic Vegetation Type.* Springer-Verlag, New York, NY, pp. 156–169.

Landsberg, J., James, C.D., Morton, S. R., Müller, W.J. and Stol, J. (2003) Abundance and composition of plant species along grazing gradients in Australian rangelands. *Journal of Applied Ecology* 40, 1008–1024.

Loiola, P.d.P., Cianciaruso, M.V., Silva, I.A. and Batalha, M.A. (2010) Functional diversity of herbaceous species under different fire frequencies in Brazilian savannas. *Flora - Morphology, Distribution, Functional Ecology of Plants,* 205, 674–681.

Luoga, E.J., Witkowski, E.T.F. and Balkwill, K. (2000) Economics of charcoal production in miombo woodlands of eastern Tanzania: some hidden costs associated with commercialization of the resources. *Ecological Economics* 35, 243–257.

Madubansi, M. and Shackleton, C.M. (2006) Changing energy profiles and consumption patterns following electrification in five rural villages, South Africa. *Energy Policy* 34, 4081–4092.

Mander, M. (1998) *Marketing of Indigenous Medicinal Plants in South Africa: a Case Study in KwaZulu-Natal*. Food and Agriculture Organisation, Rome.

Mengistu, T., Teketay, D., Hulten, H. and Yemshaw, Y. (2005) The role of enclosures in the recovery of woody vegetation in degraded dryland hillsides of central and northern Ethiopia. *Journal of Arid Environments* 60, 259–281.

Meyer, K.M., Wiegand, K. and Ward, D. (2009) Patch dynamics integrate mechanisms for savanna tree-grass coexistence. *Basic and Applied Ecology* 10, 491–499.

Milchunas, D.G., Sala, O.E. and Lauenroth, W.K. (1988) A generalized model of the effects of grazing by large herbivores on grassland community structure. *American Naturalist* 132, 87–106.

Morris, C.D. (2004) Manage the grassland not just the grass. *Bulletin of the Grassland Society of Southern Africa (Grass Roots)* 14, 16–19.

Müller, B., Frank, K. and Wissel, C. (2007) Relevance of rest periods in non-equilibrium rangeland systems: a modelling analysis. *Agricultural Systems* 92, 295-317.

Neely, C., Bunning, S. and Wilkes, A. (2009) *Review of Evidence on Dryland Pastoral Systems and Climate Change: Implications and Opportunities for Mitigation and Adaptation.* Food and Agriculture Organisation, Rome.

Noy-Meir, I. (1995) Interactive effects of fire and grazing on structure and diversity of Mediterranean grasslands. *Journal of Vegetation Science* 6, 701–710.

O'Connor, T.G. and Kuyler, P. (2009) Impact of land use on the biodiversity integrity of the moist sub-biome of the grassland biome, South Africa. *Journal of Environmental Management* 90, 384–395.

O'Connor, T.G., Martindale, G., Morris, C.D., Short, A.D., Witkowski, E.T.F. and Scott-Shaw, R. (2011) Influence of grazing management on plant diversity of Highland Sourveld grassland, KwaZulu-Natal, South Africa. *Rangeland Ecology and Management* 64, 196–207.

Olff, H. and Ritchie, M.E. (1998) Effects of herbivores on grassland plant diversity. *Trends in Ecology and Evolution* 13, 261–265.

Quinlan, T. (1995) Understanding development: different perspectives on livestock and 'range' management in Lesotho. In: Hatch, G.P. and Zacharias, P.J.K. (eds) *Proceedings of the Second Developing Areas Symposium: People, Rangelands and Development.* Grassland Society of Southern Africa, Pietermaritzburg, pp. 34–41.

Quinlan, T. and Morris, C.D. (1994) Implications of changes to the transhumance system for conservation of the mountain catchments in eastern Lesotho. *African Journal of Range and Forage Science* 11, 76–81.

Retzer, V. (2006) Impacts of grazing and rainfall variability on the dynamics of a Sahelian rangeland revisited: new insights from old data. *Journal of Arid Environments* 67, 157–164.

Richter, C.G.F., Snyman, H.A. and Smit, G.N. (2001) The influence of tree density on the grass layer of three semi-arid savanna types of southern Africa. *African Journal of Range and Forage Science* 18, 103–109.

Roques, K.G., O'Connor, T.G. and Watkinson, A.R. (2001) Dynamics of shrub encroachment in an African savanna: relative influences of fire, herbivory, rainfall and density dependence. *Journal of Applied Ecology* 38, 268–280.

Sankaran, M., Ratnam, J. and Hanan, N.P. (2004) Tree–grass coexistence in savannas revisited–insights from an examination of assumptions and mechanisms invoked in existing models. *Ecology Letters* 7, 480–490.

Scholes, R.J. and Archer, S.R. (1997) Tree-grass interactions in savannas. *Annual Review of Ecology, Evolution and Systematics* 28, 517–544.

Scoones, I., Melnyk, M. and Pretty, J.N. (1992) *The Hidden Harvest: Wild Foods and Agricultural Systems – a Literature Review and Annotated Bibliography.* Institute for Environment and Development (IIED), London.

Shackleton, C.M. (2001) Managing regrowth of an indigenous savanna tree species (*Terminalia sericea*) for fuelwood: the influence of stump dimensions and post-harvest coppice pruning. *Biomass and Bioenergy* 20, 261–270.

Shackleton, C.M., Shackleton, S.E. and Cousins, B. (2001) The role of land-based strategies in rural livelihoods: the contribution of arable production, animal husbandry and natural resource harvesting in communal areas in South Africa. *Development Southern Africa* 18, 581–604.

Shackleton, S.E. (1989) Autecology of *Cymbopogon validus* (Stapf) Stapf ex Burtt Davy in Mkambati Game Reserve, Transkei. MSc thesis, University of the Witwatersrand, South Africa.
Sher, H., Ahmad, A., Eleyemeni, M. and Fazl-i-Hadi, S. (2010) Impact of nomadic grazing on medicinal plants diversity in Miandam, Swat-Pakistan (preliminary results). *International Journal of Biodiversity and Conservation* 2, 146-154.
Smits, N.A.C., Bredenkamp, G.J., Mucina, L. and Granger, J.E. (1999) The vegetation of old-fields in the Transkei. *South African Journal of Botany* 65, 414–420.
Snyman, H.A. (1998) Dynamics and sustainable utilization of rangeland ecosystems in arid and semi-arid climates of southern Africa. *Journal of Arid Environments* 39, 645–666.
Sola, P. (2005) The community resource management plan: a tool for integrating IKS into natural resource management. *Ethnobotany Research and Applications* 3, 143–153.
Tabuti, J.R.S., Dhillion, S.S. and Lye, K.A. (2003) Firewood use in Bulamogi County, Uganda: species selection, harvesting and consumption patterns. *Biomass and Bioenergy* 25, 581–596.
Tainton, N.M. (1971) An analysis of the objectives of resting grassveld. *Proceedings of the Grassland Society of Southern Africa* 6, 50–54.
Tainton, N.M., Morris, C.D. and Hardy, M.B. (1996) Complexity and stability in grazing systems. In: Hodgson, J. and Illius, A.W. (eds) *The Ecology and Management of Grazing Systems*. CABI Publishing, Wallingford, UK, pp. 275–299.
Tsegaye, D., Moe, S.R., Vedeld, P. and Aynekulu, E. (2010) Land-use/cover dynamics in northern Afar rangelands, Ethiopia. *Agriculture, Ecosystems and Environment*, 139, 174–180.
Uys, R.G., Bond, W.J. and Everson, T.M. (2004) The effect of different fire regimes on plant diversity in southern African grasslands. *Biological Conservation* 118, 489–499.
Van Auken, O.W. (2009) Causes and consequences of woody plant encroachment into western North American grasslands. *Journal of Environmental Management* 90, 2931–2942.
Venter, A.D. and Drewes, R.H. (1969) A flexible system of management for Sourveld in Natal. *Proceedings of the Grassland Society of Southern Africa* 4, 104–107.
Visser, N., Morris, C., Hardy, M.B. and Botha, J.C. (2007) Restoring bare patches in the Nama-Karoo of South Africa. *African Journal of Range and Forage Science* 24, 1–1.
Wells, J. (1995) Population, settlements and the environment: the provision of organic materials for shelter. *Habitat International* 19, 73–90.
West, N.E. (1993) Biodiversity of rangelands. *Journal of Range Management* 46, 2–13.
Wigley, B.J., Bond, W.J. and Hoffman, M.T. (2010) Thicket expansion in a South African savanna under divergent land use: local vs. global drivers? *Global Change Biology* 16, 964–976.
Williams, V.L., Balkwill, K. and Witkowski, E.T.F. (2000) Unravelling the commercial market for medicinal plants and plant parts on the Witwatersrand, South Africa. *Economic Botany* 54, 310–327.
Wilson, K.B. (1989) Trees in fields in southern Zimbabwe. *Journal of Southern African Studies* 15, 369–383.
Zaloumis, N.P. and Bond, W.J. (2010) Grassland restoration after afforestation: no direction home? *Austral Ecology*, 36, 357–366.

22 Opportunities and Challenges for Integrating North-American Crop and Livestock Systems

Alan J. Franzluebbers, R. Mark Sulc and Michael P. Russelle

22.1 The Need to Renew Agriculture

Agriculture has become increasingly specialized in industrialized countries, in response to political and economic pressures to meet market demands of an ever-larger food and fibre processing sector (Russelle *et al.*, 2007; Hendrickson *et al.*, 2008). In the USA, specialization has been accompanied by a dramatic decline in the number of farms, from >6 million in 1920 to <2 million in 2009 (Table 22.1) and an amazing increase in productivity (Hanson and Hendrickson, 2009). Large family and non-family farms now account for >75% of agricultural sales; in 2007, the largest 2% of farms were responsible for 59% of total farm sales (NRC, 2010). However, nearly 65% of principal operators also work off the farm, and 40% work more than 200 days per year at an off-farm job, primarily to obtain sufficient income and employer-subsidized health insurance (USDA-NASS, 2009). Livestock production has changed, too. Compared with many small, diversified family farms in 1950, today there are larger specialized row-crop and/or concentrated animal feeding operations (CAFOs) (Singer *et al.*, 2009).

Kirschenmann (2009) and Steiner *et al.* (2009) expounded that the contemporary food system in the USA is entrenched in an era of cheap energy, an assumed stable climate, and a business environment in which negative environmental and social costs can be simply externalized, without

Table 22.1. Total number of farms and land in various crops of the USA (USDA-NASS, 2009).

Crop	Farms (1000s)	Mha	Crop	Farms (1000s)	Mha
Forage	870	24.9	Barley	20	1.4
Maize (grain)	348	34.9	Cotton	19	4.2
Maize (silage)	84	2.4	Tobacco	16	0.1
Soybean	279	25.9	Rice	6	1.1
Wheat	161	20.6	Sunflower	6	0.8
Orchards	116	2.0	Dry bean	6	0.6
Vegetables	69	1.9	Groundnut	6	0.5
Oat	43	0.6	Sugarbeet	4	0.5
Sorghum (grain)	26	2.7	Sugarcane	<1	0.3
Sorghum (silage)	6	0.2	Total	1685	164.5

penalty or assumed responsibility. Industrialization and concentration of agriculture have increasingly revealed how food is produced in a way that homogenizes the landscape, widens the chasm between rich and poor, fuels an epidemic of obesity, is insensitive to animal welfare, and is environmentally degrading (Pollan, 2006; Kenner, 2008; Ikerd, 2009). There is evidence that domestic markets for agricultural products are reflecting these concerns (Halloran and Archer, 2008).

There is a growing concern with specialized agricultural systems because of increasingly negative effects on the environment that are manifested in:

- Water contamination with excessive nutrients, pesticides, antibiotics, hormone mimics and pathogens.
- Sinking groundwater levels due to excessive use by a variety of stakeholders, including specialized crop production and reliance on irrigation for crops in dry regions.
- Rising atmospheric climate forcing ('greenhouse') gas (GHG) concentrations from soils being depleted in organic matter.
- Dysfunctional soils that have become exhausted from excessive tillage, wind and water erosion, salt accumulation and pesticide inputs.

In contrast, conservation systems that integrate crops and livestock could provide opportunities to capture ecological interactions, to make agricultural ecosystems more efficient at cycling nutrients, rely more on renewable natural resources, and improve the inherent functioning of soils, while achieving acceptable or improved economic returns for the farmer. Furthermore, diversifying production enterprises utilizes labour more efficiently for on-farm tasks (Hoagland *et al.*, 2010).

Although direct consumption of crops provides more protein and energy to humans than when they are processed by livestock, and although some livestock production systems have contributed to environmental degradation, livestock can utilize crops and residues not suitable as food and fibre for humans (Russelle *et al.*, 2007). In addition, conservation-oriented crop–livestock systems can provide multiple environmental benefits (Schiere *et al.*, 2002; Franzluebbers, 2007; Sulc and Tracy, 2007). Four modes of agriculture have been described (Schiere *et al.*, 2002):

- Low external input agriculture (in which demand is adjusted to resource availability and greater labour and skills are necessary to increase production).
- Expansive agriculture (where land is abundant).
- High external input agriculture (in which demand for output or profitability determines input levels, sometimes leading to environmental degradation).
- New conservation agriculture (in which production goals are matched with the resource base to achieve both profitability and environmental benefits).

Agriculture in the post-European colonization period of the USA was initially of the expansive mode – when land in the east became worn out after years of use with no inputs and declining soil organic matter, farmers simply travelled westward to explore new ground (Triplett and Dick, 2008). Congress responded to the massive degradation of land and displacement of people during the Dust Bowl of the 1930s with the Soil Conservation Act of 1935. With full implementation of the industrial revolution in agriculture following World War II, synthetic fertilizers and pest control strategies heralded an age of impressive increase in staple food production and global trade of high-value commodities (Tilman *et al.*, 2002).

Recognition of the unintended consequences of high-input agriculture on environmental quality began slowly with the publication of *Silent Spring* (Carson, 1962) and continued with the establishment of the United States Environmental Protection Agency in 1970 and passage of the Clean Water Act in 1972. Conservation tillage research and adoption in the latter part of the 20th century expanded beyond a handful of pioneers in Virginia and Ohio (Triplett and Dick, 2008), as evidenced also by the changing themes of the International Soil Tillage Research Organization conferences from ploughing, mechanization and power in the 1950s and 1960s, conservation tillage and mulch farming in the 1970s, soil compaction, strength and tilth in the 1980s, agricultural sustainability and soil quality in the 1990s, and soil organic carbon sequestration and greenhouse gas emissions in the 2000s (Lal, 2001). Several government incentive programs were developed to help overcome environmental deterioration in agricultural settings, including the Conservation Reserve Program in 1985 (land retirement), the Conservation Security Program (CSP) in 2002 (variety of stewardship incentives), and the Energy Policy Act in 2005 (biofuel production). Only the CSP views the farm within its particular context (landscape, cropping system, etc.) (Boody and DeVore, 2006). These programs contribute to an improvement in the environmental outcomes of agriculture in the USA, but are clearly insufficient in light of continuing hypoxia in the Northern Gulf of Mexico, depletion of geological water in the Ogallala aquifer, the need for regulatory approaches to reduce surface water contamination (e.g., establishment of Total Maximum Daily Loads), and high ammonia emissions from manure storage and livestock confinement facilities, among other worrisome environmental signals.

Recent emphasis in reducing fossil fuel use by substituting biomass feedstocks has resulted in specific requirements for reduction in greenhouse gas emissions and improved ecosystem services (Electronic Code of Federal Regulations, 2010). This may portend that the USA will adopt a broader view of agriculture, in general – one in which agriculture provides a wide range of beneficial ecosystem services to society. It is within this context that integrated crop–livestock systems hold enormous potential for achieving agricultural sustainability in North America.

22.2 Toward Sustainable, Integrated Agricultural Systems

Four goals of sustainable agriculture are to (NRC, 2010):

- Satisfy human food, feed and fibre needs, and contribute to biofuel needs.
- Enhance environmental quality and the resource base.
- Sustain the economic viability of agriculture.
- Enhance the quality of life for farmers, farm workers and society as a whole.

Sustainable farming systems work with ecological and biogeochemical processes and cycles to (i) maximize synergistic interactions and the beneficial use of internal resources, and (ii) minimize dependence on external inputs and use added inputs efficiently.

In its recent report assessing the scientific evidence for sustainable agricultural systems, the National Research Council recommended both incremental and transformative approaches for the future (NRC, 2010). Incremental approaches would target existing conservation practices and technologies for greater adoption, such as conservation tillage systems, cover cropping, crop diversity, traditional plant breeding, modern genetic engineering, water-use efficiency, best management practices, soil and plant tissue testing, nutrient management plans, precision agriculture, utilization of compost and green manures, integrated pest management, and livestock genetic improvement. In contrast, transformative approaches should include:

1. Developing a collaborative effort to construct an integrated vision for a future agriculture that balances and enhances the four sustainability goals.
2. Encouraging and formulating new markets and legal frameworks to advance this sustainable and integrated vision of agriculture.
3. Integrating research and extension more completely across multiple disciplines to pursue the four goals of sustainability.

4. Assessing the potential of new forms of production systems that represent a dramatic departure from the current high-input production systems.
5. Promoting system characteristics that increase resilience and adaptability with changing conditions.
6. Adjusting the mix of farming system types and practices at the landscape level to address major regional issues, such as water overdraft and environmental contamination.

Recommendation 1 might involve farmers, landowners, agribusiness, farm advisors, researchers (e.g. from soil, water, atmospheric, biological, engineering, social and economic disciplines), environmentalists, food processors, marketers, health professionals, consumers and politicians. Recommendations 4 and 6, in particular, can be adeptly addressed through developing integrated crop–livestock systems. In the following section, we outline some examples of integrated crop–livestock systems and how they might be designed for different regions of the USA, to meet various expectations.

22.3 Examples of Integrated Crop–Livestock Systems

22.3.1 Sod-based rotations

According to the US Economic Research Service, the area of cropland used for pasture peaked in the late 1960s and was about 62 million ha in 2002 (www.ers.usda.gov/data/majorlanduses/). Many dairy farmers with small- to medium-sized herds practise some form of intensive rotational grazing to gain more benefit from the pasture ('sod'). However, pastureland is often relegated to land parcels not well suited to cultivation, rather than being incorporated into rotations. This is a missed opportunity.

Sod-based rotations represent a diversity of proven systems to maintain yield, reduce pest incidence, and help protect environmental quality in a wide range of climatic and edaphic conditions (Katsvairo *et al.*, 2006; Allen *et al.*, 2007; Franzluebbers, 2007; Sulc and Tracy, 2007). Rotations may be short (as with annual cover crops), intermediate (as with biennial legumes planted with a small-grain nurse crop and harvested for forage the following year, followed by planting of grain crops), or long (as with perennial forages planted following row crops and harvested for forage by grazing, haying or silage during several years thereafter) until rotated to grain crops again. Longer rotations generally improve soil organic C and soil quality, and promote root exploration to deeper soil horizons. Improved water infiltration, with concurrent reduction in nitrate leaching, lower runoff losses, and less need for supplemental irrigation and fertilizer N are common advantages in all regions, especially when deeply rooted perennial forages are included where appropriate. Annual and perennial forage legumes further reduce purchased N requirement. These rotations frequently increase yields of the food and fibre crop phases and support higher, and often more stable, economic returns. These are significant advantages for livestock producers who diversify into crop production and for crop specialists who add pasture and livestock.

22.3.2 Grazing cover crops

Cover crops provide a viable short-rotation opportunity for almost any cropping sequence in the eastern USA, potentially improving productivity and reducing environmental threats. Despite extensive research conducted with cover crops (Hargrove, 1991; Franzluebbers, 2007), and increasingly in combination with conservation tillage systems during the past two decades, there are still too few examples of how cover crops might be successfully integrated into crop–livestock systems. Some accessible examples of farmers using cover crops in unique diversified cropping and grazing systems manners are Gabe Brown in North Dakota (www.sustainableranching.com) and Steve Groff in Pennsylvania (www.cedarmeadow farm.com).

Benefits from cover crops in annual cropping systems are numerous, including:

- Reducing water, soil and nutrient runoff.
- Improving soil tilth, structure, water infiltration and nutrient cycling.
- Modifying soil moisture, by increasing uptake and reducing evaporation at different times of the year.
- Contributing to soil organic C sequestration and soil biological diversity.
- Controlling weeds through competition, allelopathy and microclimatic alteration.
- Controlling insect and disease pressures ecologically.
- Serving as a nutrient trap in high-fertility systems.
- If leguminous, providing biologically fixed N to the cropping system.

As summarized by Gardner and Faulkner (1991), having ruminant livestock utilize cover crops in a crop production system could increase the value of cover crops, because 'planting and caring for a crop that apparently serves no immediate economic and harvestable purpose is both a foreign and unknown practice in much of the world ... details, time, and skill required to manage both crops and livestock are obvious adoption barriers to seeing cover crops as pasture'. The same authors also stated that the most basic barrier to adoption of integrated crop–livestock systems today is that many producers are reluctant to obtain or manage grazing livestock, because of a lack of experience and/or a lack of time during critical crop management periods.

A fully replicated field experiment investigating crop, animal and soil responses to three management factors was initiated in 2002 near Watkinsville, GA. Land previously in 20 years of grazed tall fescue paddocks was converted to two cropping systems (sorghum/maize grain + rye cover crop or wheat grain + pearl millet cover crop) managed under two tillage systems (no tillage or intensive tillage with initial mouldboard plough followed by disk tillage) and two cover crop scenarios (grazed by cattle or not grazed). During the first several years of production, grain yield was reduced by grazing cover crops in only one of 14 crop–tillage–year combinations (Franzluebbers and Stuedemann, 2007). Cover crop biomass, cattle performance and total gain were greater under no tillage than under conventional tillage. The integration of livestock with crops improved production from a whole-farm perspective, by utilizing cover crop biomass as forage for cattle production. The effects of grazing cover crops by cattle on soil physical properties were variable and small (Franzluebbers and Stuedemann, 2008a,b).

In Ohio, winter cover crops of annual ryegrass and oat/rye following no-till maize silage provided additional forage, of excellent nutritive value, to extend the grazing season during winter months. Good yearling dairy heifer gains were achieved without detrimental effects on subsequent maize silage productivity, and cover crops increased labile soil C (Faé *et al.*, 2009). Using conventional tillage in a crop–livestock system in Illinois, Tracy and Zhang (2008) reported that soil organic C concentration and maize grain yield were greater in a maize/cover crop pasture rotation than in continuous maize without a cover crop. Even though cattle presence on cropland may have caused soil compaction in some years, no negative effect on soil properties or maize grain yield were evident. Cold winters and multiple freeze/thaw events can cause some difficulties, especially with forage being wasted on the cropland pastures (Sulc and Tracy, 2007). Farmers also have been experimenting with planting a cover crop such as forage oat in early August, after winter wheat grain harvest, in a maize–soybean–wheat grain rotation. The oat cover crop planted after wheat is grazed in late autumn and into winter. Because freezing kills the oat crop, this practice allows early spring planting of the subsequent maize crop, helping maximize maize grain yield. Additional research and experimentation on farms is needed to identify best management practices for use of cover crop grazing within grain crop rotations in the colder regions.

22.3.3 Dual purpose use of grain crops

The most common example of this practice occurs in the southern Great Plains (Kansas, Oklahoma and Texas), where winter wheat is grown as a cash grain crop, but also as a forage. Dual-purpose wheat is an economically important strategy in the region (Redmon *et al.*, 1995), on an estimated 2.5 Mha (Taylor *et al.*, 2010). Wheat is grazed during autumn, winter and spring and can be managed to have a minimal effect on subsequent grain yield, if soil moisture at planting and subsequent precipitation during the growing season are adequate. Wheat pasture is a valuable source of high-quality forage, and stocker cattle can gain in excess of 1 kg/day. In this system, both grain and grazed forage (i.e. cattle gain) contribute significantly to total farm income. Taller wheat cultivars have the potential for extending the grazing period, while producing grain yield similar to those of the more commonly grown semi-dwarf wheat cultivars. Grazing should be terminated when wheat reaches the first hollow stem growth phase (Taylor *et al.*, 2010). The profitability of this integrated system varies with precipitation and relative value of grain versus live weight gain (Mauget *et al.*, 2009).

22.3.4 Grazing crop residues

Grazing livestock on crop residues after grain has been harvested represents one of the simplest and most economical methods for producers to integrate livestock into grain crop rotations. Crop residues are a vast feed resource available to ruminant livestock producers that can effectively reduce feed costs (Lawrence and Strohbehn, 1999). This practice has become an integral part of many cattle operations in the Midwest, but is gaining interest further east. Maize residues can provide four to five animal unit months of grazing per hectare. Grazing crop residues reduces soil surface cover by only 5 to 25% (Clark *et al.*, 2004; Lesoing *et al.*, 1996), so erosion control is not compromised. As with cover crop grazing, animal traffic on crop residues can cause soil compaction near the surface and increased soil surface roughness; however, grazing when soils are dry or frozen, or performing tillage after the winter grazing period, prevents significant detrimental effects on subsequent grain crop yields (Clark *et al.*, 2004; Lesoing *et al.*, 1996). A more intensive approach, providing feed supplementation to dairy heifers corralled on crop residues, resulted in much better capture of excreted nitrogen than spreading similar amounts of manure collected from confined animals (Powell and Russelle, 2009).

22.3.5 Integrated crop–livestock–woodlands

Most agroforestry systems involve grazing perennial forages, and there is great potential for increasing system productivity through intentional integration of trees, forages and livestock (Fike *et al.*, 2004). Spatial arrangement should be designed to optimize benefits to trees, forages and livestock while meeting management needs. With wide tree spacing, annual crops such as maize or soybean can be grown and generate income when trees are too small for grazing animals to be present. In Mississippi, maize has been grazed by cattle in such a system, rather than harvesting by machine (G.B. Triplett, personal communication, 2007). During initial research, accelerated tree growth from fertilizer applied to maize has been observed, which would shorten the tree rotation. This management system improves the potential for weed control in the tree stand with ground equipment and herbicide selection. With increasing maturity of trees, and as annual crops become shaded, perennial forages become better suited as the profitable herbaceous understorey.

Trees integrated with pasture may improve certain ecosystem services compared with treeless pastures. For example, some tree species can remove nutrients from deeper in the soil profiles that would otherwise be lost to water bodies and cause pollution. In Florida, Nair *et al.* (2007) reported that integrating trees and pasture enhanced nutrient retention in the system, thus reducing the chances for nutrient transport to surface water when compared with treeless pastures.

22.3.6 Regional integration

With increasing government regulations for nutrient management, owners of concentrated animal feeding operations (CAFOs) have been driven to develop business relationships with neighbouring farm operators, for the distribution of excess manure to cropland or pasture. Planning approaches for integrating manure management among farms are being pursued. Examples also exist where crop farms grow feed for a neighbouring CAFO and receive manure back from the CAFO for application to their croplands (Russelle *et al.*, 2007). Similar arrangements typify the large broiler chicken operations that have the need to spread floor litter (waste from broiler houses) onto crop and pasture lands of neighbouring farms (Franzluebbers, 2007). In the north-eastern USA, a number of regionally integrated potato–dairy farm operations have developed (Files and Smith, 2001). Even more sophisticated models of among-farm collaboration exist, such as in Manitoba, Canada, where manure was used to fertilize annual grain crops and pasture, grains were processed and utilized by livestock, and cow–calf pairs with replacement heifers were supported on pastures in an integrated neighbourhood of farms (Entz *et al.*, 2005).

Crop–livestock integration among farms provides opportunities to capture some of the economic and environmental benefits of a fully integrated single-farm system. The crop farmer, in particular, has the opportunity to capture the benefits of crop rotation diversity and perenniality, nutrient recycling, and perhaps improved energy efficiency. Several benefits cited by participants in the integration of potato and dairy production in Maine included: improved soil tilth and water-holding capacity, a higher proportion of marketable potatoes, and improved crop yield (Files and Smith, 2001). Development of among-farm integration is enhanced by small distances between farms, existing basic trust between individuals, and a willingness to begin slowly with modest exchanges (Files and Smith, 2001). Russelle *et al.* (2007) pointed out that an advantage of among-farm collaborations is that more people have a stake in assuring successful and mutually acceptable outcomes. However, it is not clear whether those collaborations can achieve the same range of synergies as within-farm integration.

22.4 Stepping Forward

Agricultural policy in the USA during the past century has favoured production of commodity crops for an increasingly industrial agricultural food production system. This system has increased the range of products derived from commodities, added economic value to the processing and distribution chain, and met much of the food and fibre needs of the country's growing population. It has contributed to the well-being of the world's soaring population as well, through international trade of grains (net annual USA export value of bulk agricultural products averaged US$24.6 billion over the last decade) and meat (exports of about US$11.6 billion annually). However, like many industrialized countries, agriculture now represents only a small fraction of the gross domestic product (GDP; Fig. 22.1). This diminishing contribution to the GDP and the general availability of food to most residents have now seemingly marginalized agriculture's importance to the public and to policymakers, at least in terms of funding for research and development of more sustainable agricultural approaches. Change is needed in the type and methods of agricultural production, especially given the increasing constraints of more expensive and insecure fossil fuel supplies, changing and less predictable climate, increasing need for food calories, nutrients and protein by a growing global population, and increasingly serious limits to water quality and quantity.

Future agricultural systems will probably first need to capitalize on the ingenuity, technical and economic capacity within existing agricultural organizations and businesses, while operating within the aforementioned constraints (i.e. by employing an incremental approach). Despite the difficulties of conducting this type of research (Tanaka *et al.*, 2008), a more substantial, transformative approach to increasing the sustainability of a multi-functional, robust food production system must be initiated, with a balance of tested and unique approaches across a diversity of environments (Boody and DeVore, 2006). In a forum discussion of the sustainability of grassland environments, several key messages were identified for moving forward (Steiner and Franzluebbers, 2009), including:

- Difficult times require broad vision and practical solutions.
- Agriculture can deliver environmental services.
- Sustainable agriculture can enhance economic opportunities.
- Market access and economic sustainability are possible with a diversified farming operation.
- Safe food and good health for families and communities can be obtained locally and supported with purchasing power of consumers.

A transformation of agriculture in North America is needed to increase production, mitigate past environmental damage, protect biological diversity, reduce dependence on fossil fuels, provide healthier foods, and increase economic and cultural opportunities in rural America. When incorporated into diverse agricultural systems that include livestock, perennial grasses and legumes, a wide variety of annual forages offer enhanced agro-ecosystem resilience in the face of uncertain climate and market conditions. We concur with the conclusion that 'restoring the balance between livestock density and the nutrient absorptive capacity of the surrounding environment will be central to any strategies aimed at improving the performance of any animal industry' (Powell *et al.*, 2010, p. 136). This change will require technological,

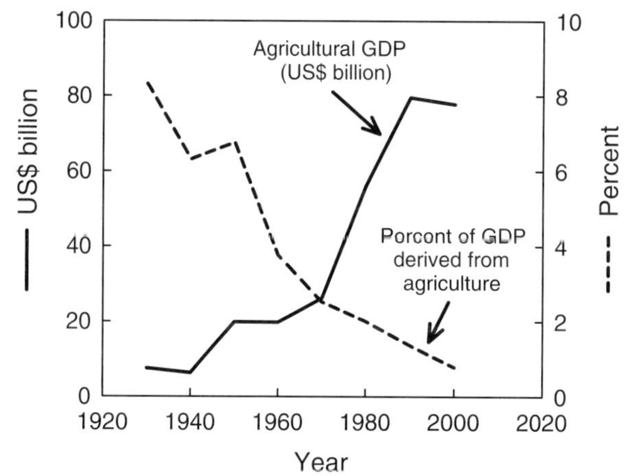

Figure 22.1. Gross domestic product (GDP; US$ billions) and percentage of the GDP derived from agriculture during the 20th century in the USA. From: Bureau of Economic Analysis, www.bea.gov/histdata/NIyear.asp.

financial, regulatory and institutional tools (Powell *et al.*, 2010), and greater skill and resilience on the part of farmers (Hendrickson *et al.*, 2008), but achieving greater integration of crops and livestock will contribute substantially towards agricultural sustainability.

References

Allen, V.G., Brown, C.P., Segarra, E., Green, C.J., Wheeler, T.A., Acosta-Martinez, V. and Zobeck, T.M. (2007) In search of sustainable agricultural systems for the Llano Estacado of the U.S. Southern High Plains. *Agriculture, Ecosystems and Environment* 124, 3–12.

Boody, G. and DeVore, B. (2006) Redesigning agriculture. *Bioscience* 56, 839–844.

Carson, R. (1962) *Silent Spring*. Houghton Mifflin, Boston, MA.

Clark, J.T., Russell, J.R., Karlen, D.L., Singleton, P.L., Busby, W.D. and Peterson, B.C. (2004) Soil surface property and soybean yield response to corn stover grazing. *Agronomy Journal* 96, 1364–1371.

Electronic Code of Federal Regulations (2010) *Title 40-Protection of the Environment. Part 80–Regulation of Fuels and Fuel Additives. Subpart M-Renewable Fuel Standard.* http://ecfr.gpoaccess.gov (accessed 31 August 2010).

Entz, M.H., Bellotti, W.D., Powell, J.M., Angadi, S.V., Chen, W., Ominski, K.H. and Boelt, B. (2005) Evolution of integrated crop–livestock production systems. In: McGilloway, D.A. (ed.) *Grassland: a Global Resource*. Wageningen Academic Publishers, Wageningen, The Netherlands, pp. 137–148.

Faé, G.S., Sulc, R.M., Barker, D.J., Dick, R.P., Eastridge, M.L. and Lorenz, N. (2009) Integrating winter annual forages into a no-till corn silage system. *Agronomy Journal* 101, 1286–1296.

Fike, J. H., Buergler, A.L., Burger, J.A. and Kallenbach, R.L. (2004) Considerations for establishing and managing silvopastures. *Forage and Grazinglands* doi:10.1094/FG-2004-1209-01-RV.

Files, A.C. and Smith, S.N. (2001) *Agricultural Integration: Systems in Action. Maine Agriculture Center Publication 002*. University of Maine, Orono, ME. http://www.umaine.edu/integratedfarming/IFAFS%20 Stuff/Integration%20Report.pdf (accessed 31 August 2010).

Franzluebbers, A.J. (2007) Integrated crop–livestock systems in the southeastern USA. *Agronomy Journal* 99, 361–372.

Franzluebbers, A.J. and Stuedemann, J.A. (2007) Crop and cattle responses to tillage systems for integrated crop–livestock production in the Southern Piedmont, USA. *Renewable Agriculture and Food Systems* 22, 168–180.

Franzluebbers, A.J. and Stuedemann, J.A. (2008a) Early response of soil organic fractions to tillage and integrated crop–livestock production. *Soil Science Society of America Journal* 72, 613–625.

Franzluebbers, A.J. and Stuedemann, J.A. (2008b) Soil physical responses to cattle grazing cover crops under conventional and no tillage in the Southern Piedmont USA. *Soil and Tillage Research* 100, 141–153.

Gardner, J.C. and Faulkner, D.B. (1991) Use of cover crops with integrated crop-livestock production systems. In: Hargrove, W.L. (ed.) *Cover Crops for Clean Water*. Soil and Water Conservation Society, Ankeny, IA, pp. 185–191.

Halloran, J.M. and Archer, D.W. (2008) External economic drivers and US agricultural production systems. *Renewable Agriculture and Food Systems* 23, 296–303.

Hanson, J.D. and Hendrickson, J.R. (2009) Toward a sustainable agriculture. In: Franzluebbers, A.J. (ed.) *Farming with Grass: Achieving Sustainable Mixed Agricultural Landscapes*. Soil and Water Conservation Society, Ankeny, IA, pp. 26–36.

Hargrove, W.L. (ed.) (1991) *Cover Crops for Clean Water*. Soil and Water Conservation Society, Ankeny, IA.

Hendrickson, J., Sassenrath, G.F., Archer, D., Hanson, J. and Halloran, J. (2008) Interactions in integrated US agricultural systems: the past, present and future. *Renewable Agricultural and Food Systems* 23, 314–324.

Hoagland, L., Hodges, L., Helmers, G.A., Brandle, J.R. and Francis, C.A. (2010) Labor availability in an integrated agricultural system. *Journal of Sustainable Agriculture* 34, 532–548.

Ikerd, J.E. (2009) Current status and future trends in American agriculture: farming with grass. In: Franzluebbers, A.J. (ed.) *Farming with Grass: Achieving Sustainable Mixed Agricultural Landscapes*. Soil and Water Conservation Society, Ankeny, IA, pp. 12–25.

Katsvairo, T.W., Wright, D.L., Marois, J.J., Hartzog, D.L., Rich, J.R. and Wiatrak, P.J. (2006) Sod–livestock integration into the peanut-cotton rotation: a systems farming approach. *Agronomy Journal* 98, 1156–1171.

Kenner, R. (2008) *Food, Inc.* (A documentary film). Perfect Meal, LLC, Los Angeles, CA.

Kirschenmann, F. (2009) Reconsidering grass: sustainable new farm designs. In: Franzluebbers, A.J. (ed.) *Farming with Grass: Achieving Sustainable Mixed Agricultural Landscapes.* Soil and Water Conservation Society, Ankeny, IA, pp. 1–11.

Lal, R. (2001) Thematic evolution of ISTRO: transition in scientific issues and research focus from 1955 to 2000. *Soil and Tillage Research* 61, 3–12.

Lawrence, J.D. and Strohbehn, D.R. (1999) *Understanding and Managing Costs in Beef Cow–Calf Herds.* Iowa Beef Center, Iowa State University, Ames, IA. www.econ.iastate.edu/faculty/lawrence/Acrobat/IRMWhitePaper.pdf (accessed 31 August 2010).

Lesoing, G., Shain, D., Klopfenstein, T., Gosey, J. and Schroeder, M. (1996) Effect of sorghum and cornstalk grazing on crop production. In: *Nebraska Beef Cattle Report,* University of Nebraska Coop. Ext. MP 66-A, University of Nebraska, Lincoln, NE. http://beef.unl.edu/beefreports/199616.shtml (accessed 31 August 2010).

Mauget, S., Zhang, J. and Ko, J. (2009) The value of ENSO forecast information to dual-purpose wheat production in the U.S. Southern High Plains. *Journal of Applied Meteorology and Climatology* 48, 2100–2117.

Nair, V.D., Nair, P.K.R., Kalmbacher, R.S. and Ezenwa, I.V. (2007) Reducing nutrient loss from farms through silvopastoral practices on coarse-textured soils of Florida, USA. *Ecological Engineering* 29, 192–199.

NRC (National Research Council) (2010) *Toward Sustainable Agricultural Systems in the 21st Century,* Committee on Twenty-First Century Systems Agriculture, Board on Agriculture and Natural Resources, National Academies Press, Washington, DC.

Pollan, M. (2006) *The Omnivore's Dilemma: a Natural History of Four Meals,* The Penguin Press, New York, NY.

Powell, J.M. and Russelle, M.P. (2009) Dairy heifer management impacts manure N collection and cycling through crops in Wisconsin, USA. *Agriculture, Ecosystems and Environment* 131, 170–177.

Powell, J.M., Russelle, M.P. and Martin, N.P. (2010) The United States. Trends in the dairy industry and their implications for producers and the environment. In: Gerber, P., Mooney, H.A., Dijkman, J., Tarawali, S. and de Haan, C. (eds) *Livestock in a Changing Landscape. Vol. 2 Experiences and Regional Perspectives.* Island Press, Washington, DC, pp. 115–139.

Redmon, L.A., Horn, G.W., Krenzer, E.G. Jr and Bernardo, D.J. (1995) A review of livestock grazing and wheat grain yield: boom or bust? *Agronomy Journal* 87, 137–147.

Russelle, M.P., Entz, M.H. and Franzluebbers, A.J. (2007) Reconsidering integrated crop–livestock systems in North America. *Agronomy Journal* 99, 325–334.

Schiere, J.B., Ibrahim, M.N.M. and van Kuelen, H. (2002) The role of livestock for sustainability in mixed farming: criteria and scenario studies under varying resource allocation. *Agriculture, Ecosystems and Environment* 90, 139–153.

Singer, J.W., Franzluebbers, A.J. and Karlen, D.L. (2009) Grass-based farming systems: soil conservation and environmental quality. In: Wedin, W.F. and Fales, S.L. (eds) *Grassland: Quietness and Strength for a New American Agriculture.* American Society of Agronomy, Crop Science Society of America, and Soil Science Society of America, Madison, WI, pp. 121–136.

Steiner, J.L. and Franzluebbers, A.J. (2009) Farming with grass – for people, for profit, for production, for protection. *Journal of Soil and Water Conservation* 64, 75A–80A.

Steiner, J.L., Franzluebbers, A.J. and Neely, C.L. (2009) Expanding horizons of farming with grass. In: Franzluebbers, A.J. (ed.) *Farming with Grass: Achieving Sustainable Mixed Agricultural Landscapes.* Soil and Water Conservation Society, Ankeny, IA, pp. 216–234.

Sulc, R.M. and Tracy, B.F. (2007) Integrated crop–livestock systems in the U.S. Corn Belt. *Agronomy Journal* 99, 335–345.

Tanaka, D.L., Karn, J.F. and Scholljegerdes, E.J. (2008) Integrated crop/livestock systems research: practical research considerations. *Renewable Agriculture and Food Systems* 23, 80–86.

Taylor, K.W., Epplin, F.M., Prorsen, B.W., Fieser, B.G. and Horn, G.W. (2010) Optimal grazing termination date for dual-purpose winter wheat production. *Journal of Agricultural and Applied Economics* 42, 87–103.

Tilman, D., Cassman, K.G., Matson, P.A., Naylor, R. and Polasky, S. (2002) Agricultural sustainability and intensive production practices. *Nature* 418, 671–677.

Tracy, B.F. and Zhang, Y. (2008) Soil compaction, corn yield response, and soil nutrient pool dynamics within an integrated crop–livestock system in Illinois. *Crop Science* 48, 1211–1218.

Triplett, G.B. Jr and Dick, W.A. (2008) No-tillage crop production: a revolution in agriculture! *Agronomy Journal* 100, S153–S165.

USDA-NASS (National Agricultural Statistics Service) (2009) *2007 Census of Agriculture Report.* United States Department of Agriculture, Washington, DC. http://www.agcensus.usda.gov/ (accessed 23 August 2010).

23 Integration of Grasslands within Crop Systems in South America

Paulo C. de Faccio Carvalho and Anibal de Moraes

23.1 Introduction

It is now recognized that intensive agriculture has led to many concerns related to loss of biodiversity, nutrient pollution and habitat fragmentation. They are the consequences of crop and livestock production systems that became increasingly specialized after the green revolution. The consumers' reaction to this has resulted in environmental regulations, which are becoming a crucial aspect of farming systems. Thus, production is no longer the sole objective, and the dilemma of production versus conservation emerges.

The world population is now demanding farming systems that are more sustainable and environmentally friendly. In this context, integrated crop–livestock systems (ICLSs) are regaining their importance, since they provide opportunities for diversification of rotations, nutrient recycling and greater energy efficiency (Entz *et al.*, 2005).

ICLSs are unique farming systems in which high levels of productivity can be reached with reasonable respect of the environmental requirements. When ICLSs are proposed within other conservation technologies (e.g. no-till), the level of synergism reached by these systems can provide ecosystem services that ensure true sustainable development (FAO, 2010). Moreover, when grasslands are integrated within crop systems, there is a reduction in the risk associated with the crop operation (Carvalho *et al.*, 2005), since livestock production seems more resilient to climate hazards and variability in economic return. While contributing to strengthening the economics of farms based on crop operations, grasslands and grazing herbivores can incorporate diversification in nutrient-cycling pathways and many other potential benefits in ICLSs (Carvalho *et al.*, 2010).

In this chapter we describe the most promising ICLS being adopted in South America, discussing the rationale and capacity to promote sustainable development. We focus on grain crops integrated with domestic grazing ruminants reared on pasture-based systems. We consider only integration at the individual farm level. While integration on an area-wide basis occurs in South America, it is rarely planned in the conceptual basis of system integration.

23.2 Integrated Crop–Livestock System Models in South America

The integration of crops and livestock is not a new technology; rather, it is a re-emerging concept. Since the domestication of plants and animals, there is

 Grassland Productivity and Ecosystem Services (eds G. Lemaire, J. Hodgson and A. Chabbi)

evidence that the ICLS was the most common pattern in the Neolithic era, when humans first gathered into small villages and farmstead groups. According to Russelle *et al.* (2007), crop production was probably first combined with animal husbandry 8 to 10 millennia ago.

In South America, ICLSs primarily involve the rotation of crops and pastures in the same area, at the individual farm level. The statistics of such systems are difficult to ascertain (estimated at 6 million ha in Brazil), but their importance and expansion are broadly recognized. ICLSs were originally used to establish pastures in a rotational sequence beginning with a grain crop, usually rice (*Oryza sativa* L.), to take advantage of the increased fertility in the short term after clearing forested land (Entz *et al.*, 2005). Recently, ICLSs have been conceived as a means for reclaiming pastures degraded by overstocking and lack of fertility, which improves productivity through land-use intensification and mitigates the clearing of native vegetation, particularly in the Cerrados and Amazon regions (Landers, 2007; Zimmer *et al.*, 2004). In those integrated systems, grain crops established on the degraded pasturelands provide the cash flow necessary for the substantial investment in lime and fertilizer needed to correct the soil fertility status. Annual (*Sorghum bicolor* L. Moench, *Pennisetum glaucum* (L.) R. Br.) and perennial (*Brachiaria* spp., *Panicum* spp.) forages are often used in rotation with soybean (*Glycine max* L.), maize (*Zea mays* L.), cotton (*Gossypium* L.), sunflower (*Helianthus annuus* L.) and bean (*Phaseolus vulgaris* L.).

The subtropical Brazilian region has around 8.0 million ha annually cultivated with soybean, maize and rice. Only around 3.5 million ha have been cultivated in rotation with winter crops such as wheat (*Triticum aestivum* L.), oat (*Avena sativa* L.), barley (*Hordeum vulgare* L.), triticale (× *Triticosecale* Witt.) and rye (*Secale cereale* L.). The remaining area represents potential income lost during winter, with soils being exposed (no cover) or simply seeded to cover crops. The cover crops used are primarily forage species, but they are rarely grazed. Hence, there is a vast area that could potentially integrate livestock grazing on winter cover crops in rotation with summer crops under no-till management (Carvalho *et al.*, 2010).

Annual crop rotations typical of southern Brazil have alternative forage species defined according to the cash crop cycle. Black oat (*Avena strigosa*) is commonly used as the preceding crop to maize, because its early maturity fits well with early planting dates required for full-season maize, whereas oat and/or Italian ryegrass (*Lolium multiflorum* Lam) are often used preceding soybean, which is planted later than maize. Italian ryegrass has the potential to perennate in those systems by annual reseeding (Carvalho *et al.*, 2005). With regard to legume utilization, *Vicia* spp. have been used with oats in rotation with maize, aiming to increase soil nitrogen availability, whilst *Trifolium* spp. and *Lotus* spp. are most commonly seen in rotation with irrigated rice, because rice yield after those species can be equivalent to, or even greater than, rice fertilized with 90 kg/ha of nitrogen (Saibro and Silva, 1999).

There is some conflict over how much residue cover is needed for no-till establishment of crops following pasture cover that is grazed. In the pasture phase, the grazing animal consumes the above-ground biomass, which is the same biomass vital to the health and functioning of the no-till crop production system. The positive linear relationship between residual biomass of the preceding cover crop and yield of the succeeding crop (Landers, 2007) in no-till systems does not encompass pastures being used by grazing animals, which has been cause for debate. This technical dilemma, together with the concern that grazing animals compact the soil, generates resistance to the adoption of grazing within no-till production areas in Brazil (Carvalho *et al.*, 2007). Probably the most common perception of farmers, who are hesitant to participate in integrated systems, is that cattle trampling has a negative effect on soil physical properties. This has proven to be a major obstacle to the adoption of ICLS, despite studies that refute this claim (Moraes *et al.*, 2002; Flores *et al.*, 2007). Quite on the contrary, an ICLS using moderate grazing intensity has proven to promote many soil chemical and physical attributes, having either no effect on, or even increasing crop yield (Carvalho *et al.*, 2010).

23.3 Benefits of Integrated Crop–Livestock Systems

Recognizing that 9 billion people will need to be provided for in a sustainable way by the year 2050, FAO (2010) promoted a debate to discuss whether ICLSs could promote food and nutritional security. A consensus was reached that ICLSs can meaningfully promote sustainable development through improved production intensification, environmental quality and livelihoods. Three main aspects provided the basis for consensus: (i) the biological perspective: ICLSs increase environmental resilience through increased biological diversity, effective/efficient nutrient cycling/recycling, improved soil health, providing ecosystem services, enhanced forest preservation, and contributing to adaptation and mitigation of climate change; (ii) the economic perspective: ICLSs enhance livelihood diversification and potential efficiency through optimization of production inputs including labour, offer resilience to economic stresses, and reduce risks; and (iii) the socio-cultural perspective: ICLSs are meant to assist farmers to diversify and meet their livelihood aspirations, ensure equitable social dynamics, particularly for elders, women and youth, and increase nutrition security and food safety while meeting consumer choice and demand. As a consequence of the broad applicability of ICLS benefits, FAO decided to host a global partnership platform and give institutional support to promote, advance and scale up the ICLS.

The success of ICLSs depends on how the different components of the system are managed. For example, grazing animals can positively affect carbon sequestration, but it depends on grazing intensity. Franzleubbers (2010) reported a 5-year study on Coastal Bermudagrass (*Cynodon dactylon* (L.) Pers.), established following conventionally tilled cropland. Soil organic carbon accumulated with moderate grazing compared with no cattle grazing, but declined at higher stocking rates. Souza *et al.* (2009) reported similar results in a 7-year soybean/black oat + Italian ryegrass succession. Annual carbon accumulation rate in the ICLS was 0.6, 1.4 and 1.3 Mg/ha for pastures managed at 10 and 20 cm height or in the absence of grazing, respectively. Higher grazing intensities not only diminished the potential of ICLS to sequester carbon, but also resulted in losses of nitrogen and particulate organic matter (Souza *et al.*, 2008; 2009).

Concerning the effect of grazing method used during the pasture phase, there are very few reports about its consequences in ICLSs. Carvalho *et al.* (2010) reported the results from an ICLS model based on a lamb-fattening operation using Italian ryegrass in the winter pasture phase in rotation with maize or soybean in summer. Classical results regarding animal production were observed, continuous stocking allowing for greater individual animal daily gains and rotational stocking yielding more gain per unit area. While the animal performance response to grazing intensity and method followed the expected classical patterns, the impact of grazing treatments on the succeeding crop was original information. Grazing intensity clearly affected soybean yield more than the grazing method (Carvalho *et al.*, 2010). The higher the herbage allowance (5 times lamb potential intake), the more biomass residue cover was left after the pasture phase for the succeeding crop, resulting in higher soybean yield compared with lower herbage allowance (2.5 times lamb potential intake), which produced only 70% of the grain yield. When pasture cover is not grazed and serves only as residue for no-till, then soybean yield is only 55% of that registered at moderate grazing

(Carvalho *et al.*, 2010). In maize, the effect of grazing intensity and method is less evident, although maize yield tended to be slightly higher with high herbage allowance and continuous stocking. No evidence was found of grazing method effects and it is noteworthy that the non-grazed treatments yielded similarly or even less than the grazed treatments (Carvalho *et al.*, 2007).

After a 9-year soybean/black oat + Italian ryegrass rotation with grazing steers (Carvalho *et al.*, unpublished results), where the pasture phase was managed with different grazing intensities corresponding to different sward heights (10, 20, 30 and 40 cm), average soybean yield in the rotation with absence of grazing was similar (2.95 Mg/ha) to that registered under light grazing (3.02 and 2.87 Mg/ha, respectively for 40 and 30 cm sward height). Soybean yield in treatments with swards managed at 20 and 10 cm was slightly lower (2.67 and 2.61 Mg/ha, respectively), but was largely compensated for by annual average animal production (495, 434, 295 and 166 kg live weight/ha, respectively for 10, 20, 30 and 40 cm sward height). These results highlight the purpose of ICLSs, aiming for higher yields per unit area, more than the highest yield in each crop rotation.

The well-managed pasture phase is recognized as being fundamental in determining the level of carbon sequestration in ICLSs. Russelle *et al.* (2007) stated that multiple agronomic and environmental benefits can be realized when land is converted from low diverse cropping systems to rotations that include forages. The longer this phase is between successions, the higher are the positive benefits (Díaz-Rosselo, 2006). Lemaire *et al.* (2003) reported that pastures have analogous effects to forests and can help agricultural systems regulate environmental fluxes to achieve multiple environmental benefits. The main positive effects of pastures include hydrological impacts and maintenance of surface- and ground-water quality, carbon sequestration, nitrogen flux regulation, gas emission regulation, organic matter stability, and soil quality maintenance, among others.

The effect of different pastures in rotation with crops on soil organic carbon content was studied in a long-term experiment in Uruguay. Four rotation systems, differing in crop sequence, pasture type (legume and non-legume) and pasture duration were monitored from 1964 to 2003 (Morón, 2003; Díaz-Rossello, 2006). The annual soil loss of continuous cropping was 0.4 Mg/ha, which represented a 46% reduction of the original soil carbon content after 40 years (Fig. 23.1).

After 3 years of a mixed pasture of white clover (*Trifolium repens*), birdsfoot trefoil (*Lotus corniculatus*) and fescue (*Festuca arundinacea*), the soil organic carbon (SOC) increased by about 0.3%, which represented 7.5 Mg/ha. Díaz-Rosselo (2006) reported how this fresh pool of SOC was so susceptible to oxidation that it decreased at an annual rate close to 2.5 Mg/ha during the following 3-year phase of annual crops. Consequently, the organic C remained in equilibrium under a 6-year complete cycle of pastures and annual crops (Díaz-Rossello, 2006).

ICLSs were fully adopted by farmers on both rain-fed crop and irrigated rice production systems during the last few decades in Uruguay (Díaz-Rossello, 2006). Survey reports indicate almost 90% of grain crops are now planted in soils in which a legume pasture has been ploughed in one of the four preceding years. The author considers it took only 20 years to change completely to this new production system. Considering the complexity of this technology in relation to the adoption of some trouble-free innovations, such as new cultivars or agrochemicals, the speed of the production change was remarkable (Díaz-Rossello, 2006).

Considering the impact of changing land use, Carvalho (2010) studied the dynamics of carbon accumulation in Amazonia and Cerrado's biomes in Brazil. The author demonstrated that even no-till systems are not sufficient to maintain soil carbon stocks if native vegetation is converted to agriculture, recording annual

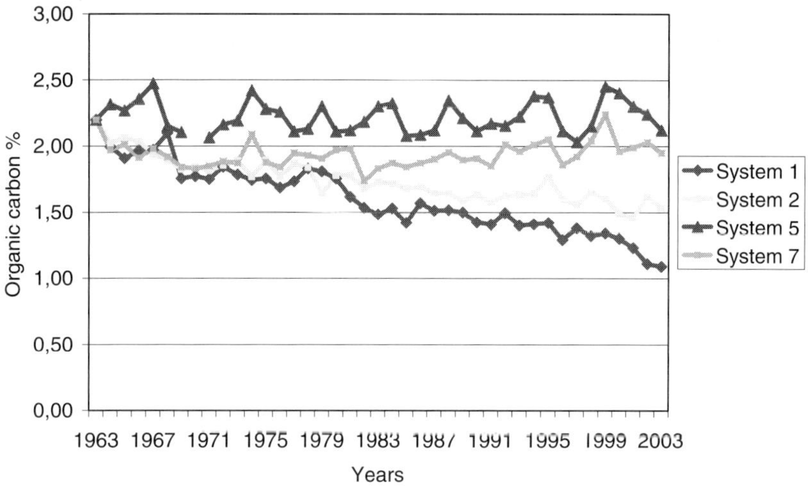

Fig. 23.1. Organic carbon dynamics in different rotation systems from 1963 to 2003 (Díaz-Rossello, 2006). System 1, continuous cropping without N and P; System 2, continuous cropping with N and P; System 5, 3-year cropping and 3-year pasture rotation; System 7, 1-year pasture and 2-year grain field crops.

carbon losses from 0.69 to 1.44 Mg/ha (Carvalho, 2010). However, the implementation of ICLSs in such areas increased annual soil carbon stocks from 0.82 to 2.58 Mg/ha. In general, no-till systems increase soil carbon stocks compared with conventional tillage. Franzluebbers (2010) reported results in the south-eastern USA comparable with those reported by Carvalho (2010) in Brazil (0.45 and 0.50 Mg/ha/year, respectively). Despite this potential of no-till technology, when it is integrated within pasture rotations under the effect of moderate grazing, the carbon sequestration and other potential benefits are much higher, since emerging properties become apparent.

23.4 Emerging Properties in Integrated Crop–Livestock Systems

Anghinoni *et al.* (2010) argued that an ICLS based on no-till is a biological system that encompasses several entities that together allow for the development of emergent properties. According to the authors, the grazing animal is the dynamic agent that produces continuous variability in the system, by the acts of grazing, trampling and excreta distribution. The soil component is the compartment that captures the consequences (+ or –) of the main evolving processes (grazing, crop rotations, residues, fertilization, machinery impact, etc.), and plants react, signalling the direction (+ or –) of those consequences.

Flores *et al.* (2008) give an example of these processes. High levels of toxic Al^{3+} constrain soybean yield, and such a situation is common in no-till systems where lime is spread over the soil. Comparing this system with another where the only difference was the presence of the grazing animal during a pasture phase, Flores *et al.* (2008) reported a lime reaction in depth of 2.5 cm/year for the non-integrated no-till system, and 5.0 cm/year in the ICLS. Ten years after lime application, soybean yield appeared to be insensitive to the increasing level of Al^{3+}, thus Anghinoni *et al.* (2010) argued that soil fertility and the classical concept of soil richness assessed by mineral concentration were not enough to characterize the yield potential of their systems. According to the authors, the classical concept is based on a reductionist, compartmentalized cause–effect relationship, so it fails to capture the holistic nature of complex ICLSs.

In this context, data from Chávez *et al.* (2010) indicate ICLS can impact ecosystem functioning. Figure 23.2 demonstrates how microbial functional diversity is enhanced by moderate grazing intensities, compared with high grazing intensity or the absence of grazing, in a soybean/oat + Italian ryegrass rotation.

The Shannon index varied from 6.9 to 7.4 after a 7-year rotation in the absence of grazing and in pastures managed at 30 cm height, respectively. The results support the hypothesis that moderate grazing in an ICLS can provide biodiversity, leading to more channels in the soil food web that stabilize the flow of energy and nutrients (Whalen and Sampedro, 2009). Since intensive farming systems in temperate regions record Shannon indexes from ~2.0 to 4.5 (Papatheodorou *et al.*, 2008; Jacinthe *et al.*, 2010), it seems ICLSs can contribute to ecological diversity and impact ecosystem functioning by determining the nature and magnitude of organic substrates' mineralization and turnover. Souza *et al.* (2010) showed that not only microbial diversity but also microbial biomass was greater in ICLS.

Tracy (2007) presented a hypothesis whereby ICLSs are a combination of developing and mature ecosystems. He developed an analogy to the principles proposed by E.P. Odum in his classical paper, *The Strategy of Ecosystem Development* (Odum, 1969). According to Tracy (2007), developing ecosystems are analogous to modern cropping systems because they are highly productive and biologically simple, but generally unstable and leaky from a mineral-cycling perspective. Because of their diversity though, mature systems are generally more stable and retentive of soil nutrients compared with developing systems. Although less productive, mature ecosystems provide many valuable ecosystem services on which we depend. Tracy (2007) argues that ICLSs contain a combination of elements of both developing and mature ecosystems, so they provide an ecological compromise between a highly productive agriculture system and an environmentally protective ecosystem. High productivity comes from annual grain crop production (developing ecosystem), while ecosystem services are provided by interactions among perennial crops, annual

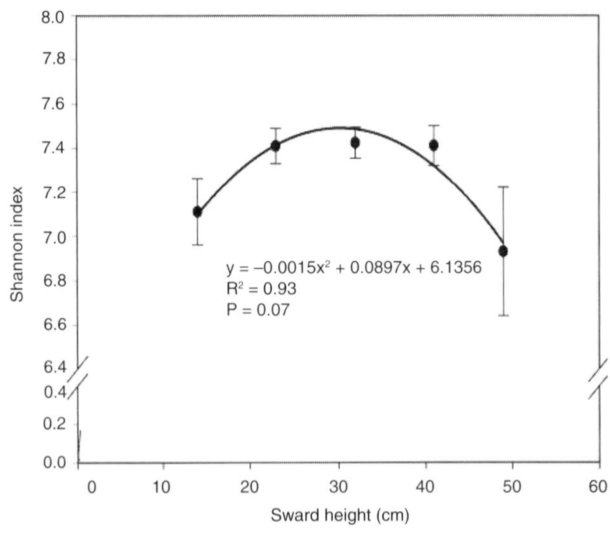

Fig. 23.2. Shannon diversity index based on the ability of microorganisms in the soil to use carbon substrates, as a function of grazing intensity in a soybean/oat + Italian ryegrass rotation. Sward heights between 10 and 40 cm were achieved by continuous variable stocking with steers. Sward height of 50 cm corresponds to the non-grazed treatment (Chávez *et al.*, 2011).

cover crops and livestock (mature ecosystem).

In essence, Tracy (2007) pointed out that modern agriculture has become diametrically opposed to the way nature, or ecosystems, develop. In this sense, Kirschenmann (2007) argued for the need for exploitation of potential biological synergies in modern agriculture, in order to face environmental challenges. Synergism is one of the main features of ICLSs, as discussed by Carvalho *et al.* (2010). Besides that, Kirschenmann (2007) argued for the development of models of farming in nature's image, evoking the concept of biomimicry. Thus, the more the farming systems reproduce nature, the more they are resilient and contribute to ecosystem services. We believe that ICLSs based on no-till, where grasslands and grazing animals are integrated with grain crops rotating in between trees, reach a level of diversity that mimics the main processes and fluxes occurring in nature (relative to specialized crop or livestock operations).

References

Anghinoni, I., Carvalho, P.C.F., Moraes A., Souza, E. D., Conte, O. and Lang C.R. (2011) Benefícios da integração lavoura-pecuária sobre a fertilidade do solo em sistema plantio direto. In: Fonseca, A.F. (ed.) *Fertilidade do Solo em Sistema Plantio Direto.* Editoria INPAG, Ponta Grossa, Brazil.

Carvalho, J.L.N. (2010) Dinâmica do carbono e fluxo de gases do efeito estufa em sistemas de integração lavoura-pecuária na Amazônia e no Cerrado. PhD thesis, Universidade de São Paulo, Brazil.

Carvalho, P.C.F., Anghinoni, I., Moraes, A., Trein, C.R., Flores, J.P.C., Cepik, C.C.T., Levien, R., Lopes, M.T., Baggio, C., Lang, C.R, Sulc, R.M. and Pelissari, A. (2005) O estado da arte em integração lavoura-pecuária. In: Gottschall, C. S., Silva, J.L.S. and Rodrigues, N.C. (eds) *Produção Animal: Mitos, Pesquisa e Adoção de Tecnologia.* Universidade Luterana do Brasil, Canoas, Brazil, pp. 7–44.

Carvalho, P.C.F., Silva, J.L.S., Moraes, A., Fontaneli, R.S., Macari, S., Bremm, C. and Trindade, J.K. (2007) Manejo de animais em pastejo em sistemas de integração lavoura–pecuária. In: Carvalho, P.C.F., Moraes, A. and Sulc, R.M. (eds.). *Proceedings of the International Symposium on Integrated Crop–Livestock Systems.* Universidade Federal do Paraná, Curitiba, Brazil. CD ROM.

Carvalho, P.C.F., Anghinoni, I., Moraes, A., Souza, E.D., Sulc, R.M., Lang, C.R., Flores, J.P.C., Lopes, M.L.T., Silva, J.L.S., Conte, O., Wesp, C.L., Levien, R., Fontaneli, R.S. and Bayer, C. (2010) Managing grazing animals to achieve nutrient cycling and soil improvement in no-till integrated systems. *Nutrient Cycling in Agroecosystems* 88, 259–273.

Chávez, L.F., Escobar L.F., Anghinoni I., Carvalho, P.C.F. and Meurer, E.J. (2011) Diversidade metabólica e atividade microbiana em sistema de integração agricultura–pecuária em plantio direto sob intensidades de pastejo. *Pesquisa Agropecuária Brasileira.*

Díaz-Rossello, R. (2006) La intensificación agrícola en el Cono Sur y los desafíos de la sostenibilidad. In: *Aportes de la Ciencia y Tecnología al Manejo Productivo y Sustentable de los Suelos del Cono Sur.* IICA/PROCISUR, pp. 11–20.

Entz, M.H., Bellotti, W.D., Powell, J.M., Angadi, S.V., Chen, W., Ominski, K.H. and Boel, B. (2005) Evolution of integrated crop-livestock production systems. In: McGilloway, D.A. (ed.) *Grassland: a Global Resource.* Wageningen Academic Publishers, Wageningen, The Netherlands, pp. 137–148.

FAO (2010) *Sete Lagoas "Consensus" on Integrated Crop Livestock Systems for Sustainable Development.* Plant Production and Protection Division Consultation Documents. www.fao.org/agriculture/crops/core-themes/spi/iclsd/outcome.

Flores, J.P.C., Anghinoni, I., Cassol, L.C., Carvalho, P.C.F., Leite, J.G.B. and Fraga, T.I. (2007) Atributos físicos e rendimento de soja em sistema plantio direto em integração lavoura–pecuária com diferentes pressões de pastejo. *Revista Brasileira de Ciência do Solo* 31, 771–780.

Flores, J.P.C., Cassol, L.C., Anghinoni, I. and Carvalho, P.C.F. (2008) Atributos químicos do solo em função da aplicação superficial de calcário em sistema de integração lavoura–pecuária submetido a pressões de pastejo. *Revista Brasileira de Ciência do Solo* 32, 2385–2396.

Franzluebbers, A.J. (2010) Achieving soil organic carbon sequestration with conservation agricultural systems in the Southeastern United States. *Soil Science Society of America Journal* 74, 347–357.

Jacinthe, P., Bills, J.S. and Tedesco, L.P. (2010) Size, activity and catabolic diversity of the soil microbial biomass in a wetland complex invaded by reed canary grass. *Plant and Soil* 329, 227–238.

Kirschenmann, F.L. (2007) Potential for a new generation of biodiversity in agroecosystems of the future. *Agronomy Journal* 99, 373–376.
Landers, J.N. (2007) Tropical crop–livestock systems in conservation agriculture: the Brazilian experience. *Integrated Crop Management* 5, 1–92.
Lemaire, G., Benoit, M. and Vertès, F. (2003) Rechercher de nouvelles organisations à l'échelle d'un territoire pour concilier autonomie protéique et préservation de l'environment. *Fourrages* 175, 303–318.
Moraes A., Pelissari A., Alves S.J., Carvalho, P.C.F. and Cassol, L.C. (2002) Integração lavoura–pecuária no Sul do Brasil. In: Mello, N.A. and Assmann, T.S. (eds). *Proceedings of Integração Lavoura–pecuária no Sul do Brasil.* Centro Federal de Educação Tecnológica, Pato Branco, Brazil, pp. 3–60.
Morón, A. (2003) *Principales Contribuciones del Experimento de Rotaciones Cultivos-pasturas de INIA La Estanzuela en el Area de Fertilidad de Suelos (1963–2003).* Serie Técnica Nº134. Instituto Nacional de Investigación Agropecuaria, Colonia, Uruguay.
Odum, E.P. (1969) The strategy of ecosystem development. *Science* 164, 262–270.
Papatheodorou, E.M., Efthimiadou, E. and Stamou, G.P. (2008) Functional diversity of soil bacteria as affected by management practices and phenological stage of *Phaseolus vulgaris. European Journal of Soil Biology* 44, 429–436.
Russelle, M.P., Entz, M.H. and Franzluebbers, A.J. (2007) Reconsidering integrated crop–livestock systems in North America. *Agronomy Journal* 99, 323–324.
Saibro, J.C. and Silva, J.L.S. (1999) Integração sustentável do sistema arroz x pastagens utilizando misturas forrageiras de estação fria no litoral norte do Rio Grande do Sul. In: Gottschall, C., Silva, J.L.S. and Rodrigues, N.C. (eds) *IV Ciclo de Palestras em Produção e Manejo de Bovinos de Corte.* Editoria da ULBRA, Canoas, Brazil, pp. 27–56.
Souza, E.D., Costa, S.E.V.G.A., Lima, C.V.S., Anghinoni, I., Meurer, E.J. and Carvalho, P.C.F. (2008) Carbono orgânico e fósforo microbiano em sistemas de integração agricultura–pecuária submetidos a intensidades de pastejo em plantio direto. *Revista Brasileira de Ciência do Solo* 32, 1273–1282.
Souza, E.D., Costa, S.E.V.G.A., Anghinoni, I., Carvalho, P.C.F., Andrigueti, M. and Cao, E. (2009) Estoques de carbono orgânico e de nitrogênio no solo em sistema de integração lavoura–pecuária em plantio direto submetido a intensidades de pastejo. *Revista Brasileira de Ciência do Solo* 33, 1829–1836.
Souza, E.D., Costa, S.E.V.G.A., Anghinoni, I., Lima, C.V.S., Carvalho, P.C.F. and Martins, A.P. (2010) Biomassa microbiana do solo em sistema de integração lavoura–pecuária em plantio direto, submetido a intensidades de pastejo. *Revista Brasileira de Ciência do Solo* 34, 79–88.
Tracy, B. (2007) The ecology of integrated crop–livestock systems. In: Carvalho, P.C.F., Moraes, A. and Sulc, R.M. (eds.). *Proceedings of the International Symposium on Integrated Crop–Livestock Systems.* Universidade Federal do Paraná, Curitiba, Brazil. CD ROM.
Whalen, J.K. and Sampedro, L. (2009) *Soil Ecology and Management.* CAB International, Wallingford, UK.
Zimmer, A.H., Macedo, M.C.M. and Kichel, A.N. (2004) Integrated Agropastoral Production Systems. In: Guimarães E.P. (ed.) *Agropastoral Systems for the Tropical Savannas of Latin America.* CIAT, Cali, Colombia, pp. 253–290.

Part V

Grasslands as Components of Landscape Ecology and Agricultural Systems at the Regional Level

Editor: Gilles Lemaire

Most of the environmental and ecological impacts of agriculture systems can be characterized and evaluated only at spatial scales larger than the individual farm area. But most of the decisions for management of these systems are driven by economic and social conditions relevant for each individual farm. This mismatch makes the optimization of some environmental and ecological services very difficult to obtain at the level of landscape and region.

The overall objectives of Part V are: (i) to analyse spatial interactions between grassland areas with other land-use and management systems, such as forests and arable crops, and their consequences for regulating environmental fluxes and biodiversity at the landscape level; and (ii) to analyse the importance of the diversity of land-use systems at the landscape and regional level for regulating environmental fluxes and biodiversity, and the role of grasslands as a component of this diversity. The role and impact of grassland areas for regulating water flows and quality of resource at catchment level, in relation to other types of land-use system is analysed in Chapter 24. Interactions of grasslands with forests and other woodlands is analysed in Chapter 25, with emphasis on the use of agro–sylvo–pastoralism as a system for landscape management. Interactions of grasslands with arable cropping systems are analysed, as a means of enhancing the restoration of biodiversity at the landscape level within intensive cereal cropping systems, in Chapter 26. Examples of the integration of livestock systems using grassland areas with cropping systems, for a more sustainable agricultural system at the regional level, are developed in Chapter 27 for arid and semi-arid Mediteranean countries and in Chapter 28 for tropical conditions, with an emphasis on peri-urban systems. These examples demonstrate the necessity for maintaining diversity in land-use systems at the regional level, through associations and synergies between different agriculture production systems.

24 Temperate Grasslands in Catchment Systems: the Role of Scale, Connectivity and Thresholds in the Provision and Regulation of Water Quality and Quantity

Christopher J.A. Macleod and Robert C. Ferrier

24.1 Introduction

There is an increasing need to understand the linkages between biotic and abiotic processes and structures, their resultant patterns and functions at single and multiple scales, and the services that they provide to society regarding the provision and regulation of water in catchment systems (Brauman *et al.*, 2007). Ecological and economic systems are complex sets of components linked by dynamic processes and functions. Ecosystem services provide a framework to assess the benefits to humans from ecological systems (Daily *et al.*, 2000). There is a need to identify and quantify scaling rules that describe the production and delivery of ecosystem services (Limburg *et al.*, 2002) and improve our understanding of the trade-offs between ecosystem services (Bennett *et al.*, 2009).

24.2 Management of Grassland Systems and their Integration at Catchment Level

The management of grassland areas varies from very low levels of stocking in nutrient-poor but species-rich pastures to improved grassland areas that support higher numbers of beef cattle, dairy cattle, sheep, horses and other livestock. The type and management of grasslands are influenced by precipitation, temperature, topography, geology, soil texture, location, and social and macro-economic factors. A common approach is to classify grasslands based on the pH of the soil (acid, neutral or calcareous), their level of intensity (low or high) and their water status (flood meadows/wetlands) (Table 24.1). The wide range of land-use systems allied to grassland land cover leads to non-linear relationships between grassland land cover and its functions (Verburg *et al.*, 2009), resulting in the need for land-use activity data to supplement remotely sensed land-cover data.

Linkages between terrestrial and aquatic ecosystems have long been recognized (Likens and Bormann, 1974) and remain a vibrant area of research (e.g. Wasson *et al.*, 2010). It is increasingly realized that our understanding, observations and models of the land and water components of landscapes and riverscapes need to be more integrated (as reviewed by Allan, 2004). In terrestrial ecology, early work on pattern and process of plant communities has advanced through viewing

 Grassland Productivity and Ecosystem Services (eds G. Lemaire, J. Hodgson and A. Chabbi)

Table 24.1. Examples of the broad-scale service provision by different temperate grassland types.

	Semi-natural grassland				
Service	Acid grassland[a]	Neutral grassland	Calcareous grassland	Intensive grassland	Flood meadow
Provisioning					
Food					
Meat	+	+	++	+++	++
Milk	0	+	+	+++	++
Fibre	+	+	++	+++	+
Regulating					
Water quality	+	+	+	- - -	++
Soil infiltration	+	+	+	+/-	+
Groundwater recharge	0	0	+	++	+++
Flood attenuation	+/-	+	+	+	+++
Supporting					
C Sequestration	++	+	+	+	++
GHG emissions	+/-	+/-		- - -	--
Biodiversity	+/-	+/-	+++	--	+++
Pollination	+/-	+/-	++	-	++
Cultural					
Traditional practices	+/-	+/-	+++	- - -	+++
Recreation	+	+	++	-	+/-

[a] Service provision as judged by the authors. Weak positive = +, medium positive = ++, strong positive = +++, weak negative = -, medium negative = --, strong negative = ---, no clear service provision = 0, +/- = mixed relationships.

these as patches with hierarchical dynamics (Wu and Loucks, 1995). In parallel to these developments in landscape ecology, conceptual models of rivers as linear continuums have evolved to consider the hierarchy of patch dynamics (Thorp *et al.*, 2006). By considering both landscapes and riverscapes as hierarchical patch dynamics, it is possible to construct a new approach to how we understand and manage catchment systems and the components they contain in a more integrated manner. Landscape ecology and catchment hydrology both deal with patterns and processes, as well as their interactions and functional implications, at a variety of scales (Turner and Chapin, 2005). Landscapes are increasingly being referred to as complex adaptive systems in which patterns at higher levels emerge from smaller-scale local interactions (Levin *et al.*, 1997). This complexity arises from abiotic constraints and disturbances interacting with interspecific and intraspecific biotic processes within a hierarchy of spatial and temporal scales (Schroder, 2006). Interdisciplinary approaches with complementary perspectives are required to enable a greater understanding of these complex interactions between biotic and abiotic processes and patterns occurring across temporal and spatial scales (Schroder, 2006).

In this chapter we present our conceptual understanding of the role of multifunctional grassland areas in hierarchical patch dynamics of catchment systems. Biotic and abiotic processes, structures and patterns influence functions at individual and multiple spatial and temporal scales and the cross-scale interactions between them. To understand how we can manage multifunctional grassland areas in catchments requires integration across multiple disciplinary perspectives. We propose a systems perspective that integrates the key concepts of scale, connectivity and thresholds, and their indicators.

24.3 Application of the Ecosystem Service Framework: Grassland Areas in Catchment Systems

The ecosystem service framework enables natural and social scientists to work together and communicate their results with wider stakeholders. Although the MA (Millennium Ecosystem Assessment, 2003) notes broad scale functions such as water quality, hydrological regulation, nutrient cycling and soil erosion regulation, our current knowledge of how grassland management influences these provisioning and regulating services in catchments is skewed towards an understanding at the plant to plot scales.

24.3.1 Importance of a systems approach

There is a growing interest in understanding complex environmental systems from interdisciplinary perspectives, e.g. landscape ecology (Schroder, 2006), systems ecology (Jorgensen *et al.*, 2007) and ecohydrology (Rodriguez-Iturbe, 2000) to name a few. Though there have been large advances in systems-based approaches, e.g. systems dynamics (Forrester, 2007) and ecosystem ecology (Odum, 1953), reductionist scientific methods have dominated during the past 50 years due to the overwhelming complexity of the natural world. This has resulted in the specialization of science and the development of a wide range of scientific disciplines with their own methods, language and norms. These are able to understand individual components of complex hierarchical systems, but have provided little progress on the goal of understanding larger temporal and spatial interactions and system functioning. Preceding chapters in this book set out the detailed disciplinary progress that has been achieved in expanding the depth and breadth of our knowledge of grasslands, from the gene to field scale.

Hydrological catchments are complex systems due to nonlinear and dynamic processes involving biogeochemical, ecological, hydrological and human systems, with interactions and feedbacks resulting in unexpected behaviour at larger scales (Gordon *et al.*, 2008). There is a need to understand how interacting processes have resulted in the structure and function of hydrological systems and how these will change (Wagener, 2007). To understand and manage grasslands as components of catchment systems, there is a need to integrate our understanding from disparate disciplines of molecular science, plant science, ecology, soil science, biogeochemistry, hydrology, landscape ecology and social sciences. To aid this integration, there is a need to consider the scale, connectivity and thresholds of these hierarchical complex adaptive catchment systems, and their indicators.

24.3.2 Importance of scale

In hydrology, the importance of scale and scale-dependent dominant processes is well established (Bloschl and Sivapalan, 1995). Processes important at one scale are not always important at other scales (Sivapalan *et al.*, 2003), and the dominance of processes alters with changing scales (Grayson and Bloschl, 2000). In ecology, it is realized that the observer creates a perceptual bias when making observations at a particular scale. Levin (1992) stated ‘the key to prediction and understanding lies in the elucidation of mechanisms underlying observed patterns. Typically these mechanisms operate at different scales than those on which the patterns are observed.’ Understanding ecosystems in part involves learning how loosely or tightly coupled are processes at different scales (O’Neill *et al.*, 1986). Complex systems contain processes that function at different, and, at times, overlapping space and time scales (Limburg *et al.*, 2002). Stakeholders at diverse spatial scales can have different interests in the ecosystem services provided and their valuations, and that needs to be explicitly captured in ecosystem management plans (Hein *et al.*, 2006).

24.3.3 Importance of connectivity

Cross-scale interactions are processes that occur at one temporal or spatial scale and interact with processes at a different scale, leading to non-linear dynamics with sudden threshold responses (Holling, 1992). Peters *et al.* (2008) set out a conceptual connectivity framework to understand the hierarchy of interacting scales. In hydrology, connectivity is fundamental, since water and the soluble and particulate substances entrained by it move from one location to another based on physical gradients, creating hydrological flow paths in streams and rivers (Amoros and Bornette, 2002) or changes in spatial patterns of contributing source areas (Western *et al.*, 2001). In ecology, the connectivity between organisms and habitat patches is a key to understanding processes and patterns across multiple scales (Cottenie and De Meester, 2003).

24.3.4 Importance of thresholds and their indicators

The importance of thresholds in understanding and managing ecological systems has been clearly demonstrated (Scheffer *et al.*, 2001). Zehe and Sivapalan (2009) suggested there were three types of threshold behaviour in hydrological systems. The first is at the process level, controlled by local forcings and states. The second is the response of systems of intermediate complexity that were controlled by catchment topography interacting with systems states and boundary conditions. The third type includes changes in the 'architecture' of human geo-ecosystems. There is a need to improve and develop indicators that will warn us of thresholds of large declines in ecosystem services (Limburg *et al.*, 2002).

24.3.5 Integration of scale, connectivity and thresholds in the provision of ecosystem services

The services identified in Table 24.1 highlight the important role of grasslands in regulating water provision and quality (in the widest fit-for-use context). Biotic and abiotic processes, structures and their resultant spatial and temporal patterns regulate and provide timely and clean water. In Table 24.2 we set out examples of processes and structures that operate at different spatial scales, leading to spatial and temporal patterns and functions that

Table 24.2. Examples of processes, structures and function for: (A) flood regulation water service and (B) water quality services associated with grassland systems at different spatial scales.

	Increasing spatial scale →				
	Plant	Pedon/plot	Field	Reach	Catchment
A					
Functions	Infiltration	Runoff generation	Drainage	Flood regulation	Water provision and flood regulation
Structures	Roots, soil structure	Botanical composition	Field drains	Channel morphology	Dams
Processes	Infiltration	Infiltration, evaporation		Ground–surface water interactions	Flow routing, abstraction
B					
Functions	Decomposition	Nutrient cycling	Filter, production	Clean water	Water quality
Structures	Soil aggregates	Soil structure and horizons		Connectivity to reach	River network
Processes	Denitrification		Crop production	De-gassing	Routing, nutrient spiralling

provide beneficial services. There are many more processes, structures, patterns and functions that occur in and between grassland patches in catchment systems.

One way to understand the importance of scale, connectivity and thresholds for the provision and regulation of ecosystem services is to imagine two catchments with increasing complexity:

1. *A catchment system with a low level of complexity.* The simplest conceptual catchment system comprises one type of geology, one soil type with one horizon, constant topography, one climate with constant solar inputs, rainfall and temperature, one grass species within a single land-management approach, with regular and equal-sized fields and farms. Within this system there would be a wide range of biotic and abiotic ecological processes, structures and resultant patterns and functions providing intermediate and final services of human benefit. In this simple system, scale interactions would be fundamental to understanding the contribution of the complete grassland coverage on the provision of clean and timely water at the catchment outlet. The dominance of certain processes and the resultant functions at different spatial scales would control these outputs. The reach and catchment functions, and the services they provide, would be less sensitive to the connectivity and thresholds of the system, e.g. grassland areas to first-order streams, since this would be constant in space and time.
2. *A catchment system with a higher level of complexity.* Consider a similar catchment system, but with a range of grassland use intensities and grass species present. The dynamic steady-state conditions reached in the simpler system would now be perturbed by anthropogenic activities that would vary in time and space. These would include, for example, the addition of macro-nutrients, the presence of livestock, and tillage operations. This would result in the rates of individual biological, chemical and physical processes and their associated structures being altered. The dominance of processes and their resultant functions at individual scales may change, and the scale of these disturbances will be key to the delivery of reach- and catchment-scale services. Connectivity becomes important as the catchment land use is no longer homogeneous in space and time and the thresholds of biological, chemical and physical processes control the functioning of the system.

In reality our catchment systems have many more dimensions of heterogeneity of biotic and abiotic processes and structures, at multiple spatial and temporal scales, leading to complex patterns and functional responses that control the delivery of ecosystem services.

24.4 Grassland Multiple Functions from the Plant to Catchment Scale

Increasing attention in scientific domains is being directed to understanding the functioning of individual components of grasslands, be it from a soil perspective (e.g. Robinson *et al.*, 2009) or plant ecology perspective (e.g. Tilman *et al.*, 1996). The majority of scientific studies to date have examined single functions at a particular spatial scale, e.g. biodiversity on carbon stocks at the plot scale. However, functions occur at a range of temporal (Bardgett *et al.*, 2005) and spatial (Ettema and Wardle, 2002) scales. Current knowledge of biotic and abiotic processes and structures related to controlling water quality and quantity tends to be at the plant to plot scale, due to the ease of observations at a small spatial scale. In this section we briefly highlight key aspects of our understanding of how grassland areas can regulate and provide clean and timely water from the plant to catchment scales.

24.4.1 Water quantity regulation functions

Plant to pedon/plot scale functions

The presence and type of vegetation are key to hydrological regulation at small spatial scales. In many natural dry grassland ecosystems, infiltration is increased over non-vegetated areas, and the loss or change in vegetation cover has increased the amount of degraded ecotypes and altered runoff characteristics (Dregne, 2002). Grass species may alter hydrological regulation at the plant to plot scale through altering the ability and rate of water to infiltrate soil, the storage capacity of surface soils, different water-use efficiencies, and differing levels of surface roughness that may affect surface runoff (Macleod *et al.*, 2007).

Field to catchment scale functions

At a field scale, first-order hydrological controls are influenced by rainfall and intensity, hydraulic conductivity and soil water storage. At larger scales, preferential flow pathways, subsurface flow and landscape connectivity influences are increasingly important (Lohse *et al.*, 2009). Intensively managed grazing systems can result in reductions in pasture yield, soil compaction and increased runoff generation. These effects are likely to occur in temperate climates under moisture-excess conditions (Drewry *et al.*, 2008). Temperate grasslands generally need little or no irrigation; therefore water-use efficiency is high.

Water 'farming' or the use of grasslands to store episodic floodwater can reduce downstream risk and build resilience. From an economic perspective, the use of grassland areas is preferable to the potential loss of arable or horticultural crops, with less entrainment of sediment from bare soils. Although many of these practices show tangible benefit at the local scale, their impact is lessened at large catchment scales (Parrott *et al.*, 2009). At the catchment scale, the balance between grasslands and other land-use types and their management has a direct consequence on discharge, although the inherent properties of each catchment (topography, geology, etc.) are also important.

24.4.2 Water quality regulation functions

Plant to pedon/plot scale functions

Important studies have demonstrated that there is a link between grass species biodiversity and soil functions (Tilman *et al.*, 1996). The majority of studies on the impact of grassland management on water quality have been carried out at the plant to plot scale due to the difficulties in establishing replicated experiments on a larger spatial scale (e.g. Sharpley *et al.*, 1981; Scholefield and Stone, 1995). These studies have found that inputs of inorganic and organic fertilizers and hydrological regimes control losses of N and P.

Field to catchment scale functions

At the field to catchment scales there are additional structures and associated processes (e.g. field drains and farm buildings) that influence the provision of clean water. At the field scale, lysimeters have demonstrated the linkage between land management, field drains and losses of N (Scholefield *et al.*, 1993) and P (Hawkins and Scholefield, 1996), sediment (Bilotta and Brazier, 2008) and microbial hazards (Oliver *et al.*, 2010). At the farm/headwater catchment scale, fluxes of P (Page *et al.*, 2005), N (Granger *et al.*, 2010), sediment (Bilotta *et al.*, 2010) and microbial hazards (Oliver *et al.*, 2010) have been linked to catchment structures and processes.

In lakes and river channels, exchanges with groundwater through the hyporheic zone influence hydrochemical and biogeochemical processes. Responses are damped down in lower-order river channels, where additional in-stream processes such as bed scour, deposition, re-suspension and nutrient spiralling, influence observed water quality (Fisher *et al.*, 1998). Interrogating existing databases of water quality and relating these to catchment characteristics can help understand relationships between land-based activities and their impact on water quality. In Ireland, Donohue *et al.* (2006) have used water-quality data based on the Irish water quality rating system for 797 sites and compared these to catchment characteristics. They found that the presence of urban areas (up to 10% of the contributing area) reduced the probability to approximately 50% of 'good' ecological status; with arable areas up to 20% coverage was linked to a similar decrease in ecological status; and catchments with 100% pasture land cover were still able to deliver an approximately 50% estimated probability of attaining 'good' ecological status. A recent modelling study of the role of riparian and catchment land use on water quality across several European countries found positive relationships between riparian grassland and water quality (Wasson *et al.*, 2010).

24.5 Role of Grasslands in Catchment Rehabilitation

Current and emerging legislation, such as the EU Water Framework Directive (2000/60/EC) and EU Floods Directive (2007/60/EC), are encouraging the rehabilitation of regulated and degraded inland water bodies. This can be complex due to the dynamic nature of rivers and the fact that rehabilitation strategies require a whole-systems approach (structure, function and dynamics; Poudevigne *et al.*, 2002) Hansen and Baattrup-Pedersen (2006) highlight the importance to move from small-scale to whole-catchment-scale interventions. This has widespread implications for the mosaic of land management that is sustainable within a given catchment and one that will be catchment specific. It also provides a direct link to thinking about riverine dynamics and ecological process as an inherent part of catchment dynamics, not independent from them. Similarly, Death and Collier (2010) have recently shown that riparian management is likely to be more successful when directed at rehabilitating or protecting headwater catchments than short reaches of a river. There has been increased awareness that current land-use practices and management have enhanced surface-water generation at the local scale (O'Connell *et al.*, 2007), and national initiatives are promoting Natural Flood Management practices (e.g. 'Making Space for Water' (Defra, 2005)). The nature and intensity of grassland management is critical when developing strategies for riparian rehabilitation. Multifunctional river restoration strategies are now looking to decrease the amount of riparian margin under intensive use, to alter water-table profiles to encourage inundation, and the temporal establishment of wetland and wet-grass areas.

24.6 Perspectives and Future Research Needs

There is an increasing need to understand the linkages between processes, structure and their resultant functions at single and multiple scales, and the services that they provide regarding the provision and regulation of water in catchment systems. Understanding how grassland management increases beneficial functions from the plant to catchment scales is complicated by the range of possible grassland land-use intensities (Verburg *et al.*, 2009). In this chapter we build on advances in the hydrological and landscape ecology literatures, to highlight the need to consider scale, connectivity and thresholds if we are to link our management of grassland patches to catchment-scale services in complex hierarchical catchment systems. This will involve interdisciplinary research across scales linked to societal needs. Grasslands can play a beneficial role in catchment rehabilitation, but the complex interactions of land-use intensity with cross-scale interactions means there is a greater need for the integration of empirical data collection and modelling activities. Hydrological and grassland scientists need to embrace interdisciplinary systems based approaches to understand and predict environment change (Schroder, 2006; Macleod *et al.*, 2007; Wagener *et al.*, 2010). The ecosystem service framework, alongside other frameworks, e.g. system dynamics (Forrester, 2007), and establishment of long-term socio-ecological systems (Anderson *et al.*, 2008) can provide frameworks and experimental platforms for the required interdisciplinary research.

Acknowledgement

The authors acknowledge the support of the Scottish Government's Rural and Environment Research and Advisory Directorate.

References

Allan, J.D. (2004) Landscapes and riverscapes: the influence of land use on stream ecosystems. *Annual Review of Ecology Evolution and Systematics* 35, 257–284.

Amoros, C. and Bornette, G. (2002) Connectivity and biocomplexity in waterbodies of riverine floodplains. *Freshwater Biology* 47, 761–776.

Anderson, C.B., Likens, G.E., Rozzi, R., Gutierrez, J.R., Armesto, J.J. and Poole, A. (2008) Integrating science and society through long-term socio-ecological research. *Environmental Ethics* 30, 295–312.

Bardgett, R.D., Bowman, W.D., Kaufmann, R. and Schmidt, S.K. (2005) A temporal approach to linking aboveground and belowground ecology. *Trends in Ecology and Evolution* 20, 634–641.

Bennett, E.M., Peterson, G.D. and Gordon, L.J. (2009) Understanding relationships among multiple ecosystem services. *Ecology Letters* 12, 1394–1404.

Bilotta, G.S. and Brazier, R.E. (2008) Understanding the influence of suspended solids on water quality and aquatic biota. *Water Research* 42, 2849–2861.

Bilotta, G.S., Krueger, T., Brazier, R.E., Butler, P., Freer, J., Hawkins, J.M.B., Haygarth, P.M., Macleod, C.J.A. and Quinton, J.N. (2010) Assessing catchment-scale erosion and yields of suspended solids from improved temperate grassland. *Journal of Environmental Monitoring* 12, 731–739.

Bloschl, G. and Sivapalan, M. (1995) Scale issues in hydrological modeling – a review. *Hydrological Processes* 9, 251–290.

Brauman, K.A., Daily, G.C., Duarte, T.K. and Mooney, H.A. (2007) The nature and value of ecosystem services: an overview highlighting hydrologic services. *Annual Review of Environment and Resources* 32, 67–98.

Cottenie, K. and De Meester, L. (2003) Connectivity and cladoceran species richness in a metacommunity of shallow lakes. *Freshwater Biology* 48, 823–832.

Daily, G.C., Soderqvist, T., Aniyar, S., Arrow, K., Dasgupta, P., Ehrlich, P.R., Folke, C., Jansson, A., Jansson, B.O., Kautsky, N., Levin, S., Lubchenco, J., Maler, K.G., Simpson, D., Starrett, D., Tilman, D. and Walker, B. (2000) Ecology – the value of nature and the nature of value. *Science* 289, 395–396.

Death, R.G. and Collier, K.J. (2010) Measuring stream macroinvertebrate responses to gradients of vegetation cover: when is enough enough? *Freshwater Biology* 55, 1447–1464.

Defra (2005) *Making Space for Water, Taking Forward a New Government Strategy for Flood Control and Coastal Erosion Risk Management in England.* Department for Environment, Food and Rural Affairs, London.

Donohue, I., McGarrigle, M.L. and Mills, P. (2006) Linking catchment characteristics and water chemistry with the ecological status of Irish rivers. *Water Research* 40, 91–98.

Dregne, H.E. (2002) Land degradation in the drylands. *Arid Land Research and Management* 16, 99–132.

Drewry, J.J., Cameron, K.C. and Buchan, G.D. (2008) Pasture yield and soil physical property responses to soil compaction from treading and grazing - a review. *Australian Journal of Soil Research* 46, 237–256.

Ettema, C.H. and Wardle, D.A. (2002) Spatial soil ecology. *Trends in Ecology and Evolution* 17, 177–183.

Fisher, T.R., Lee, K.Y., Berndt, H., Benitez, J.A. and Norton, M.M. (1998) Hydrology and chemistry of the Choptank River basin. *Water Air and Soil Pollution* 105, 387–397.

Forrester, J.W. (2007) System dynamics - a personal view of the first fifty years. *System Dynamics Review* 23, 345-358.

Gordon, L.J., Peterson, G.D. and Bennett, E.M. (2008) Agricultural modifications of hydrological flows create ecological surprises. *Trends in Ecology and Evolution* 23, 211–219.

Granger, S.J., Hawkins, J.M.B., Bol, R., White, S.M., Naden, P., Old, G., Bilotta, G.S., Brazier, R.E., Macleod, C.J.A. and Haygarth, P.M. (2010) High temporal resolution monitoring of multiple pollutant responses in drainage from an intensively managed grassland catchment caused by a summer storm. *Water Air and Soil Pollution* 205, 377–393.

Grayson, R. and Bloschl, G. (2000) Spatial Patterns in Catchment Hydrology, Observations and Modelling. Cambridge University Press, Cambridge.

Hansen, H.O. and Baattrup-Pedersen, A. (2006) A new development: stream restoration In: Sand-Jensen, K., Friberg, N. and Murphy, J. (eds), *Running Waters.* National Environmental Research Institute, Silkeborg, Denmark.

Hawkins, J.M.B. and Scholefield, D. (1996) Molybdate-reactive phosphorus losses in surface and drainage waters from permanent grassland. *Journal of Environmental Quality* 25, 727–732.

Hein, L., van Koppen, K., de Groot, R.S. and van Ierland, E.C. (2006) Spatial scales, stakeholders and the valuation of ecosystem services. *Ecological Economics* 57, 209–228.

Holling, C.S. (1992) Cross-scale morphology, geometry, and dynamics of ecosystems. *Ecological Monographs* 62, 447–502.

Jorgensen, S.E., Fath, B.D., Bastianoni, S., Marques, J.C., Muller, F., Nielsen, S.N., Patten, B.C., Tiezzi, E. and Ulanowicz, R.E. (2007) *A New Ecology Systems Perspective*. Elsevier, Oxford.

Levin, S.A. (1992) The problem of pattern and scale in ecology. *Ecology* 73, 1943-1967.

Levin, S.A., Grenfell, B., Hastings, A. and Perelson, A.S. (1997) Mathematical and computational challenges in population biology and ecosystems science. *Science* 275, 334–343.

Likens, G.E. and Bormann, F.H. (1974) Linkages between terrestrial and aquatic ecosystems. *Bioscience* 24, 447–456.

Limburg, K.E., O'Neill, R.V., Costanza, R. and Farber, S. (2002) Complex systems and valuation. *Ecological Economics* 41, 409–420.

Lohse, K.A., Brooks, P.D., McIntosh, J.C., Meixner, T. and Huxman, T.E. (2009) Interactions between biogeochemistry and hydrologic systems. *Annual Review of Environment and Resources* 34, 65–96.

Macleod, C.J.A., Binley, A., Hawkins, S.L., Humphreys, M.W., Turner, L.B., Whalley, W.R. and Haygarth, P.M. (2007) Genetically modified hydrographs: what can grass genetics do for temperate catchment hydrology? *Hydrological Processes* 21, 2217–2221.

Millennium Ecosystem Assessment (2003) *Ecosystems and Human Well-being: a Framework for Assessment*. Island Press, Washington DC.

O'Connell, E., Ewen, J., O'Donnell, G. and Quinn, P. (2007) Is there a link between agricultural land-use management and flooding? *Hydrology and Earth System Sciences* 11, 96–107.

O'Neill, R.V., DeAngelis, D.L., Waide, J.B. and Allen, T.F.H. (1986) *A Hierarchical Framework for the Analysis of Scale*. Princeton University Press, Princeton, NJ.

Odum, E.P. (1953) *Fundamentals of Ecology*. Saunders, Philadelphia, PA.

Oliver, D.M., Page, T., Hodgson, C.J., Heathwaite, A.L., Chadwick, D.R., Fish, R.D. and Winter, M. (2010) Development and testing of a risk indexing framework to determine field-scale critical source areas of faecal bacteria on grassland. *Environmental Modelling and Software* 25, 503–512.

Page, T., Haygarth, P.M., Beven, K., Joynes, A., Butler, P., Keeler, C., Freer, J., Owens, P. and Woods, G. (2005) Spatial variability of soil phosphorus in relation to the topographic Index and critical cource areas: sampling for assessing risk to water quality. Journal of *Environmental Quality* 34, 2263–2277.

Parrott, A., Brooks, W., Harmar, O. and Pygott, K. (2009) Role of rural land use management in flood and coastal risk management. *Journal of Flood Risk Management* 2, 272–284.

Peters, D.P.C., Groffman, P.M., Nadelhoffer, K.J., Grimm, N.B., Coffins, S.L., Michener, W.K. and Huston, M.A. (2008) Living in an increasingly connected world: a framework for continental-scale environmental science. *Frontiers in Ecology and the Environment* 6, 229–237.

Poudevigne, I., Alard, D., Leuven, R. and Nienhuis, P.H. (2002) A systems approach to river restoration: a case study in the Lower Seine Valley, France. *River Research and Applications* 18, 239–247.

Robinson, D.A., Lebron, I. and Vereecken, H. (2009) On the definition of the natural capital of soils: a framework for description, evaluation, and monitoring. *Soil Science Society of America Journal* 73, 1904–1911.

Rodriguez-Iturbe, I. (2000) Ecohydrology: a hydrologic perspective of climate-soil-vegetation dynamics. *Water Resources Research* 36, 3–9.

Scheffer, M., Carpenter, S., Foley, J.A., Folke, C. and Walker, B. (2001) Catastrophic shifts in ecosystems. *Nature* 413, 591–596.

Scholefield, D. and Stone, A.C. (1995) Nutrient losses in runoff water following application of different fertilisers to grassland cut for silage. *Agriculture, Ecosystems and Environment* 55, 181–191.

Scholefield, D., Tyson, K.C., Garwood, E.A., Armstrong, A.C., Hawkins, J. and Stone, A.C. (1993) Nitrate leaching from grazed grassland lysimeters - effects of fertilizer input, field drainage, age of sward and patterns of weather. *Journal of Soil Science* 44, 601–613.

Schroder, B. (2006) Pattern, process, and function in landscape ecology and catchment hydrology - how can quantitative landscape ecology support predictions in ungauged basins? *Hydrology and Earth System Sciences* 10, 967–979.

Sharpley, A.N., Ahuja, L.R., Yamamoto, M. and Menzel, R.G. (1981) The kinetics of phosphorus desorption from soil. *Soil Science Society of America Journal* 45, 493–496.

Sivapalan, M., Bloschl, G., Zhang, L. and Vertessy, R. (2003) Downward approach to hydrological prediction. *Hydrological Processes* 17, 2101–2111.

Thorp, J.H., Thoms, M.C. and Delong, M.D. (2006) The riverine ecosystem synthesis: biocomplexity in river networks across space and time. *River Research and Applications* 22, 123–147.

Tilman, D., Wedin, D. and Knops, J. (1996) Productivity and sustainability influenced by biodiversity in grassland ecosystems. *Nature* 379, 718–720.

Turner, M.G. and Chapin, F.S. (2005) Causes and consequences of spatial heterogeneity in ecosystem function. *Ecosystem Function in Heterogeneous Landscapes*, 9–30.

Verburg, P.H., van de Steeg, J., Veldkamp, A. and Willemen, L. (2009) From land cover change to land function dynamics: A major challenge to improve land characterization. *Journal of Environmental Management* 90, 1327–1335.

Wagener, T. (2007) Can we model the hydrological impacts of environmental change? *Hydrological Processes* 21, 3233–3236.

Wagener, T., Sivapalan, M., Troch, P.A., McGlynn, B.L., Harman, C.J., Gupta, H.V., Kumar, P., Rao, P.S.C., Basu, N.B. and Wilson, J.S. (2010) The future of hydrology: an evolving science for a changing world. *Water Resources Research* 46.

Wasson, J.G., Villeneuve, B., Iital, A., Murray-Bligh, J., Dobiasova, M., Bacikova, S., Timm, H., Pella, H., Mengin, N. and Chandesris, A. (2010) Large-scale relationships between basin and riparian land cover and the ecological status of European rivers. *Freshwater Biology* 55, 1465–1482.

Western, A.W., Bloschl, G. and Grayson, R.B. (2001) Toward capturing hydrologically significant connectivity in spatial patterns. *Water Resources Research* 37, 83–97.

Wu, J.G. and Loucks, O.L. (1995) From balance of nature to hierarchical patch dynamics: a paradigm shift in ecology. *Quarterly Review of Biology* 70, 439–466.

Zehe, E. and Sivapalan, M. (2009) Threshold behaviour in hydrological systems as (human) geo-ecosystems: manifestations, controls, implications. *Hydrology and Earth System Sciences* 13, 1273–1297.

25 Interactions of Grasslands with Forest at the Landscape Level: Topical Issues and Challenges Attached to Land Management in Agro–Silvo–Pastoral Landscapes

Gérard Balent and Annick Gibon

25.1 Introduction

At the regional level, grasslands are most often components of rural landscapes made of spatial associations of different types of ecosystems, which are grouped together within a local agro–silvo–pastoral system. The ways grassland, forest and/or scrub ecosystems are spatially arranged within landscape, and their use combined within agro–silvo–pastoral systems, are important factors that determine the nature of the usually wide range of ecological, social and economic services that these agroecosystems provide. Ecosystem services at the regional level do not result only from the intrinsic properties of single ecosystems, but also from their interactions with other ecosystems that have different properties. This is even true for primary production services, as we will show later on in this chapter. It is well known, for example, that grasslands in watershed bottoms play an important role in flood limitation and that hedges in steep cropland help to prevent wind and water erosion. Many studies in landscape ecology also stress the major role of the habitat mosaic in biodiversity, especially in fauna communities (Bretagnolle *et al.*, Chapter 26, this volume). For example, forest and grassland fields in agricultural landscapes act as reservoirs and also as corridors for many species, including insects that are useful for pollination and crop-pest control. Additionally, societal changes have increased interest in the visual quality and aesthetics of rural landscapes, their cultural and heritage value, and their economic value for rural development through tourism and other activities.

The ecosystems with which grasslands are associated are very diverse in agro–silvo–pastoral landscapes throughout the world. We briefly review them in the first section of the chapter, before providing an insight into the main types of agro–silvo–pastoral system in the world and the challenges they currently face. Due to the complex and multiscale interactions between the natural and human activities that agro–silvo pastoral systems encapsulate, from the field to the landscape level, it is not easy for either research or society to get a comprehensive picture of the challenges currently associated with the future of their grasslands, landscapes and ecosystem services.

In the second section of the chapter, we review some research advances that support an enhanced understanding of the inter-linkages between human societies and ecosystems in agro–silvo–pastoral systems. The views we discuss rely on theories recently developed in sustainability science and landscape ecology (Gunderson and Holling, 2002; Wu, 2006). We illustrate our ideas with examples of agro–silvo–pastoral systems in southern-Europe mountain and Mediterranean conditions.

In the last section we discuss how current knowledge can provide new prospects for meeting the tricky challenges currently associated with the sustainable management of agro–silvo–pastoral landscapes, and the conservation or enhancement of their ecosystem services. These challenges are many indeed, in these times of huge, rapid and uncertain change at local and global levels. We concentrate in this chapter on sustainable land management at the landscape level.

25.2 Ecosystem Mosaics and Land Management in Agro–Silvo–Pastoral Landscapes

25.2.1 An overview of ecosystem variety in agro–silvo–pastoral landscapes

Agro–silvo–pastoral systems are associated with a wide range of types of ecosystems (Table 25.1), from short-lived single-species vegetation stands to native and very ancient cultural ecosystems of outstanding biodiversity and patrimonial value, and they have an essential role in the provision of varied ecosystem services.

There is a widely accepted categorization of grasslands into three main types (e.g. Lemaire *et al.*, 2005): (i) **natural grasslands**, in areas where climate or other natural environment conditions cannot support forests or allow tree establishment; (ii) **semi-natural grasslands**, which represent long-term naturalized vegetation maintained by domestic herbivore exploitation; and (iii) **cultivated grasslands**, included in crop rotations in cropland areas. There is no widely accepted categorization, using a simple terminology, for ecosystems other than grasslands in the continuum between treeless and fully treed (Putz and Redford, 2010). The concept of the English word 'forest', understood as open or closed tree cover managed for timber or/and other wood production, which is widely accepted in the western world, is restrictive and relatively recent. In Old English, the term was used to designate all non-cultivated areas including trees, where a variety of goods were extensively harvested.

A main basic reason why non-grassland ecosystems are not easily classified on a scientific basis is that the usual classification systems, based on tree cover only, lack recognition of the various sorts of scrub ecosystems, i.e. ecosystems dominated by multiple stemmed woody plants that can exceed 5 m height, but may never get much taller, such as cerrado, chaparral, fynbos, thicket, garrigue, mallee, maquis and sand heath. However, the FAO and others agencies define shrubs as 'woody perennial plants, generally > 0.5m and < 5m in height at maturity without a definite crown'. 'Shrublands', which figure prominently in many maps, are often combined with grasslands and savannahs in a single category for analysis, a fact that hampers consideration of the specificities of each land type's characteristics, use and ecological dynamics. Consequently, Putz and Redford (2010) recommend using the tradition of designating anthropogenic non-forest vegetation types as derived woodland, derived savannahs, and derived

Table 25.1. Categorization of the world's major ecosystem/vegetation/land cover types along the continuum from treeless to fully treed (from Putz and Redford, 2010).

Grasslands	Dominated by graminoids, i.e. grasses and grass-like plants	<5% tree-crown cover; a tree being defined as a woody plant > 5 m tall
Savannahs	Graminoids dominated with scattered trees	5–20% tree-crown cover
Woodlands	Graminoids in the understory of trees	20–40% tree-crown cover
Open forests	Dominated by trees	40–70% tree-crown cover
Closed forests	Dominated by trees	>70% tree-crown cover

scrub, so as not to confuse them with structurally similar but generally much more species-rich communities. Indeed, in all categories of anthropogenic ecosystem, as well as grasslands, some are recently settled and generally have a vegetation of a low biodiversity, while others are ancient, have long-term naturalized ('semi-natural') vegetation, and are extremely rich in biodiversity.

For the purpose of this chapter, we therefore use a categorization of the whole range of ecosystems in agro–silvo–pastoral landscapes based on a dichotomy between natural and semi-natural ecosystems with long-term naturalized vegetation on the one hand, and anthropogenic ecosystems with similar physiognomic characteristics but of recent creation on the other (Table 25.2). This categorization appears to allow definition of ecological functions at the landscape level, the roles and values of each ecosystem for agriculture and society, and each ecosystem's contribution to the adaptive capacity of agro–silvo–pastoral systems in adjusting to global changes and environmental hazards.

25.3 The Diversity in Agro–Silvo–Pastoral Systems and Landscapes and the Current Challenges for their Future

The variety in agro–silvo–pastoral landscapes and systems throughout the world can be broadly categorized into three main types, in reference to the topical research and society issues raised by their future. These follow, in particular, from the ways local land-use and livestock-grazing practices impact the structure and function of ecosystems and landscapes, and the drivers of change in local livestock production and land-use systems.

25.3.1 Cultural landscapes and tradition-based agro–silvo–pastoral systems

Located in temperate and tropical regions, cultural landscapes result from a slow

Table 25.2. Overall categorization of grasslands, shrublands and treed ecosystems (from Lemaire *et al.*, 2005; Putz and Redford, 2010).

Natural grasslands	< 5% tree-crown cover	Natural graminoid vegetation or long-term naturalized graminoid vegetation anthropogenically determined
Cultivated grasslands		Seeded graminoid vegetation of a short lifetime in cropping areas
Shrublands	< 5% tree-crown cover	Natural and semi-natural ecosystems dominated by woody plants that can exceed 5 m height but never much taller
Derived shrublands		Recent anthropogenic shrubland ecosystem derived from other ecosystems
Savannahs	5–20% tree-crown cover	Natural or semi-natural graminoid-dominated vegetation with scattered trees
Derived savannahs		Recent anthropogenic savannah ecosystem derived from other ecosystems
Woodlands	20–40% tree-crown cover	Natural or semi-natural vegetation with graminoids in the understory of trees
Derived woodlands		Recent anthropogenic woodland ecosystem derived from other ecosystems
Open forests	40–70% tree-crown cover	Natural or semi-natural vegetation dominated by trees
Derived open forests		Recent anthropogenic open forest derived from other ecosystems
Closed forests	>70% tree-crown cover	Natural or semi-natural vegetation fully dominated by trees
Cultivated forest		Recent anthropogenic forest derived from other ecosystems

co-evolution between natural and social systems (Antrop, 2005). These landscapes present a wide diversity in their composition and patterns, depending on the natural environmental conditions, and the ways rural societies have shaped and managed the land. They usually comprise mosaics of (a few) croplands and ecosystems with semi-natural vegetation of a site-specific character (e.g. savannahs and woodlands in the Sahel; non-irrigated and irrigated semi-natural grasslands in mountains, shrublands and woodlands in Mediterranean hills). In addition to an outstanding biodiversity of major conservation value, they are acknowledged as offering an increasing range of ecosystem services of important local and global value.

Although some of these cultural landscapes have a relatively short history, most were shaped by a long tradition of agro–silvo–pastoralism dating back to prehistoric times. Regional land-use systems most often include a mix of privately used land units and commons. Semi-natural ecosystems and landscapes result from a long co-adaptive process driven by the rural community's objectives to ensure both their social reproduction and the sustainability of the natural resources of their territory (e.g. Bourbouze and Gibon, 1999).

Tradition-based agro–silvo–pastoral systems and landscapes have become increasingly challenged by changes in societies and policies, especially since the mid-20th century due to the spread of the 'green revolution' policy for agricultural development (e.g. McDonald *et al.*, 2000). Those that survived up to now are currently imperilled throughout the world. In the north, landscapes experience spontaneous encroachment with shrubs and trees and/or reforestation with plantations, in relation to agricultural population decline, and remaining agricultural land is subject to land-use competition for urbanization, often in relation to tourism pressure (e.g. Gibon *et al.*, 2010). In the south, agro–silvo–pastoral areas experience an increasing drought occurrence that leads to desertification, as the result of climate change, and are subject to irrigation plans and land-use competition for crops, as the result of national population growth and policies for food self-sufficiency (e.g. Bah *et al.*, 2010).

25.3.2 Recently created agropastoral landscapes and land-use systems

Agricultural systems were recently developed in the world's 'land reserves', to enhance the livelihood of poor people in growing national populations, such as in the tropical 'native' forests of the Amazon Basin, Indonesia and central Africa, and also in temperate areas such as New Zealand. Agro–silvo–pastoral areas were established as natural forest by colonialist farmers practising a family-based agriculture, whose installation was most often supported by public policies (e.g. Bommel *et al.*, 2010). Farm populations in these areas evolved, over time, towards a mix of small-scale farms and large-scale farms practising a capitalistic, or even speculative, agriculture that tended to base its development on a continuous land expansion, to the detriment of pristine or natural forest ecosystems that provided major environmental services on a global scale. Topical issues in these 'new' agro–silvo–pastoral landscapes concern the search for innovative crop–livestock management systems able to secure land fertility and enhance agroecosystem biodiversity and services, as well as meet the local-to-global challenge of stopping (or at least limiting) the deforestation process (Muchagata and Brown, 2003).

25.3.3 Emerging 'agroforestal' landscapes

Based on research-coined innovative technologies, loose plantations of trees in otherwise treeless ecosystems are increasingly regarded, in science and policy, as a solution to help mitigate the global effects of deforestation on carbon emission. Currently promoted by public policies or private agencies, such plantations are being created in cropland and grassland ecosystems in temperate areas (e.g. Farley, 2007). In most places, the application of these novel land-use practices appears, for now, to be superimposed on, rather than to fit into, a local co-adaptation process between ecosystems and land-use systems. The questions raised by science- and policy-designed agroforestal landscapes concern the assessment of the effectiveness of ecosystem–service enhancement they currently achieve, and the way they can be locally incorporated into sustainable socio-ecological systems.

25.4 A Framework for a Multiscale Assessment of Agro–Silvo–Pastoral Systems and their Changes

The objective of managing grassland and forest landscapes so that they can provide a variety of ecosystem services calls for an in-depth understanding of the interactions between ecological and management processes. This is a tricky challenge indeed, due to the multi-scale and nested inter-relationships encapsulated in each landscape, from the individual land unit, domestic animal and society member, up to the landscape and the whole local society. A first basic difficulty, extensively reported in the literature, follows from the scale mismatches between ecological and management processes.

25.4.1 Scale mismatches between management and ecological processes

A scale mismatch between the levels of organization of land management and the levels of organization of ecological processes could affect a clear understanding of the relationships existing between the two hierarchies of levels, as well as the effectiveness of policies intended to limit any resulting undesirable effects (Cumming *et al.*, 2006; Table 25.3).

In a recent work reviewing the ways authors approach spatial scale mismatch in the literature and the solutions they propose to resolve it, Pelosi *et al.* (2010) conclude that the major importance attached to scale mismatches in literature arises from the fact that authors generally do not use a systemic approach for the integrated assessment of social and ecological processes and their relationships. Such an approach is required to appraise the spatial, temporal and/or functional inter-linkages between management and ecological processes. Since many researchers consider ecological and management processes separately, the scale/level terminology and theoretical frameworks used cannot match. We consider that, by using a hierarchical conceptual model of the ecological systems, the social systems and their inter-linkages, as recently suggested in landscape ecology and sustainability science (Wu, 2006), a better match between the representations of the ecological processes and of the management processes could be achieved. Additionally, spatially explicit modelling is necessary to help efficiently manage the interactions between social systems and ecological processes, at their relevant levels of organization, in agro–silvo–pastoral landscapes.

Table 25.3. The different types of mismatch between spatial, temporal and functional scales and ecological and social processes (adapted from Cumming *et al.*, 2006).

Consequences of mismatch		Fine scale ecological	Large scale ecological
Spatial scales	Large scale social	Too many managers, 'micromanager' syndrome	
	Fine scale social		No solutions for global problems, unmanaged essentials
Temporal scales	Large scale social	Social response time too slow	
	Fine scale social		Management decisions lack continuity and consistency
Functional scales	Large scale social	Ecosystem over-exploitation	
	Fine scale social		Potential disruption of social system

25.4.2 The challenge of creating a reference model to understand the system

As shown by Balent and Stafford-Smith (1993; Fig. 25.1), the management of the interactions between animals and resources in agro–silvo–pastoral systems can be analysed using an integrated hierarchical framework. This implies focus on a particular management process, differences in the rate of behaviour of which will define the different levels of organization and the associated relevant entities and scales (Allen and Starr, 1982). Such an approach also allows assessment of how interactions at one level relate to those at other levels, an issue which is too often forgotten in research.

This conceptual model proved particularly useful in analysing the apparent inconsistency between different management objectives that generally occurs in agro–silvo–pastoral systems, e.g. maximizing grazing intake by animals in the short term versus sustainability of grassland production in the long term. Using this model, Gibon *et al.* (2010) focused on eliciting the logic of the different stakeholders and simulating the consequences of their interactions, in a transdisciplinary research project on sustainable land management. The aims were sustainability of pastoral activities and the long-term landscape provision of various amenities and functions (Gibon *et al.*, 2010). An agroecological characterization of the pastoral resources could be made independently of the management practices and goals of the land users, although some consider that pastoral resources have to match the functions assigned by the end users, quasi-independently of local ecological conditions. These two points of view only differ because they address two different levels of organization within the pastoral system. The agroecological approach deals with the management of grasslands at the field level, with long-term sustainability as

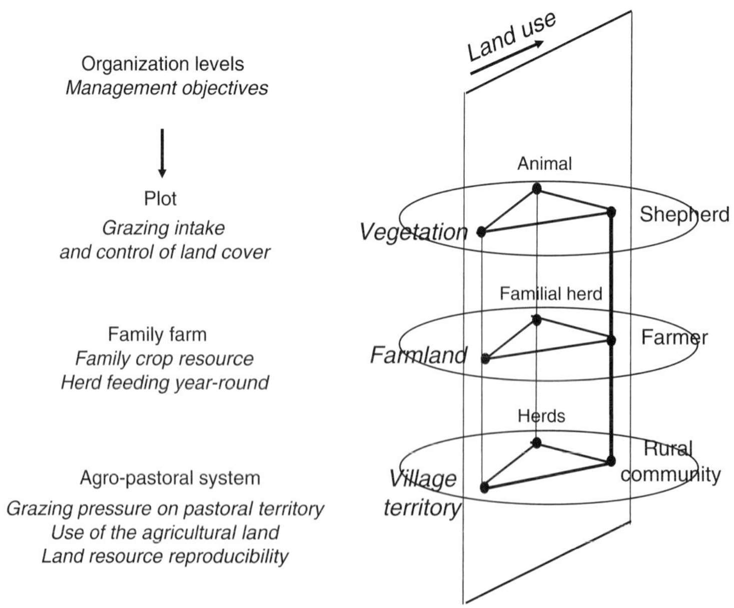

Fig. 25.1. A conceptual model of the organization of a tradition-based agropastoral system (adapted from Balent and Stafford-Smith, 1993). The different levels are defined by differences in the rate of decisions in animal–vegetation interactions (daily, seasonal and yearly). Scales of space and time vary in the opposite direction to the rate of decision.

the objective, whereas the second approach deals with the annual balance between herbage on offer and the requirement of the animals at the farm level.

25.4.3 Changes in pastoral system organization: the key factor for understanding grasslands and forest dynamics at a landscape level

In pastoral systems the fundamental basis for organization (Kolasa and Pickett, 1989) is the patrimonial management of the resources of the territory to ensure the socio-economic sustainability of the community. This is often guaranteed by the tight social control exerted by the community over individuals. A general process of change in the institutional organization of the governance of the land and its ecosystems took place in European mountain landscapes during the second half of the 20th century. It led to the progressive abandonment of collective strategies in local land-use systems and their gradual replacement by uncoordinated individual ones.

A model built for the Pyrenean agro-pastoral systems, to assess multi-scale changes in land management organization and landscape dynamics (Fig. 25.2), can be widely adapted to other pastoral systems, as shown by Balent and Stafford-Smith (1993). Such a model of the concomitant patterns of change in land-use systems and landscape ecological traits is helpful in assessing current challenges to landscape sustainability. The change in landscape grain was the main visible pattern resulting from the management system dynamics explored in the model. As a consequence, underlying ecological processes regarding vegetation dynamics, such as seed bank and seed rain, are more difficult to control and important changes can then occur at landscape and field levels, namely the development of shrubs at the field level and of forest patches at the landscape level (e.g. Jurena and Archer, 2003).

25.4.4 Traditional ecological knowledge: a major source of knowledge for the multi-purpose and multi-scale management of heterogeneous ecosystems

Although some ecologists attempted many years ago to raise awareness of the very intensive character of land use in some tradition-based agro–silvo–pastoral systems, even with severe natural environmental limitations (e.g. for the Serengeti, see Sinclair and Norton-Griffiths, 1979), little consideration was given to the management skills of the land users. An awareness recently developed, in natural resource management (NRM) and sustainability science, that people in traditional socio-ecological systems often develop a great diversity of technological know-how and practices for ecosystem management as a result of the empirical expertise they gain in their resource-use history (Berkes *et al.*, 2000). Various studies of pastoralists' information systems and indigenous knowledge strongly support the view that the traditional practices of these people include special skills with respect to multipurpose management of multiple species ecosystems, and mosaic landscapes, deep knowledge of local ecosystem successional processes in relation to anthropogenic pressures, and also 'ways of responding to and managing pulses and ecological surprises' (Berkes *et al.*, 2000). Researchers in NRM, agriculture and forestry increasingly recognize the sophisticated ecological rationality underlying the practice of farmers in the tradition-based management of complex ecosystems. In a review of management challenges of grasslands for production, the environment and the landscape, Gibon (2005), stressed that a major feature in tradition-based land-management practices of semi-natural grassland ecosystems, from the land unit to the landscape level, was the capacity not only to deal with, but also to maintain and benefit from, heterogeneity within and between the ecosystems.

A study of Mediterranean farmers' management practices in tradition-based

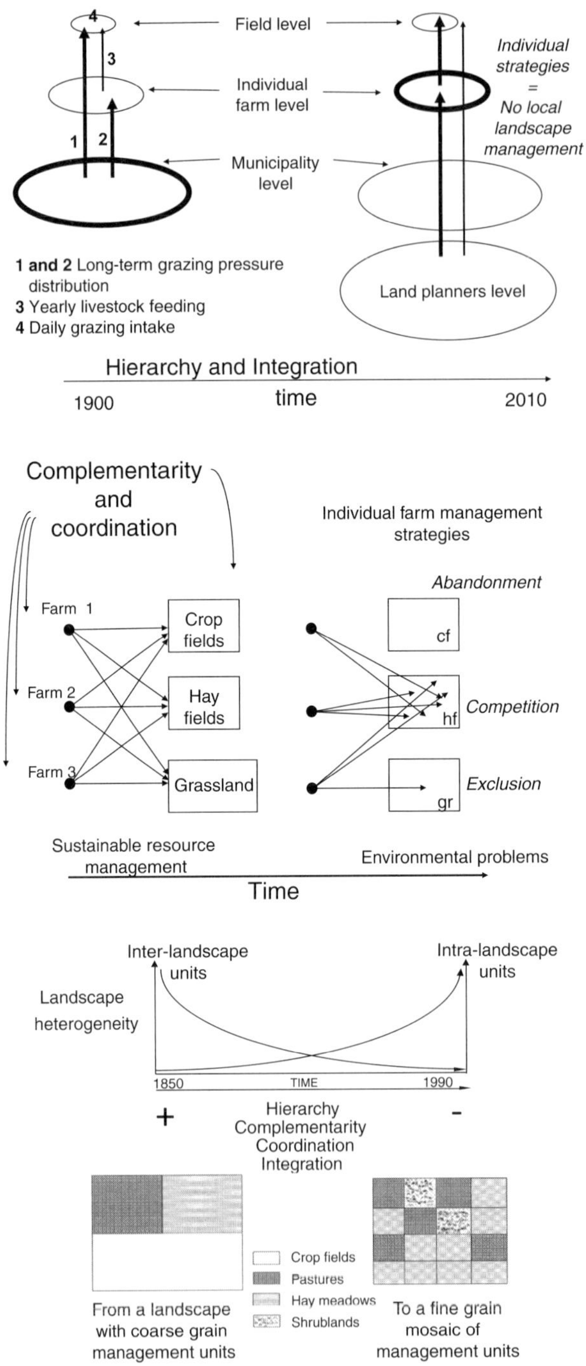

Figure 25.2 Changes in organization of pastoral systems result in changes in landscape grain and heterogeneity (from Gibon and Balent, 2005).The progressive abandonment of collective control of land-use management and the correlative emergence of individual strategies progressively overcame coordination and complementarity in resource management. There is a shift in landscape grain that increases the risk of shrub encroachment at the field level.

agropastoral systems provided outstanding illustrations of similar knowledge and skills in the multi-purpose management of the intra- and inter-ecosystem variety in landscape mosaics made of croplands, grasslands and shrublands, at a variety of temporal horizons (Meuret *et al.*, 2006). The skills included management of the elaboration of goat or/and sheep daily diets from the combination of various types of ecosystem. As illustrated in Fig. 25.3, the conceptual model 'Menu', built by Meuret (Meuret, 2000) from combined field studies of animal grazing behaviour and shepherd management practices, supports an understanding of the biological processes underpinning animal ingestive behaviour and animal–vegetation interactions incorporated into shepherding practices. The model also takes account of the genetic and cultural skills of domestic herbivores, in conditioning their capacity to make their own living in local ecosystem and landscape conditions, and their inter-generational transmission within herds. Educational techniques of grazing livestock are part of the tradition-based management

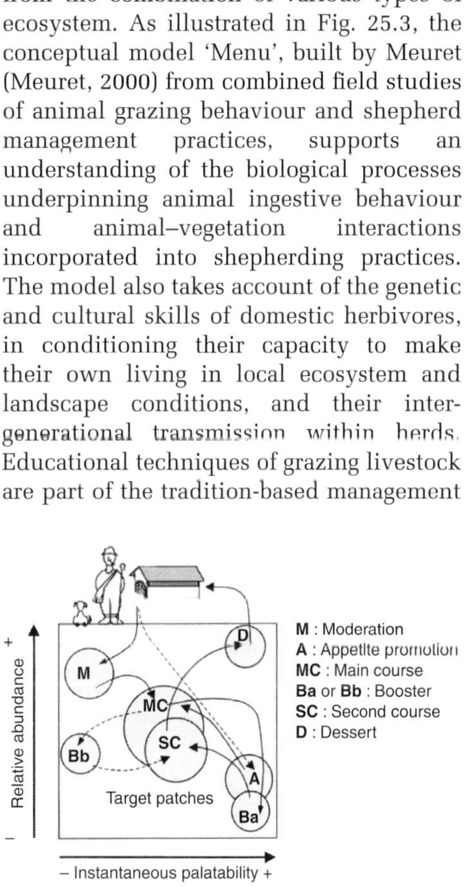

Figure 25.3 The 'Menu' conceptual model of the daily grazing circuit of a shepherded flock. The daily grazing circuit appears as an ordered sequence of offered patches during one main meal, the shepherd objective being to constantly revive the animal's motivation to eat on resources of medium or low quality ('target patches') in the course of the day. A booster patch can be of low plant abundance but high instantaneous palatability (Ba); or conversely of medium abundance and very low instantaneous palatability (Bb). In addition a booster action can be watering the flock or offering salt licks (from Meuret, in Baumont *et al.*, 2000.)

practices of pastoralists in many areas (e.g. Sahelian semi-natural savannahs). There is a recent regenerated interest in such techniques in Europe, in relation to agri-environmental policies.

25.5 The Challenge Raised by the Management of Complex Ecosystems and Ecosystem Mosaics for Sustainability

The emergence of global environmental concerns, and the growing awareness of the basic requirement to account for a wide range of services when evaluating ecosystems, deeply questioned conventional research views on agro–silvo–pastoral ecosystems and land-management systems in the last two decades.

25.5.1 An ongoing change in paradigm for building appropriate scientific advances for enhancing landscape management for ecosystem services

Until a few years ago, mainstream research on the improved management of ecosystems and landscapes mainly relied on the experimental search for new technologies and management packages, aimed at enhancing a single production function (e.g. timber from forest, as mentioned in the first section of the chapter). This approach neglected the direct and indirect contributions of the ecosystem to the provision of other goods and services to rural communities, the requirements for ecosystem sustainability (e.g. resilience capacity), and the continuation of production functions (e.g. resistance to climatic hazards, flexibility in harvest dates).

There was indeed, over several decades, a widening gap between the mainstream views in agricultural research, and consequently in advisory and development services, about appropriate technologies and management for enhancing the benefits of agro–silvo–pastoral ecosystems, and the views and practices of farmers in tradition-based agro–silvo–pastoral systems.

It would, however, be a mistake to consider indigenous knowledge as 'old-fashioned' or 'static' (Berkes *et al.*, 2000), and a novel challenge for scientists is to help account for this traditional ecological knowledge (TEK) in research and development programmes on pastoral resource management. Awareness begins to spread among scientists that mitigation of the current knowledge gap may enhance the chances for success in research and development programmes for sustainability of ecosystem and landscape services. Traditional knowledge, tuned into a continuously evolving environment, and constantly incorporating outside influences and inside innovations, can be highly adaptive. Even FAO and the World Bank have recently developed a shared vision for a new integrated approach to agricultural education, research and extension, i.e. the Agricultural Knowledge and Information System (AKIS) (Niamir-Fuller, 1998). In terms of pastoralism, AKIS should include pastoralists, educators, researchers, extensionists, input suppliers, marketing agents and others who have knowledge and information that can be used to improve livestock production in a way that ensures food security and sustainable livelihoods for pastoral people. It is essential to have a sound understanding of the nature of pastoralists' TEK, the way it is shared and transferred, and the factors that influence its reproduction. Incorporating TEK into projects can contribute to local empowerment and provide valuable input for alternative NRM strategies. Furthermore, it appears to us that not only the sustainable development of NRM and land-use systems but also the survival and renewal of cultures and social identity make it essential to emphasize traditional/local knowledge and information systems and enhance their transfer.

25.5.2 Adaptive co-management of complex social-ecological systems

Traditional knowledge and resource management systems were characterized by the building of local ecological knowledge from observation, and interpreting and responding to feedback from the environment, in order to adapt to change and to guide future developments carefully. Indeed, traditional systems relied on principles similar to the current recommendations for adaptive management of social-ecological systems, as listed by Folke *et al.* (2005) in a recent literature review:

- Ecological knowledge and understanding of ecosystem processes and functions are required to understand resource and ecosystem dynamics that could contribute to resilience of resource management in a changing environment. Different sources of knowledge and understanding are needed for managing complex adaptive systems, including social processes of knowledge creation and transmission.
- Successful management is characterized by continuous testing, monitoring and re-evaluation to enhance adaptive responses, acknowledging the inherent uncertainty in complex systems. It is increasingly proposed that knowledge generation of ecosystem dynamics should be explicitly integrated with adaptive management practices, rather than striving for optimization based on past records.
- Adaptive management deals with perturbations, uncertainty and surprises. It is not sufficient to consider the ecological dynamics of the ecosystems under management. We need to take into account changes in climatic conditions, disease outbreaks, market trends, subsidies and governmental policies.

As these authors state: 'The challenge for (any) social-ecological system is to accept uncertainty, be prepared for change and surprise, and enhance the adaptive capacity to deal with disturbance. Non-resilient social-ecological systems are vulnerable to external change, whereas a resilient system may even make use of disturbances as opportunities to transform into more desired states' (Folke *et al.*, 2005).

25.6 Conclusions

There is increasing evidence for the need to move towards more integrative research and dialogue with land managers to deal with the complexity of agro–silvo–pastoral systems. In agreement with Wu (2006), we consider that the sustainable management of complex mosaic landscapes including grasslands, shrublands, forests and cultivated areas requires transdisciplinary research, i.e. integration of inter-disciplinary research with participation of all of the stakeholders. Integrative landscape ecology and socio-ecological theories provide a hierarchical framework for holistic approaches to real-world agro–silvo–pastoral systems. They also allow researchers to develop theories and study methods for assessing the spatial heterogeneity, relating processes to observed patterns, scaling management and ecological processes and accounting for global change and uncertainty. However, the shift in research practices towards such an integrated approach has a long and risky way to go: firstly, for scientific reasons, because it assumes decreasing generality in research findings, increasing emphasis on corroboration/scenario analysis, rather than on falsification/validation hypotheses, and recognition of subjectivity and uncertainty in system description and prediction, etc. There are also institutional problems, because if transdisciplinary research is recognized as a trigger for scientific innovation, 'the major funding research agencies have so far failed to fully appreciate the opportunity and the competitive advantage it brings' (Jahn, 2009). This results, in particular, in a lack of commonly accepted quality criteria and weak evaluation methods for this type of research.

To end on a more promising note, we consider that the current development of landscape ecology, as a science aimed at the integrated assessment and modelling at meso-scales of both the ecological processes and their management, and its recognition as a sustainability science will help to enhance scientific knowledge of the dynamics of complex mosaic landscapes associating grasslands with various other ecosystems, and their multi-purpose management. Integrated studies of local agro–silvo–pastoral systems as complex adaptive social-ecological systems appear to us to represent the high road in helping to integrate scientific and traditional ecological knowledge in sustainable resource management tools and policies.

References

Allen, T.F.H. and Starr, T.B. (1982) *Hierarchy. Perspectives for Ecological Complexity.* University of Chicago Press, Chicago.

Antrop, M. (2005) Landscape change and the urbanization process in Europe. *Landscape and Urban Planning* 67, 9–26.

Bah, A., Toure, I., Fourage, C., Gaye, I., Leclerc, G., Soumare, A., Ickowicz, A. and Diop, A. (2010) An agent-based model to study land allocation policies and their effect on pastoral and territorial dynamics in the Ferlo (Senegal). *Cahiers Agricultures* 19, 14–20.

Balent, G. and Stafford-Smith, D.M. (1993) A conceptual model for evaluating the consequences of management practices on the use of pastoral resources. In: *Proceedings 4th International Rangeland Congress.* CIRAD Editions, Montpellier, pp. 1158–1164.

Berkes, F., Colding, J. and Folke, C. (2000) Rediscovery of traditional ecological knowledge as adaptive management. *Ecological Applications* 10, 1251–1262.

Bommel, P., Bonaudo, T., Barbosa, T., Bastos da Veiga, J., Vieira Pak, M. and Tourrand, J.F. (2010) La relation complexe entre l'élevage et la forêt en Amazonie brésilienne: une approche par la modélisation multi-agents. *Cahiers Agricultures* 19, 104–111.

Bourbouze, A. and Gibon, A. (1999) Ressources individuelles ou ressources collectives? L'impact du statut des ressources sur la gestion des systèmes d'élevage des régions du pourtour de la Méditerranée. *Options Méditerranéennes, Série A, Séminaires Méditerranéens* 38, 289–309

Cumming, G., Cumming, D. and Redman, C. (2006). Scale mismatches in social–ecological systems: causes, consequences, and solutions. *Ecology and Society* 11, 14.

Farley, K.A. (2007) Grasslands to tree plantations: forest transition in the Andes of Ecuador. *Annals of the Association of American Geographers* 97, 755–771.

Folke, C., Hahn, T., Olsson, P. and Norberg, J. (2005) Adaptive governance of social–ecological systems. *Annual Review of Environment and Resources*, 30, 441–473.

Gibon, A. (2005) Managing grassland for production, the environment and the landscape. Challenges at the farm and the landscape level. *Livestock Production Science*, 96, 11-31.

Gibon, A. and Balent, G. (2005) Landscapes on the French side of the Western and Central Pyrenees. In: Pinto-Correia, T., Bunce, R.G.H. and Howard, D.C. (eds) *Landscape Ecology and Management of Atlantic Mountains.* Agricultural Productivity Enhancement Program, Wageningen, The Netherlands, pp. 65–74.

Gibon, A., Sheeren, D., Monteil, C., Ladet, S. and Balent, G. (2010) Modelling and simulating change in reforesting mountain landscapes using a social-ecological framework. *Landscape Ecology* 25, 267–285.

Gunderson, L.H. and Holling, C.S. (2002) *Panarchy: Understanding Transformations in Human and Natural Systems.* Island Press, Washington DC.

Jahn, T. (2009) Understanding socio-ecological systems: frontier research for sustainable development. Implications for European research policy. In: *Conference: Sustainable Development: Implications for European Research*, 26–28 May 2009. Instutute for Social–Ecological Research, Brussels.

Jurena, P.N. and Archer, S. (2003) Woody plant establishment and spatial heterogeneity in grasslands. *Ecology*, 84, 907–919.

Kolasa, J. and Pickett, S.T.A. (1989) Ecological systems and the concept of biological organization. *Proceedings of the National Academy of Sciences USA*, 86, 8837–8841.

Lemaire, G., Wilkins, R. and Hodgson, J. (2005) Challenges for grassland science: managing research priorities. *Agriculture, Ecosystems and Environment* 108, 99–108.

McDonald, A., Crabtree, J.R., Wiesinger, G., Dax, T., Stamou, N., Fleury, P., Gutierrez Lazpita, J. and Gibon A. (2000) Agricultural abandonment in mountain areas of Europe: Environmental consequences and policy response. *Journal of Environmental Management* 59, 47–69.

Meuret, M. (2010) Modèle MENU: le berger vu comme un chef cuisinier. In: Meuret, M. (ed.) *Un Savolr-faire de Bergers.* Editions Quae, Versailles, France, pp. 167–190.

Meuret, M., Debi,t S., Agreil, C. and Osty, P.L. (2006) Educating the herd's calves and heifers: an empirical knowledge usable in mountain agro-environmental policies? *Natures Sciences Sociétés* 14, 343–352 (in French).

Muchagata, M. and Brown, K. (2003) Cows, colonists and trees: rethinking cattle and environmental degradation in Brazilian Amazonia. *Agricultural Systems* 76, 797–816.

Niamir-Fuller, M. (1998) The resilience of pastoral herding in Sahelian Africa. In: Berkes, F. and Folke, C. (eds) *Linking Social and Ecological Systems: Management Practices and Social Mechanisms for Building Resilience.* Cambridge University Press, Cambridge, pp. 250–284

Pelosi, C., Goulard, M. and Balent, G. (2010) The spatial scale mismatch between ecological processes and agricultural management: do difficulties come from underlying theoretical frameworks? *Agriculture, Ecosystems and Environment* 139, 455–462.

Putz, F.E. and Redford, K.H. (2010) The importance of defining 'forest': tropical forest degradation, deforestation, long-term phase shifts, and further transitions. *Biotropica* 42, 10–20.

Sinclair, A.R.E. and Norton-Griffiths, M. (1979) *Serengeti: Dynamics of an Ecosystem.* University of Chicago Press, Chicago.

Wu, J.G. (2006) Landscape ecology, cross-disciplinarity, and sustainability science. *Landscape Ecology* 21, 1–4.

26 The Role of Grassland Areas within Arable Cropping Systems for the Conservation of Biodiversity at the Regional Level

Vincent Bretagnolle, Bertrand Gauffre, Helmut Meiss and Isabelle Badenhausser

26.1 Introduction

Farmland is the most common habitat in France as well as in Europe as a whole (56% and 52% of the land surface, respectively). As a likely consequence, farmland habitat harbours a major part of European biodiversity. For example, 50% of European bird species live in rural landscapes (Potts, 1997; Tucker, 1997). In Europe as well as in other parts of the world, the process of agricultural intensification has led to the simplification and specialization of farmed landscapes, increased uses of inputs such as pesticides and fertilizers, and the mechanization of agricultural practices. At the landscape level, agricultural intensification has led to a loss of natural or semi-natural habitats and a decrease in habitat heterogeneity. The effects of agricultural intensification on biodiversity are well demonstrated: many studies and reviews document the negative effects of intensification on plants, insects, birds and mammals, in various countries (e.g. Krebs *et al.*, 1999; Robinson and Sutherland, 2002; Benton *et al.*, 2003, Inchausti and Bretagnolle, 2005; Julliard *et al.*, 2004; Fried *et al.*, 2009). Recent empirical evidence strongly supports the view that plant and animal diversity decreases with increasing crop yield, a good proxy of agricultural intensification (see Geiger *et al.*, 2010 and references therein). However, the evidence that agricultural intensification has led to biodiversity loss is mainly based on correlative studies carried out on large spatial scales (regions or countries), whereas the ecological processes involved are not well identified. Moreover, most studies dealt only with few taxa, and there is a lack of studies investigating the effects of biodiversity loss on ecosystem services, e.g. soil conservation, nutrient cycling, groundwater purification, pollination and biological control (but see Geiger *et al.*, 2010). Lastly, it is currently uncertain whether biodiversity loss at the regional scale results more from cropping intensification at the field scale or from the loss of natural elements from the landscape.

In France, mixed farming systems (integrating crop and livestock production) have strongly declined during the past 60 years, which was mainly due to the European Common Agricultural Policy (CAP) and other macro-economic constraints. Some regions specialized in intensive cereal production (e.g. La Beauce), others on intensive livestock production (e.g. Brittany), generating uniform patterns of land use across the landscape on large spatial scales. At a landscape level, cereal agroecosystems are characterized by high spatial variability (i.e. the fields generate strong spatial discontinuity) and temporal variability (due to crop rotations, harvesting, mowing and ploughing). Cereal landscapes, being highly fragmented in space, provide natural populations with a spatial patchwork of habitats of temporally varying quality, depending on species characteristics. Ploughing, by introducing temporally asynchronous alterations of habitat quality, is probably the strongest human disturbance in these systems. For many types of organisms, it may lead to extinction of the local population within the patch/field. Hence, a metapopulation dynamics framework appears appropriate to study biodiversity in such agro-ecosystems. In landscapes dominated by intensive cereal cropping, multi-annual forage crops such as meadows, mown grasslands and forage crops can be considered as 'perennial habitats', since they are usually kept for 3–4 years (or even longer). They differ radically from all annual crops in terms of mechanical disturbance (soil tillage, sowing, cutting for harvest) but also often receive fewer pesticide and fertilizer inputs.

Patches of perennial crops, as well as field margins and other non-crop habitats, may act as shelters for wildlife within the 'matrix' of less favourable annual crops. In regards to biodiversity and trophic network, the presence, abundance and distribution of these perennial habitats may have strong impacts on the metapopulation and meta-community dynamics of various organisms (Hanski, 1999). To investigate the role of a particular habitat in biodiversity, particularly in cereal systems, it is therefore important to address the metapopulation dynamics of the organisms in a spatially explicit way and at large spatial scales, since dispersal and colonization, two major processes of metapopulation dynamics, are obviously spatial phenomena. They mainly depend on the distance between source populations and on the structure of the landscape that affects the movement of individuals between habitat patches (Akçakaya, 2000).

In regions now dominated by cereal crop production, the gradual disappearance of grasslands and multi-annual forage crops negatively impacted fauna and flora but, more generally, presumably led to the reduction or disappearance of ecosystem services, such as beneficial functions for ground-water quality, sequestration of greenhouse gases, and functional biodiversity. It may be hypothesized that the simultaneous presence, at the landscape scale, of surfaces with herbaceous perennial habitats and annual crops, may maintain high levels of agricultural production while limiting negative impacts on the environment and on biodiversity. In the following chapter, we provide empirical results from our long-term studies on voles, plants, grasshoppers, birds and trophic interactions between these groups carried out in the Chizé study region (Zone Atelier 'Plaine et Val de Sèvre') in central western France (see Fig. 26.1 and Bretagnolle *et al.*, 2011, for details of the study site). We show that trophic interactions and biodiversity management, as well as conservation, require the presence of grasslands at the landscape scale.

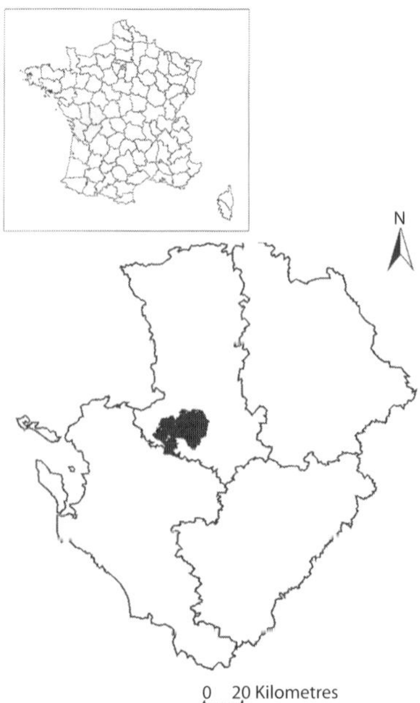

Fig. 26.1. A map showing the study site (light grey) and its location in France.

26.2 Metapopulation Dynamics of Voles at the Landscape Level

The common vole *Microtus arvalis* is a small-sized rodent abundant in Western European agroecosystems (Mitchell-Jones *et al.*, 1999; Salamolard *et al.*, 2000), where it shows large cyclic variation in abundance (Lambin *et al.*, 2006). Common voles are not restricted to grasslands, perennial crops and edge habitats (e.g. field margin strips) in intensively cultivated areas but also show high densities in cereal or rapeseed fields (see Salamolard *et al.*, 2000). Short-term effects of farming practices on common vole population dynamics and demographic parameters have already been investigated (Jacob, 2003; Jacob and Hempel, 2003). Harvesting, mowing and mulching (occurring in both perennial and annual habitats) do not destroy nest sites (subterraneous burrows), but potentially eliminate a substantial proportion of the local population by direct killing or increased risk of predation due to reduced vegetation height and cover. No dispersive movement appears to be associated with these activities (Jacob and Hempel, 2003). Ploughing (occurring solely in annual crops), which destroys common vole burrows, seems to be the only agricultural practice to eradicate the common vole (Jacob, 2003).

Although common voles are not restricted to grasslands, the maintenance of perennial surfaces at the landscape level has often been viewed as enhancing the common vole's cycles, which, in turn, is supposed to affect the trophic network, in particular the many carnivores that rely on voles as preys. As support for this idea, Butot and Leroux (2001) found a negative correlation between the amplitude of cycles of common vole in the marshes of Western France and the degree of agriculture intensification (conversion of grasslands to cereals). Such a result suggests that common vole metapopulations in agricultural mosaics may follow a source–sink dynamic between permanent and temporary habitats. We addressed this question in a landscape genetic study, and showed that common vole gene flow in our study site (450 km^2) was not sufficiently limited to result in a strong genetic structure (Gauffre *et al.*, 2008). Rather, we observed a slight, but significant, isolation by distance pattern that shaped the roughly homogeneous genetic structure of the studied metapopulation. This result does not contradict a source–sink dynamic model between permanent and temporary habitats but suggests that, if such dynamics exist, they occur in the context of large effective population sizes and high movement rates among the different habitat types constituting the agricultural matrix. In another study, we demonstrated that dispersal rates are high and strongly male biased at short distances, whereas long-distance dispersal is not rare and affects males and females in similar proportions (Gauffre *et al.*, 2009). Genders also differ in the timing of dispersal. Males migrate continuously from colonies to colonies, whereas females may disperse just once and thus are involved in the foundation of new colonies.

If gene flow is not affected by habitat fragmentation, it may be because the common vole life cycle (voles reproduce at few weeks of age and have several generations per year) and metapopulation dynamics (through intense dispersal and colonization) are faster than the perturbation dynamics of the landscape, hence mitigating the short temporary suitability of most habitats (i.e. annual crops). Furthermore, the spatio–temporal variation in habitat quality resulting from agricultural land use could select for increased dispersal propensity (Ronce, 2007). The persistence of the common vole in the agricultural landscape seems to be conditioned by the animal's capacity to exploit perennial habitats (e.g. grasslands, always favourable but representing a minor part of land use) as well as ephemeral habitats (annual crops that are not available all year round).

26.3 Plants and Weeds in Cereal Systems: Spatial and Temporal Interdependences between Annual Crops and Grasslands

Plants, including both cultivated crop grassland species as well as wild 'weed' species, are the basis of food webs in agroecosystems and are important determinants of habitat quality for animals (Badenhausser *et al.*, 2009b). Plants also show strong interactions with other plants. The abundance of wild 'weed' species is, for instance, strongly determined by that of crop plants, due to competition (and facilitation). Each crop may favour some types of weed species while suppressing others. This may be due to characteristics of the crop plant (such as its growth dynamic and competitive ability) as well as crop-specific field management practices (including soil tillage, sowing dates, fertilization, harvesting dates, etc.) (Doucet *et al.*, 1999). Therefore, the succession of crops grown year after year on a field may have strong impacts on weed community dynamics (e.g. see the review of Liebman and Dyck, 1993). Due to the separation of crop and livestock production (see above), today's crop rotations are often only composed of annual crops. The inclusion of temporary grasslands (perennial forage crops) into crop rotations may have particularly strong impacts on weed communities, as these types of perennial crops differ in many biotic and abiotic aspects from annual crops (see Sebillotte, 1980; Viaux *et al.*, 1999; and Meiss *et al.*, 2010a for a review). Recent weed surveys in our study site, as well as field and greenhouse experiments carried out in eastern France, showed that some plant species will profit from the conditions in these 'perennial crops', including biennial and perennial species and annual species with rosettes. At the same time, other species will be suppressed, including some annual species, especially those with an upright or climbing morphology, that are often dominant (and most problematic) in annual cereal crops (Meiss *et al.*, 2008, 2010b).

These changes in weed species composition may be due to several mechanisms acting on several stages of the plant life cycle. The absence of soil tillage in temporary grasslands may reduce weed germination. The survival of established plants may be favoured by the absence of soil tillage but will be negatively affected by the competition of the perennial vegetation cover and by the frequent hay cuttings (Meiss *et al.*, 2009). Finally, weed seed survival may be reduced in perennial crops, as seeds stay on the soil surface where they are more accessible to seed predators than seeds in tilled annual crops. Moreover, experimental studies indicated that post-dispersal weed seed predation (consumption of seeds) by vertebrates and invertebrates increases with vegetation cover in perennial crops (Meiss *et al.*, 2010c).

By modifying the plant species composition, temporary grasslands devoted to forage (or fibre or energy) production may thus both contribute to the regulation of weed populations that are most harmful in annual crops and favour other plant species that are less problematic for the following annual crops. Moreover, specific fields, such as overwinter stubble fields that favour the access of farmland birds to seed resources during winter (Moorcroft *et al.*, 2002), may also be integrated into such long crop rotations at the moment when seed bank densities are highest, thus at the end of the period of annual crops. By diversifying crop rotations dominated by annual cash crops with perennial crops and overwinter stubble fields, the functions of crop production, weed management and biodiversity conservation may thus be wisely combined on the scale of the dynamic landscape mosaic.

26.4 Grasshopper Spatial Dynamics and the Role of Perennial Crops

Grasshoppers (Orthoptera: Acrididae) play a major trophic role, being the primary herbivores in grassland habitats, and also because they are prey for other invertebrates, e.g. spiders and vertebrates, notably farmland birds that use them as food for chicks (Baker, 2004). Like many invertebrates, grasshoppers need perennial habitats due to the fact that their survival from one year to the other is achieved by eggs that are deposited in the soil. As ploughing destroys eggs, annual crops are sinks for grasshoopper populations. Thus, maintaining grasshopper populations in the landscape can be achieved only through perennial habitats, such as grasslands or field boundaries. However, these areas are rare in intensive agrosystems, and temporary grasslands are subject to frequent destruction due to crop rotation. Local extinctions at the field scale can be compensated only if new habitats are available and if species can colonize them, i.e. have sufficient dispersal abilities (Baur *et al.*, 2005). For example, the distance between favourable habitats could be more restrictive for species like the wingless *Pezotettix giornae* than for the winged species *Calliptamus italicus*, which may be more affected by habitat loss than by habitat fragmentation (Tscharntke and Brandl, 2004).

In that context, grassland availability and distribution at the landscape level have a major impact on grasshopper populations. At the field scale, grassland vegetation type and the agricultural practices have strong impacts on grasshopper abundance (Wingerden *et al.*, 1992; Guido and Gianelle, 2001). For example, grasshopper densities in our study site, during the period 2003–2006, were rather low in lucerne crops (about 0.8 individuals per m^2) compared with other temporary grasslands (about 3.1 individuals per m^2). In lucerne, the lower numbers of grasshoppers could be explained by the agricultural practices on this crop. Surveys that compared extensive lucerne fields without any input and with no cutting from June to August and conventional fields with at least one herbicide and one insecticide treatment showed that grasshopper densities may be seven to ten times higher in the extensive fields (Badenhausser *et al.*, 2008). This response may also be caused by the number of plant species, which was nearly twice as high in the extensive plots as in the conventional plots (32 plant species; Badenhausser *et al.*, 2008). This is indicated by the positive relationship between grasshopper abundance and plant diversity shown in Fig. 26.2. Increased plant diversity could diversify the offer of different refuges and microclimates to species having different habitat preferences. Moreover, plant diversity may enhance grasshopper diversity, as grasshopper species could differ in their food preferences (Specht *et al.*, 2008). Thus, management practices such as the amount of inputs and the number of cuttings may be responsible directly or indirectly for the high observed spatial variability in grasshopper abundance at the landscape scale (Badenhausser *et al.*, 2009a). Moreover, during the period 1999–2010 in our study site, grasshopper densities exhibited strong temporal variability at the scale of the season but also at the scale of the year (Badenhausser *et al.*, 2009a). The range of grasshopper abundances at the peak period of adult stage, i.e. at the beginning of August, during the time period 1999–2010 varied from 0.35 grasshopper/m^2 in 2001 to 8.5 grasshoppers/m^2 in 2004 (Badenhausser *et al.*, 2009a).

26.5 Dependence of Threatened Top Predator Species on their Prey in Cereal Systems

Insects and small mammals make up the bulk of biomass in food supply for higher trophic levels in agroecosystems. For example, the common vole often represents a key resource in the trophic chain (Lambin *et al.*, 2006). In Western France,

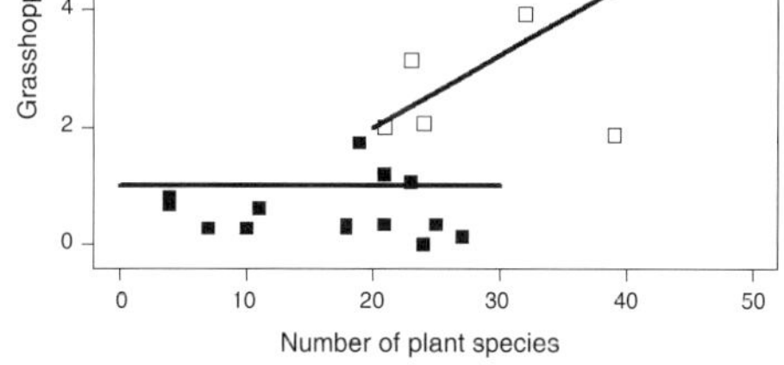

Fig. 26.2. The relationship between grasshopper density and plant richness in lucerne crops. AES, agri-environment scheme.

Montagu's Harrier (*Circus pygargus*) appears to be highly dependent on common vole abundance, since its population density displays a numerical response to the cyclic dynamic of this prey (Salamolard *et al.*, 2000; Millon and Bretagnolle, 2008). Thus, the persistence of this flagship raptor species is conditioned by the maintenance of common vole populations. Similarly, differences in grasshopper availability appear to be critical to the Little Bustard (*Tetrax tetrax*) productivity. Though the populations of Little Bustard in southern France, which are not migratory, are relatively stable (Jolivet, 1997), the population of western France, which winters in Spain (Villers *et al.*, 2010), has undergone one of the steepest declines ever documented to date for a bird species in Europe: from 7800 males in 1978 to 390 in 1996 (a decrease of 95% in 18 years; Inchausti and Bretagnolle, 2005; Bretagnolle *et al.*, 2011), and 300 in 2008. There is a strong positive relationship between yearly average grasshopper abundance (calculated as the mean abundance over the surveyed grasslands for a given year) and total annual productivity of Little Bustard on our study site (Fig. 26.3), as estimated by the number of fledglings counted in post-nuptial groups (see Bretagnolle *et al.*, 2011 for methods). The case is not unique, as most bird chicks in cereal systems feed primarily on insects (e.g. Rands, 1986; Baines *et al.*, 1996; Panek, 1997) and especially on Orthoptera.

26.6 The Conservation of Biodiversity in Cereal-based Cropping Systems

Many species, particularly birds, though often still being extremely common, have fallen sharply in abundance across the agricultural plains of Europe (Fuller *et al.*, 1995; Potts, 1997). In Western Europe, about 1% of the avifauna of lowland landscapes disappears annually (Donald *et al.*, 2001; Julliard *et al.*, 2004). In France, between 1989 and 2003, bird populations (all species combined) declined by 3%, while birds using agricultural habitats declined by 25% (Julliard *et al.*, 2004). Paradoxically, however, agroecosystems, though they dominate the land surface of

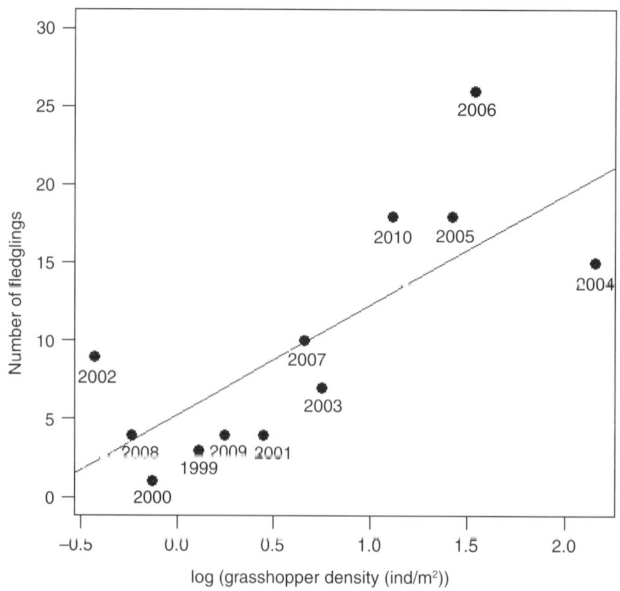

Fig. 26.3. Numerical response of Little Bustard fecundity (chick productivity) and grasshopper density (each point represents a year).

Europe, have been poorly studied until recently in regard to trophic interactions and community ecology. Likewise, almost no conservation effort has yet been undertaken on these areas because of private land ownership and exploitation that forbids the creation of nature reserves, the most-used conservation strategy.

Instead, agri-environment schemes (AESs) and NATURA 2000 are the two main (if not the only) tools available in order to mitigate the devastating effects of intensive agriculture on biodiversity. In intensive cereal systems, AESs should be mainly targeted at the conservation and management of permanent and temporary grasslands due to (i) their key role as habitats for different taxa, (ii) their potential for weed management when inserted into the rotations and (iii) other ecosystem services. However, several studies have shown that AESs (and organic farming, which is often included in AESs, for instance in France) do not always have the expected positive effects on biodiversity (e.g. Kleijn and Sutherland, 2003).

On our study site, we implemented different AESs in order to protect the Little Bustard. Most of these schemes were based on grassland restoration and modification in grassland management. The strong decline in Little Bustard has been linked to the lowering in surfaces of perennial habitats for breeding, but also to strong decreases in insects (particularly grasshoppers) for feeding (Bretagnolle and Inchausti, 2005; Bretagnolle *et al.*, 2011). The latter resulted simultaneously from a decrease in grasslands as the breeding habitat of grasshoppers and the intense use of insecticides and herbicides (grasshoppers are herbivorous). Our conservation strategy for the Little Bustard was therefore to counteract the loss of habitat and the low availability of food resources. We developed protection measures providing favourable nesting plots (decrease of agricultural activities to minimize the risk of destruction of nests and females) and resources, and more generally to encourage farmers to restore perennial vegetation covers. The Little Bustard population, which showed an initial decrease by a factor of five in just 8 years (about 13% per year since 1996), has now almost completely recovered, in no more than 5 years (Bretagnolle *et al.*, 2011).

26.7 Conclusion

Our results indicate that, in intensive cereal systems, grasslands have a critical role in shaping the distribution and abundance of organisms of different trophic levels, including plants, grasshoppers, small mammals and birds. In addition, we show that the population dynamics of top level predators such as raptors (e.g. Montagu's Harrier) or bustards, which are both of high conservation concern, are driven by the abundances of their prey (insects and small mammals) which ultimately depend to a large extent on grasslands. The management of grassland habitat in such ecosystems is therefore critical not only for the maintenance of ecosystem services, such as those depending on functional biodiversity, but also for the conservation of threatened species. For the latter, grassland must be managed at the regional rather than the local scale, because these species forage on vast areas and occur at rather low density, so that their population dynamic arises at the regional scale. The implementation of AESs, when specifically targeted at mitigating the underlying causes of declining numbers of top predators, can help to maintain or conserve biodiversity and threatened species. In addition, the implementation of AESs in our study area has also benefitted other organisms, such as grasshoppers and passerine birds such as the Corn Bunting (*Milaria calandra*). Currently, we are testing whether the AESs dedicated to harrier and bustard conservation also have positive impacts on other taxa, including plants, grasshoppers, beetles and other passerine birds.

Acknowledgements

We would like to thank the many people that contributed to our field work. Paul Amouroux, Carole Attié, Laurent Bruneteau,

Nadine Guillon, Marilyn Vandier, Marilyn Roncoroni and other students helped in sampling grasshoppers; Steve Augiron, Alexandre Villers, Maxime Galan, Camille Fontaine, Adrien Pinot and many other students helped in bird surveys and trapping voles; Safia Médiène, Laurent Grelet, Anne-Caroline Denis, Damien Charbonnier, Luc Bianchi, Dominique Le Floch, Bruno Chauvel, Fabrice Dessaint, François Bretagnolle, Émilie Cadet, Jacques Gasquez, Florence Strbik, and Solène Bellanger and others helped in weed surveys. Part of this work was supported by INRA (ECOGER national program), ANR BIODIVAGRIM, ADVHERB and ECOCYCLES. B. Gauffre received a grant from ACI ECCO, H. Meiss from the French research ministry.

References

Akçakaya, H.R. (2000) Conservation and management for multiple species: integrating field research and modeling into management decisions. *Environmental Management,* 26, 75–83.

Badenhausser, I., Mediene, S., Paris Le Clerc, N. and Bretagnolle, V. (2008) Effects of agri-environnemental agreements on acridids and plant species richness in alfalfa crops. In: Hopkins, A. *et al.* (eds) *22st International Symposium of the European Grassland Federation.* Swedish University of Agricultural Sciences (SLU), Uppsala, Sweden, pp. 54–56.

Badenhausser, I., Amouroux, P., Lerin, J. and Bretagnolle, V. (2009a) Acridid (Orthoptera: Acrididae) abundance in Western European grasslands: sampling methodology and temporal fluctuations. *Journal of Applied Entomology* 133, 720–732.

Badenhausser, I., Chauvel, B., Cosson, J.F., Darmency, H., Houte, S., Lemaire, G., Lazrak, E.G., Médiène, S. and Bretagnolle, V. (2009b) Réconcilier agriculture et environnement dans les paysages céréaliers. *Biofutur* 305, 27–30.

Baines, D., Wilson, I.A. and Beeley, G. (1996) Timing of breeding in black grouse *Tetrao tetrix* and capercaillie *Tetrao urogallus* and distribution of insect food for the chicks. *Ibis* 138, 181–187.

Baker, A.M. (2004) Insects as food for farmland birds – is there a problem? In: Van Emden, H. and Rothschild, M. (eds) *Insect and Bird Interactions.* Intercept Ltd, Andover, UK, pp. 37–50.

Baur, B., Coray, A., Minoretti, N. and Zschokke, S. (2005) Dispersal of the endangered flightless beetle Dorcadion fuliginator (Coleoptera: Cerambycidae) in spatially realistic landscapes. *Biological Conservation* 124, 49–61.

Benton, T.G., Vickery, J.A. and Wilson, J.D. (2003) Farmland biodiversity: is habitat heterogeneity the key? *Trends in Ecology and Evolution* 18, 182–188.

Bretagnolle, V. and Inchausti, P. (2005) Modelling population reinforcement at a large spatial scale as a conservation strategy for the declining Little Bustard (*Tetrax tetrax*) in agricultural habitats. *Animal Conservation* 8, 59–68.

Bretagnolle, V., Villers, A., Denonfoux, L., Cornulier, T., Inchausti P. and Badenhausser, I. (2011) Rapid recovery of a depleted population of Little Bustards *Tetrax tetrax* following provision of alfalfa through an agri-environment scheme. *Ibis* 153, 4–13.

Butet, A. and Leroux, A. B. A. (2001) Effects of agriculture development on vole dynamics and conservation of Montagu's harrier in western French wetlands. *Biological Conservation* 100, 289–295.

Donald P.F., Green, R.E. and Heath, M.F. (2001) Agricultural intensification and the collapse of Europe's farmland bird populations. *Proceedings of the Royal Society of London B* 268, 25–29.

Doucet, C., Weaver, S.E., Hamill, A.S. and Zhang, J.H. (1999) Separating the effects of crop rotation from weed management on weed density and diversity. *Weed Sciences* 47, 729–735.

Fried G., Petit, S., Dessaint, F. and Reboud. X. (2009) Arable weed decline in Northern France: crop edges as refugia for weed conservation? *Biological Conservation* 142, 238–243.

Fuller, R.J., Gregory, R.D., Gibbons, D.W., Marchant, J.H., Wilson, J.D., Baillie, S.R. and Carter, N. (1995) Population declines and range contractions among lowland farmland birds in Britain. *Conservation Biology* 9, 1425–1441.

Gauffre, B., Estoup, A., Bretagnolle, V. and Cosson, J.F. (2008) Spatial genetic structure of a small rodent in a heterogeneous landscape. *Molecular Ecology* 17, 4619–4629.

Gauffre, B., Petit, E., Brodier, S., Bretagnolle, V. and Cosson, J.F. (2009) Sex-biased dispersal patterns depend on the spatial scale in a social rodent. *Proceedings of the Royal Society of London B* 276, 3487–3494.

Geiger F., Bengtsson, J., Berendse, F., Weisser, W.W., Emmerson, M., Morales, M.B., Ceryngier, P., Liira, J., Tscharntke, T., Winqvist, C., Eggers, S., Bommarco, R., Pärt, T., Bretagnolle, V., Plantegenest, M., Clement, L.W., Dennis, C., Palmer, C., Oñate, J.J., Guerrero, I., Hawro, V., Aavik, T., Thies, C., Flohre, A., Hänke, S., Fischer, C., Goedhart, P.W. and Inchausti, P. (2010) Persistent negative effects of pesticides on biodiversity and biological control potential on European farmland. *Basic and Applied Ecology* 11 97–105.

Guido, M. and Gianelle, D. (2001) Distribution patterns of four Orthoptera species in relation to microhabitat heterogeneity in an ecotonal area. *Acta Oecologica* 22,175–185.

Hanski, I. (1999) *Metapopulation Dynamics.* Oxford University Press. Oxford.

Inchausti, P. and Bretagnolle, V. (2005) Predicting short-term extinction risk for the declining Little Bustard (*Tetrax tetrax)* in intensive agricultural habitats. *Biological Conservation* 122, 375–384.

Jacob, J. (2003) Short-term effects of farming practices on populations of common voles. *Agriculture, Ecosystems and Environment* 95, 321–325.

Jacob, J., and Hempel, N. (2003) Effects of farming practices on spatial behaviour of common voles. *Journal of Ethology* 21, 45–50.

Jolivet, C. (1997) L'Outarde canepetière *Tetrax tetrax*: le déclin s'accentue. *Ornithos* 4, 63–72.

Julliard, R., Jiguet, F. and Couvet, D. (2004) Common birds facing global changes: what makes a species at risk? *Global Change Biology* 10, 148–154.

Kleijn, D. and Sutherland, W. (2003) How effective are European agri-environment schemes in conserving and promoting biodiversity? *Journal of Applied Ecology* 40, 947–969.

Krebs, J.R., Wilson, J.D., Bradbury, R.B. and Siriwardena, G.M. (1999) The second silent spring? *Nature* 400, 611–612.

Lambin, X., Bretagnolle, V. and Yoccoz, N.G. (2006) Vole population cycles in northern and southern Europe: Is there a need for different explanations for single pattern? *Journal of Animal Ecology* 75, 340–349.

Liebman, M. and Dyck, E. (1993) Crop-rotation and intercropping strategies for weed management. *Ecological Application*, 3(1), 92-122.

Meiss, H., Munier-Jolain, N., Henriot, F. and Caneill, J. (2008) Effects of biomass, age and functional traits on regrowth of arable weeds after cutting. *Journal of Plant Disease Protection* Special Issue XXI, 493–499.

Meiss, H., Bonnot, R., Strbik, F., Waldhardt, R., Caneill, J. and Munier-Jolain, N. (2009) Cutting and competition reduce weed growth: additive or interactive effects? In: *XIIIth International Conference on Weed Biology*. Association Francaise de Protection des Plantes, Dijon, France, pp. 28–37.

Meiss, H., Médiène, S., Waldhardt, R., Caneill, J. and Munier-Jolain, N. (2010a) Contrasting weed species composition in perennial alfalfas and six annual crops: implications for Integrated Weed Management. *Agronomy Sustainable Development* 30, 657–666.

Meiss, H., Médiène, S., Waldhardt, R., Caneill, J., Bretagnolle, V., Reboud, X. and Munier-Jolain, N. (2010b) Perennial alfalfa affects weed community trajectories in grain crop rotations. *Weed Research* 50, 331–340.

Meiss, H., Lagadec, L.L., Munier-Jolain, N., Waldhardt, R. and Petit, S. (2010c) Weed seed predation increases with vegetation cover in perennial forage crops. *Agriculture, Ecosystem and Environment* 138, 10–16.

Millon, A. and Bretagnolle, V. (2008) Predator population dynamics under a cyclic prey regime: numerical responses, demographic parameters and growth rates. *Oikos* 117, 1500–1510.

Mitchell-Jones, A.J., Amori, G. and Bogdanowicz, W. (1999) *The Atlas of European Mammals.* T. & A.D. Poyser, London.

Moorcroft, D., Whittingham, M.J., Bradbury, R.B. and Wilson, J.D. (2002) The selection of stubble fields by wintering granivorous birds reflects vegetation cover and food abundance. *Journal of Applied Ecology*, 39, 535–547.

Panek, M. (1997) The effect of agricultural landscape structure on food resources and survival of grey partridge *Perdrix perdrix* chicks in Poland. *Journal of Applied Ecology* 34, 787–792.

Potts, D. (1997) Cereal farming, pesticides and grey partridges, In: Pain, D.J. and Pienkowski, M.W. (eds) *Farming and Birds in Europe. The Common Agricultural Policy and its Implications for Bird Conservation.* Academic Press, London, pp. 150–177.

Rands, M.R.W. (1986) The survival of gamebird (Galliformes) chicks in relation to pesticide use on cereals. *Ibis* 128, 57–64.

Robinson, R. and Sutherland, W. (2002) Post-war changes in arable farming and biodiversity in Great Britain. *Journal of Applied Ecology* 39, 157–176.

Ronce, O. (2007) How does it feel to be like a rolling stone? Ten questions about dispersal evolution. *Annual Review of Ecology, Evolution and Systematics* 38, 231–253.

Salamolard, M., Butet, A., Leroux, A. and Bretagnolle, V. (2000) Responses of an avian predator to variations in prey density at a temperate latitude. *Ecology* 81, 2428–2441.

Sebillotte, M. (1980) The role of grasslands in crop rotations. *Fourrage* 83, 79–124.

Specht, J., Scherber, C., Unsicker, S.B., Koehler, G. and Weisser, W.W. (2008) Diversity and beyond: plant functional identity determines herbivore performance. *Journal of Animal Ecology* 77, 1047–1055.

Tscharntke, T. and Brandl, R. (2004) Plant–insect interactions in fragmented landscapes. *Annual Review of Entomology* 49, 405–430.

Tucker, G. (1997) Priorities for bird conservation in Europe: the importance of the farmed landscape. In: Pain, D. and Pienkowski, M.W. (eds), *Farming and Birds in Europe. The Common Agricultural Policy and its Implications for Bird Conservation.* Academic Press, San Diego, CA, pp. 79–116.

Viaux, P., Bodet, J.M. and Le Gall, A. (1999) Complémentarité herbe-cultures dans les rotations. *Fourrages* 160, 345–358.

Villers, A. Millon, A., Jiguet, F., Lett, J.M., Attie, C., Morales, M.B. and Bretagnolle, V. (2010) Migration of wild and captive-bred Little Bustards: releasing birds from Spain threatens attempts to conserve declining French populations. *Ibis* 152, 254–261.

Wingerden, W., Kreveld, A. and Bongers. W. (1992) Analysis of species composition and abundance of grasshoppers (Orth., Acrididae) in natural and fertilized grasslands. *Journal of Applied Entomology* 113, 138–152.

27 Interactions between Cereal Cropping Systems and Pastoral Areas as the Basis for Sustainable Agriculture Development in Mediterranean Countries

Aïssa Abdelguerfi and Tayeb Ameziane El Hassani

27.1 Introduction

Mediterranean countries are located around the Mediterranean Sea in a region bordered by mountains that limit the geographical extension of olive trees northward and palm dates southward. Typically, the Mediterranean climate is characterized by an erratic rainfall distribution pattern with alternating periods of dry/hot summer and humid/cool winter[1]. The natural vegetation consists of forests, woody plants, shrublands (matorral, maquis and garrigue), grasslands and pastures, rangelands and steppes. The Mediterranean flora is rich and diversified, with more than 500 range species (Talamucci and Chaulet, 1989). Legume species often represent the basic formation of open and forest pastures, prairie and fallow land. Fallow cultivation is practised widely in the southern Mediterranean, where it makes a substantial contribution to animal feeding. Animal species also present great diversity in their genotypic characters, which is reflected in the wide range of local breeds seen around the Mediterranean.

In the Euro-Mediterranean sub-region, the main agroecological zones are characterized by humid/sub-humid climates, whilst those in North Africa and the Middle East are arid/semi-arid. Aridity increases vulnerability to drought, which is a normal feature of climate and a recurrent phenomenon in all Mediterranean countries (Ameziane, 2002). In fact, recent drought episodes have had devastating effects on crops, livestock, populations and the environment in most countries from Morocco to Iran (FAO, 2002). Irrigated agriculture is well developed in some countries but, with the exception of Egypt, the countries still depend heavily on rainfed, cereal-based and extensive livestock production systems. Most soils are of low fertility and some, such as the vertisols, present management problems. Land degradation and desertification are major concerns and threaten the sustainability of the farming systems in the whole region. Prevailing conditions of aridity, drought and desertification have negative impacts on the availability of water resources and those result in water scarcity. The challenges and threats of water scarcity among dryland populations of the entire region are expected to increase in magnitude and scope, particularly in those countries that will be hit harder by climate change. Projected climate change trends for the region (e.g. Brown and Crawford, 2009) indicate further negative impacts on water availability and desertification, which are the main threats to food security and ecosystem services in the Mediterranean. Comprehensive accounts of the desertification and associated security issues in the region, as well as the state of natural resources degradation in some countries and proposed policy measures for action, have been given elsewhere (Kepner *et al.*, 2006; Ouassou *et al.*, 2006). This chapter examines the existing interactions between cereal cropping systems and pastoral areas, as the base for sustainable agriculture development in Mediterranean countries. Ways to reconcile delivery of ecosystems services to farmers and herders with biodiversity conservation, for a better integration of agriculture and livestock in the region, are also proposed.

27.2 Agriculture and Livestock Integration in Northern and Southern Parts of the Mediterranean

Agriculture in the region is characterized by diverse and complex production systems, which reflect the wide range of biophysical environments of the Mediterranean. These systems can be grouped into four main categories: (i) cultivation of cereal crops (wheat, barley and maize) and food legumes (broad beans, chickpeas and lentils) in rain-fed areas; (ii) olive, vine and other tree crops in dry areas; (iii) fruits and vegetables in irrigated perimeters, often associated with industrial crops (sugarbeet, sugarcane, cotton and sunflower) and irrigated forages (lucerne, berseem, maize and sorghum); and (iv) pasture-based animal production systems, which maintain complementary relationships with the three other types of production systems. Complementarities between the production systems in southern Mediterranean countries differ from those prevailing in the northern countries. In fact, cropping systems vary widely between north and south of the Mediterranean, particularly in terms of cereal cultivation and animal husbandry. In southern countries, only limited acreages of forage crops are cultivated and animal feeding relies heavily on natural pastures, on cereals in different forms (green, grain, straw and crop residues, and industry by-products), on fallow land, and, to some, extent on commercial feed. There are strong interactions between cereal-cropping and livestock-feeding systems. Integrated cereal–livestock production systems mainly prevail in arid and semi-arid zones of the southern part of the Mediterranean. In the northern part, there are also interactions between cereal cropping and livestock but with peculiar differences that could be analysed in terms of:

- Fallow lands in relation to integrated cereal–livestock production systems.
- Prairie grass in relation to animal husbandry and livestock production systems.
- Pastures in marginal lands and forest pastures in relation to animal husbandry.
- Complementarities between regions in relation to land use for crops and livestock.

27.2.1 Importance and role of fallow lands in cereal–livestock integration

Fallow land practices in the Mediterranean can be traced back to earlier rain-fed cropping by the Romans (Sagne, 1950). Since then, some practices have disappeared from temperate zones, due to intensive cropping, but many forms of fallow cultivation still prevail in the Mediterranean region. Fallow practices have different functions in the southern and northern countries of the region. In the north, evolution of the Common Agricultural Policy (CAP) has led to using fallow land as a tool to regulate production and market prices of cereals, oil and protein crops. When such commodities are overproduced, farmers are encouraged, through subsidies, to leave their land fallow; conversely, in the case of low production, they would cultivate their fallow lands[2]. In southern countries where cereal cultivation dominates the cropping systems as a result of limited rainwater potentials, different forms of fallow land are still used and millions of hectares are left fallow annually. Here, integrated rain-fed cereal–small ruminant production systems cannot be replaced by other types of sustainable land use. This is because livestock is able to provide valuable services to large populations and communities, through the use of extended areas and diverse agroecosystems. The cereal–livestock production systems also provide additional revenues, through income-generating activities, for local communities, while ensuring environmental conservation (Abbas and Abdelguerfi, 2005). Such utilities and services of the arid and semi-arid ecosystems of the Mediterranean may be quantified and valued (e.g. Carrière and Toutain, 1995). Other forms of crop–livestock interactions could also exist even on arable farms, like, for instance, renting fallow land as a grazing pasture to other farmers and herders, particularly in the event of a pronounced drought (Benniou and Aubry, 2009). During dry years, the pastoral communities move their livestock from drought-hit zones to zones with normal rain, which defines the transhumance system. In integrated cereal–livestock systems of North Africa, cereal crops contribute to animal feeding in three different forms: direct use as grains and straw (barley) or as hay (oats), in addition to industrial by-products (all cereals), making a total contribution to forage ration energy of about 75% in Algeria and 45% in Morocco and Tunisia (Jouve *et al.*, 1995); winter grazing as green forage, when grain production is compromised by lack of early spring precipitation (dual purpose barley, wheat and triticale); and grazing fallow and crop residues (all cereals), as part of the cereal–fallow rotation.

Previous studies on fallow management (Oussible and Bourarach, 1998) and recent reports (Benniou and Aubry, 2009) identified two periods for soil tillage depending on bioclimatic zone and farm size. For large farms, spring cultivation of fallow land accounted for 30–50%, while the remaining fallow area was allocated for grazing prior to late autumn soil cultivation when early rains occur. Small-scale farmers usually delay the cultivation period of their fallow land, so that livestock can benefit both from early grazing of cereal residues after harvest, and from late grazing of forage resources during the fallow period. Farmers with less intensive livestock production prefer to delay fallow cultivation further, at the expense of expected higher yields from early cereal cultivation. This is because they greatly appreciate the 'free' contribution of fallow grazing to livestock feeding, indicating again the strong relationships between cereal cropping and animal husbandry in the Mediterranean. On the other hand, farmers clearly prefer to use traditional cereal varieties, with their relatively high biomass-to-grain ratios, for feeding to their livestock. Thus, cereal cultivation in the Mediterranean is greatly associated with animal husbandry in terms of land use and technical interventions such as soil tillage in spring and autumn periods, choice of cultivars, and resource allocation to grazing.

In North Africa, fallow cultivation has always been viewed by policy makers as a constraint to intensification, which could be alleviated through drastic reduction of the fallow acreages. Planned agricultural development measures, however, were not successful in reducing fallow land, despite the proposed incentives (Abbas *et al.*, 2006). The first measure was to encourage adoption of an *intensive technical package for cereal production* and early cultivation of fallow land in rain-fed zones with deep soils and average annual rainfall *above 400 mm*; the second measure was to replace cereal–fallow rotation by the Australian *ley farming system* in cereal production zones with average rainfall *below 400 mm annually*[3]. The proposed ley farming system consisted of introducing a self-regenerating forage legume, such as annual medics and sub-clovers, or a food legume crop, during the fallow phase of the rotation. The ley farming options for cereal and forage intensification were encouraged by regional applied research centres such as the International Centre for Agricultural Research in Dry Areas (ICARDA), and were also thoroughly investigated by the National Research and Extension Systems (NARES) in most Mediterranean countries. The potential benefits of the ley farming system – providing valuable forage biomass, increasing cereal yields, building up soil fertility and preserving biodiversity – were extensively investigated in experimental plots and on-farm trials in most southern countries of the region (Ameziane *et al.*, 1989; Abdelguerfi, 1993) and in the north (Masson and Gintzburger, 2000; Crespo, 2006). Despite encouraging research findings, technology transfer of this system to Mediterranean countries did not achieve the expected results at the farm level. Reeves (1993) confirmed this assessment and questioned whether ley farming in Mediterranean zones was just a passing phase. Halse (1993) summarized Australian attempts to introduce ley farming to North Africa; he also assessed major problems encountered and important constraints to introducing the ley farming system in the Mediterranean basin.

27.2.2 Importance of grasslands for development of livestock systems and environmental conservation

Grasslands and native pastures form the basis of forage resources available to ruminants while providing valuable environmental services to communities (see MAP France, 2002). Forage production not only contributes to raising animal production and agricultural productivity but also to ensuring good quality products, protecting the environment, keeping landscape quality, and ensuring viable economic turnover in rural areas. Despite these advantages, however, much acreage of prairies has disappeared from both sides of the Mediterranean. In the north, progressive intensification of agricultural production since the 1960s, and the associated economic and social constraints, have led to more specialized agriculture, with a clear separation between livestock farming, which relies on grasslands, and field crops such as cereals. This move created specialized territories, with intensive cereal production plains on the one hand and concentrations of intensive livestock production systems on the other (Lemaire, 2007). In the southern part, natural grasslands also regressed as a result of cereal 'extensification'. The case of Algeria provides a good illustration, with acreages of natural prairies being around 1 million ha early in the 20th century but only some 20,000 ha in 2000. Continuous destruction of native species has occurred as a result of a lack of development planning and also a lack of incentives (Abbas *et al.*, 2006).

Recent analysis of the multiple roles and functions of French grasslands over two centuries indicated that the main changes in grassland dynamics over this period essentially occurred during the last few decades. In this review, Huyghe (2008) observed an increased acreage of annual forage crops, a strong decline of perennial forage legumes and a reduction in permanent grasslands. In parallel, numbers of sheep and dairy cows decreased while the number of suckling cows increased, leading to modifications in the types of grassland and forage crop and in their management regimes. Changes in the location of the animal production and a specialization of the territories were also induced. Using systemic analysis of the supply chain in France, Huyghe (2009) identified three functions as being major for all stakeholders. The first is the contribution of livestock farming to economic performance, and, more specifically, to the balance between farmer income and workload. The second function is the contribution to quality of animal products, which secures their access to markets and satisfies consumer expectations. The third function is environmental conservation, given the fact that, under adequate management, grasslands significantly reduce soil erosion, limit nitrogen loss and reduce fossil energy consumption through reducing use of nitrogen fertilizers and feed concentrates. Species diversity and plant life cycle duration in grasslands may also preserve flora and fauna biodiversity. Therefore, the contribution of grasslands to farm economic performance and environmental conservation function should be reconciled and rehabilitated in the Mediterranean.

27.2.3 Importance of marginal land and forest pasture in providing support services to livestock husbandry

In the Mediterranean, livestock feeding has always benefited from non-cultivated lands, shrublands and forests. Hence, agro–silvo–pastoral systems greatly dominate the Mediterranean landscapes. This has resulted from a combination of factors, including soil and climate characteristics, land-use patterns and the social/cultural/economic setup of human societies across the region (Joffre *et al.*, 1987). Forage productivity of marginal land, shrubs and forest pastures varies from region to region and mainly depends on rainfall, soil type, botanical composition of plant covers and stocking rate. Amount of precipitation and overgrazing greatly affect productivity and sustainability of the pastures. The silvo-pastoral systems generally suffer from low nutrient (N, P, K) status, since animal manure produced locally is often exported to cultivated land (Qarro, 1991). Several studies have shown the benefits of agro–silvo–pastoral systems, such as the Spanish dehesa (Joffre *et al.*, 1987) and the espinal agroforestry system of the Mediterranean-type climate region of Chile (Ovalle *et al.*, 1990), for soil fertility conservation. The benefits of such production systems for controlling fire hazards are also well documented (Abbas *et al.*, 2006). Mediterranean pastures under forest cover, like in Morocco, greatly contribute to livestock feeding, particularly during hard times of cold winters or long periods of drought. Qarro *et al.* (1996) found that forest management seeking 60–70% of oak tree cover was ideal for both timber and livestock production. Also, very rich herbaceous strata were found under forest cover in the natural pine ecosystem (*Pinus pinaster* var. *maghrebiana*) of the central Mid-Atlas of Morocco. Belghazi *et al.* (1996) identified a large number of useful species including grasses (six perennials, three annuals), legumes (seven perennials, ten annuals) and composites (eight perennials, three annuals), indicating the important role of forest pastures in biodiversity conservation.

In a recent study on modelling relationships between land-use dynamics and livestock farming systems in the Luberon area of Mediterranean France, Lasseur *et al.* (2010) provided a clear illustration of the importance of taking the diversity of present forms of sheep farming (structural characteristics, farming practices and animal husbandry) into account to forecast future relationships between agro–silvo–pastoral activities and landscape dynamics. They also referred to complementarities between regions for livestock feeding, which enable animals to contribute significantly to increased production, income generation, and improved sustainability of diverse annual and perennial cropping systems in the Mediterranean.

27.2.4 Diversity of land use and complementarities between regions for livestock feeding

The integration of crop and animal production is well developed in the farming systems of the Mediterranean, particularly in small-scale agriculture. This is because it is common to invest in livestock as a source of savings to buffer against crop failure. There are also marked complementarities in resource use in these systems, with inputs from one system being supplied to others, or feeding resources of one region being used in other regions. These interactions reflect the interdependence of regions for livestock feeding but also the secular roles of nomadism and transhumance in the Mediterranean. At higher altitude and in marginal steppe zones, it is not possible to feed livestock locally all the year round. During winter (October to March), cropped cereals and the fallow areas of the vast plains generally provide shelter for livestock, conserved forage hay, green plants from the fallow and cut or grazed forage crops, particularly dual purpose barley up to late spring/early summer (April–May). Cereal stubble left after harvest is then grazed before moving the livestock to high-altitude pastures for grazing during the hot summer months, usually until around the first snow in December. Agro–silvo–pastoral and pastoral systems still represent important forms of land use in the southern Mediterranean, although these systems are undergoing profound transformations of a social, economic and environmental nature. Compared with northern Mediterranean systems, the pastoral production systems of North Africa rely heavily on the mobility of herds and herders and on the existence of vast collective pastoral zones with different agroecological attributes. These territories greatly contribute to livestock feeding, and their use is governed by an arsenal of legislative and cultural regulations, including religious as well as traditional and modern laws to resolve conflicts within and/or between communities arising from the mobility of herders and their livestock (Bourbouze, 2000). Attempts to improve agriculture–livestock integration and conserve biodiversity of these fragile ecosystems should consider the cultural and regulatory values of local communities.

27.3 Proposed Actions for Improving Agriculture–Livestock Integration, Preserving Biodiversity and Sustaining Ecosystem Services in the Mediterranean

27.3.1 Paying farmers for environmental services

Farmers manage biodiversity at different scales (spatial, temporal and institutional) but their decisions are primarily driven by the benefits they can get in the short term, particularly in terms of provisioning services. McIvor (Chapter 5, this volume) discussed the importance of spatial and temporal scales in the management of grassland resources. The institutional dimension of biodiversity management was recently discussed by Jackson *et al.* (2010), who argued that, institutionally, the concept of ecosystem services has helped to bring ecosystem issues to the level of policymakers. However, 'current institutional policies rarely offer adequate incentives to promote management systems that support biodiversity as a public good, for example, for its option value for improving provisioning and regulating services in the future'. For the temporal scales of biodiversity management, Jackson *et al.* (2010) discussed three groups of concerns that should be addressed in relation to delivery of ecosystem services (Table 27.1): 'first, the short term flow of goods and services on which financial returns and *efficiency* are focused; second, *persistence* and continuity through investment in assets and stocks, upon which sustainability is focused; and third, *human capacity* to deal with *change*, to ensure future agroecosystem functioning in ways that include current likely options as well as those that are still unknown'. Therefore, improving delivery of ecosystem services would require paying farmers for environmental services, in the form of incentives, as was advocated by FAO (2007) based on case studies of the relationship between biodiversity, natural resource degradation and poverty alleviation.

27.3.2 Revisiting the ley farming system

The dynamics of landscapes during the past decades has been characterized by an ongoing intensification of land use, and an increasing number of former meadows and pastures becoming either fallow land in the northern part of the Mediterranean or cropped to cereals in the southern countries, as a result of changing social and economic conditions. Among the proposed actions that could have reshaped agricultural landscapes of the

Table 27.1. Perspective on ecosystem services in agricultural landscapes and their association with different types of agricultural time frames (Jackson *et al.*, 2010).

	Ecosystem service			
Temporal scale	Provisioning	Regulating	Cultural	Supporting[a]
Efficiency	+++	+	+	( )
Persistence	+	++	++	++
Change	( )	+	+	+++

[a] Supporting services underpin other types of ecosystem services, such as the genetic basis for evolutionary change.
( ) Unknown.

Mediterranean was the introduction of the ley farming system to replace cereal–fallow rotations with cereal–forage crops, for better crop–livestock integration. In Australia, self-regenerating forage legumes like annual medics and subterranean clovers are widely used in rotation with cereals and form the basis of an integrated and sustainable cereal–livestock production system. This innovative approach was tested on a large scale in North Africa and West Asia. Although adaptation of this system to the Mediterranean basin did not achieve the expected development results, a wealth of related research findings already exists in the region. These findings need to be revisited and recommendations made on how to design viable management options for adoption of the ley farming benefits. With the available genetic resources and appropriate management practices, the development of ley farming could improve forage production and quality, increase cereal yields, build up soil fertility and preserve local biodiversity. This would require substantial investment in a new set of accommodations, especially in developing technical packages that are economically feasible, socially acceptable and environmentally in favour of biodiversity conservation.

27.3.3 Rehabilitating the function of grassland and fallow development

In the northern Mediterranean, interactions between animal and grain-producing farms must be reassessed for biodiversity conservation purposes, including the role of the grassland and fallow land, in view of the re-shaping of territories already fully specialized in either animal or grain production. The contribution of grasslands to farm economic performance and their implication in product quality as well as their environmental conservation function should be underlined and rehabilitated. The fallow area strategy initiated by the European CAP might be an opportunity for agriculture development and biodiversity conservation, including the conservation of pollinator insects. Decourtye *et al.* (2007) discussed advantages for farmers and beekeepers of introducing flowering fallows in field crop zones, which could offer various sources of nectar and pollen to bees and other pollinators. They rightly argued that European rules and regulations, as well as agricultural practices regarding fallow areas, must evolve in order to help insect populations to develop. Sowing fallow areas with melliferous and polleniferous species would be a good contribution to promotion of beekeeping and protection of biodiversity in the Mediterranean. Thus, sustaining the advantages for farmers and beekeepers resulting from fallow development in cropped zones of Mediterranean Europe is an important policy issue.

27.3.4 Strengthening agro–silvo–pastoralism and complementarities between regions

In the Mediterranean, three major agricultural zones could be considered: the higher rainfall (rainfall >600 mm), cereal (rainfall 350–600 mm) and steppe zones (rainfall 200–350 mm). Animals are moved from the pastoral steppe lands to the higher rainfall zones to allow fattening before slaughter, or to rear another lamb under more favourable conditions with a more regular feed supply. In fact, decision making in semi-arid areas is based upon obtaining reasonable levels of production with a minimum of risk. Livestock and cereal production are integrated as buffering factors to control the consequences of variable rainfall, particularly in the southern Mediterranean where traditional farming systems were much diversified. Nowadays, farming systems are becoming market-oriented mixed crop–livestock farming, with more specialized production systems in the northern Mediterranean. A recent review indicated that, in several countries, large-scale, nature-adapted pastoral systems could offer an alternative to industrial livestock raising and keep a high biodiversity on the landscape level (Plachter and Hampicke, 2010). Many international organizations and financial agencies, with the United Nations development Programme (UNDP) as the implementing agency, are advocating the World Initiative for Sustainable Pastoralism (WISP; www.iucn.org/wisp). This seeks a greater recognition of the importance of sustainable pastoral development for both poverty reduction and environmental management. Support of Mediterranean agro–silvo–pastoralism, in a way that is similar to WISP methodology and incentives, could form the basis for reconciling livestock development and biodiversity conservation in the region.

27.4 Conclusions

The principal characteristics of the Mediterranean climate are extreme seasonal and inter-annual variability of rainfall distribution, and pronounced aridity, which increases vulnerability to drought. Irrigated agriculture is well developed in some countries but most still depend heavily on rain-fed cereal-based farming and extensive livestock production systems. Land degradation, drought and desertification are major concerns and threaten the sustainability of the farming systems under severe water-scarcity conditions. Against this background, however, integration of crop and animal production is well developed in the farming systems, particularly those in the small-scale agriculture of southern countries. Marked complementarities in resource use in these systems are identified, with inputs from one system being supplied to others, or feeding resources of one region being used in other regions. Livestock feeding has always benefited from cereal-cropped land and non-cultivated areas, grassland and rangeland, shrubs and forests. These interactions reflect the secular roles of the agro–silvo–pastoralism, nomadism and transhumance that dominate the Mediterranean landscapes. Although integrated crop–livestock production systems in the region are complex and country specific, they are generally tending towards intensification of crop and livestock production, with cropping systems tending to monoculture, particularly in the northern part of the Mediterranean.

For a better delivery of ecosystem services and sustainable cereal/pastoral-based agricultural development in the Mediterranean countries, decision makers need to readjust policies and take actions aimed at: (i) paying farmers and herders for their environmental services; (ii) revisiting the ley farming system for its adoption in the Mediterranean basin; (iii) rehabilitating the function of grasslands and fallow land; and (iv) strengthening agro–silvo–pastoralism and complementarities between regions. Globally, the Mediterranean countries need to design and implement a comprehensive programme for agricultural and rural development adapted to each of the agroecological zones of the region. This would require investment in a new set of accommodations, especially in developing new crop varieties and technical packages that are environmentally and economically safe and socially acceptable, broadening the forage resources base, developing efficient feeding systems, tailoring research and extension knowledge base, scaling up research findings where appropriate, and developing value-added products and easing their access to markets. Strengthening national research and extension institutions and capacity building of civil society should be part of the investment, in order to cope with the continually evolving production systems and sustain ecosystem services in the context of the water scarcity and changing climate of the Mediterranean region.

Notes

1 According to Le Houérou (1971), the Mediterranean basin enjoys climatic similarities with other regions of the world (California, South Africa, Chile and parts of Australia), but also shows important climatic differences from these regions.

2 Unfavourable climatic conditions and related drought during 2010 led to reduced forage availability, which convinced the French authorities to authorize the grazing of fallow lands, where applicable.

3 Subsidies were allocated for purchasing seeds and soil tillage equipment for the ley farming system.

References

Abbas, K. and Abdelguerfi, A. (2005) Perspectives d'avenir de la jachère pâturée dans les zones céréalières semi-arides. *Fourrages* 184, 533–546.

Abbas, K., Abdelguerfi-Laouar, M., Madani, T., M'Hammedi Bouzina, M. and Abdelguerfi, A. (2006) Place des légumineuses dans la valorisation de l'espace agricole et pastoral en régions Nord d'Algérie. In: Abdelguerfi, A. (ed.) *Proceedings of International Workshop "Diversity of Fodder Fabaceae and their Symbionts: Biotechnological, Agronomic, and Environmental Applications"*. Algiers, Algeria, pp. 309–320.

Abdelguerfi, A. (1993) The use of annual medics in pasture systems in Algeria. In: Christiensen, S., Materon, L., Falcinelli, M. and Cooks, P. (eds) *Introducing Ley farming to the Mediterranean Basin.* ICARDA, Aleppo, Syria, pp.135–143.

Ameziane, T.E. (2002) Drought preparedness and risk management in the Mediterranean Region. In: International Union of Nature Conservation (ed.) *Proceedings of International Symposium "The Mediterranean Water Week"*. Athens, Greece.

Ameziane, T.E., Mazhar, M. and Berkat, O. (1989) Seed reserve and self-regeneration of annual medic pastures in a Mediterranean environment. In: *Proceedings XVI International Grassland Congress, Montpellier, France,* Vol II. Association Français pour la Production Fourragère, Versailles, France, pp. 1545–1546.

Belghazi, B., Ezzahiri, M., Qarro, M., Sabir, M. and Didi, H.Z. (1996) Relations entre les potentialités ligneuses et pastorales dans les peuplements naturels de Pin Maritime; cas de la pinède de Tamrabta. *Annales de la Recherche Forestière du Maroc,* 47-57.

Benniou, R. and Aubry, C. (2009) Place et rôle de l'élevage dans les systèmes de production agricole en régions semi-arides de l'est de l'Algérie. *Fourrages* 198, 239–251.

Bourbouze, A. (2000) Pastoralisme au Maghreb: la révolution silencieuse. *Fourrages* 161, 3–21.

Brown, O. and Crawford, A. (2009). *Rising Temperatures, Rising Tensions: Climate Change and the Risk of Violent Conflict in the Middle East.* International Institute for Sustainable Development, Toronto, Canada.

Carrière, M. and Toutain, B. (1995) *Utilisation des Terres de Parcours par l'Élevage et Interaction avec l'Environnement. Outils d'Évaluation et Indicateurs.* CIRAD/IEMVT, Paris, France.

Crespo, D.G. (2006) The role of legumes on the improvement of the grazing resources and the conservation of the "montado/dehesa". In: Abdelguerfi, A. (ed.) *Proceedings of International Workshop "Diversity of Fodder Fabaceae and their Symbionts: Biotechnological, Agronomic, and Environmental Applications"*. Algiers, Algeria, pp. 298–308.

Decourtye, A., Lecompte, P., Pierre, J., Chauzat, M. P., and Thiebeau, P. (2007). Introduction de jachères florales en zones de grandes cultures: atouts pour agriculteurs et apiculteurs. *Cahiers Agricultures* 16, 213–218.

FAO (2007) *The State of Food and Agriculture: Paying Farmers for Environmental Services.* Food and Agriculture Organisation, Rome, Italy.

FAO (2002). *The State of Food and Agriculture 2002: Agriculture and Global Public Goods Ten Years After the Summit.* Food and Agriculture Organisation, Rome, Italy.

Halse, N.J. (1993) Australian attempts to introduce the ley farming system to West Asia and North Africa. In: Christiensen, S., Materon, L., Falcinelli, M. and Cooks, P. (eds) *Introducing Ley farming to the Mediterranean Basin.* ICARDA, Aleppo, Syria, pp.1–14.

Huyghe, C. (2008) La multifonctionnalité des prairies en France: I. Les fonctions de production. *Cahiers Agricultures* 17, 427–435.
Huyghe, C. (2009) La multifonctionnalité des prairies en France: II. Conciliation des fonctions de production et de préservation de l'environnement. *Cahiers Agricultures* 18(1), 7–16.
Jackson , L., Noordwijk van M., Bengtsson, J., William, F., Lipper, L., Pulleman, M., Said, M., Snaddon, J. and Vodouhe, R. (2010) Biodiversity and agricultural sustainagility: from assessment to adaptive management. *Current Opinion in Environmental Sustainability* 2, 80–87.
Joffre, R., Vacher, J., de Los Llanos, C. and Long, G. (1987) The dehesa: an agrosylvopastoral system of the mediterranean region with special reference to the Sierra Morena area of Spain. *Agroforestry Systems* 6, 71–96.
Jouve, A.M., Belghazi, S. and Kheffache, Y. (1995) La filière des céréales dans les pays du Maghreb: constante des enjeux, évolution des politiques. Les agriculteurs maghrébins à l'aube de l'an 2000. *Options Méditerranéennes* 14, 170–192.
Kepner, W.G., Rubio, J.L., Mouat, D.A. and Pedrazzini, F. (eds) (2006) *Desertification in the Mediterranean Region: a Security Issue.* Springer, Amsterdam.
Lasseur, J., Bataille, J-F., Beylier B, Etienne, M., Legeard, J-P. and Luccioni, J. (2010) Modélisation des relations entre dynamiques des territories et des systèmes d'élevage dans le massif du Luberon. *Cahiers Agricultures* 19, 90–96.
Le Houérou, H.N. (1971) Le rôle de l'écologie végétale dans les études de mises en valeur de la région méditerranéenne. *Bulletin de la Recherche Agronomique Gembloux,* Vol. Ext., 68–87.
Lemaire, G. (2007) Interactions entre systèmes fourragers et systèmes de grandes cultures à l'échelle d'un territoire. Intérêts pour l'environnement. *Fourrages* 189, 19–32.
Masson, P. and Gintzburger, G. (2000) Les légumineuses fourragères dans les systèmes de production méditerranéens: utilisations alternatives. *Options Méditerranéennes,* 395–406.
MAP France (2002) Vers une intégration de la diversité biologique et paysagère pour une agriculture durable. In: *Conférence Paneuropéenne à Haut Niveau sur l'Agriculture et la Biodiversité, Strasbourg, 4 March 2002.* La Documentation Française, Paris, France, pp. 94–186.
Ouassou, A., Ameziane, T.E. and Lajouad, L. (2006) State of natural resources degradation in Morocco and plan of action for desertification and drought control. In: Kepner, W.G. *et al.* (eds) *Desertification in the Mediterranean Region: a Security Issue.* Springer,, Amsterdam, pp 251–258.
Oussible, M. and Bourarach, E.H. (1998) *Installation Précoce des Cultures d'Automne. Cas des Céréales.* Bulletin de liaison du programme national de transfert de technologie en agriculture (PNTTA). Institut Agronomique et Vetérinaire. Hassan II, Rabat. http://www.vulgarisation.net/01-49.htm.
Ovalle, C., Aronson, J., Del Pozo, A. and Avendano, J. (1990) The espinal agroforestry systems of the Mediterranean-type climate region of Chile: state of the art and prospects for improvement. *Agroforestry Systems* 10(3), 213-239.
Plachter, H. and Hampicke, U. (eds) (2010) *Large-scale Livestock Grazing: a Management Tool for Nature Conservation.* Springer, Berlin.
Qarro, M. (1991) Facteurs de variation des teneurs en azote du sol des parcours de la région d'Aïn Leuh (Maroc). In: 4th *International Rangeland Congress.* Association Française de Pastoralalisme, Montpellier, France, pp. 226–228.
Qarro, M., Sabir, M., Belghazi, B. and Ezzahiri, M. (1996) Effets des traitements sylvicoles sur le développement des potentialités herbagères dans les taillis de chêne vert (*Quercus ilex* L.). *Annales de la Recherche Forestière du Maroc,* 57–59.
Reeves, E.M.A. (1993) Is ley farming in Mediterranean zones just a passing phase? In: *Proceedings of the XVII International Grassland Congress, Auckland, New Zealand.* New Zealand Grassland Association, Dunedin, New Zealand, pp. 2169–2177.
Sagne, J. (1950) *L'Algérie pastorale : ses Origines, sa Fonction, son Passé, son Présent,* son Avenir. Fontana, Algiers.
Talamucci, P. and Chaulet, C. (1989) Contraintes et évolution des ressources fourragères dans le Bassin Méditerranéen. In: *Proceedings XVI International Grassland Congress, Montpellier, France,* Vol. III. Association Français pour la Production Fourragére, Versailles, France, pp. 1731–1740.

28 A Way for Developing an Integrated System at the Landscape Level in Tropical Areas

René Poccard-Chapuis and Philippe Lecomte

28.1 Landscape Changes

Extending from the boundaries of the equatorial rainforests to the temperate highlands or drier biomes, 'savannahs', 'llanos', 'cerrados' and other grasslands are a central component of tropical landscapes. After a long period of unperturbed climax, large parts of tropical grasslands enter a succession (sere), and such landscapes are also highly affected by increasing human activities.

The ways in which landscapes are affected and respond to these trends are highly variable. Land-use change is complex and dragged along by a range of drivers that are regionally specific. In many tropical countries, land abundance has for some time slowed the introduction of new technologies that can raise productivity (Rosegrant *et al.*, 2009). The basic resource endowments (pedo-climate, land tenure, labour and capital), the role of agricultural and livestock sectors, and the socio-economic and institutional frameworks, are major determinants of the differing evolution processes. Where forest, crop and/or livestock activities interact, significant changes are also taking place in the organization of supplies procurement, processing and products distribution (Steinfeld *et al.*, 2010).

Human populations in these tropical ecosystems are predominantly rural, and a significant proportion of the active people are engaged in agriculture. In absolute terms FAO (2006) estimates that the agricultural population will continue to increase, although the proportion of the total population involved in agriculture is declining.

Arable and grazing lands in these tropical landscapes have expanded considerably (Steinfeld *et al.*, 2006). During the past 20 years, large forest conversions to crop and grassland have occurred in the Amazon Basin, Central and West Africa and South-east Asia, while forest area has increased, as a result of agricultural land abandonment, in other non-tropical regions. Considerable expansion of maize croplands for energy and of soybean as a protein source in animal feed has occurred in Latin America. Actually, confined livestock production systems in industrializing countries are the source of much of the world's short-cycle meat production, and such systems are very rapidly expanding in tropical landscapes. To meet the increasing demand, Bruinsma (2003) estimates that at least 75% of total production growth until 2030 will be in Asian and Latin American systems, with far less growth of these confined systems in Africa. In these intensifying mixed systems, forage, food–feed crops and livestock are vital.

Many grass and croplands have been converted to other uses, including urban development around many major cities.

28.2 System Integration at Landscape Level

In the future, grazing systems will also increasingly provide ecosystem goods and services (Millenium Ecosystem Assessment, 2005) that will be traded. How tropical grassland areas and systems may be affected is not clear. The mixed crop–livestock systems will continue to be critical to future food security, as two-thirds of the global population live in these systems. Some of the higher potential mixed systems in Africa are already facing resource pressures. Various responses are possible, including efficiency gains and intensification options (Herrero *et al.*, 2010). Increasing competition for land also comes from large private sector investments in land for industrial cropping, for food or biofuels. In tropical grassland regions, future scenarios on bioenergy use raise difficult questions (van Vuuren *et al.*, 2009), and there are large gaps concerning the likely trade-offs between food, feed and fuel in production systems, particularly when one considers second-generation bioenergy technologies.

Driven by sustained concerns about climate change, food or energy security and speculative income sources, such large investments tend to be detrimental to small agricultural households' livelihoods. Compared with an average expansion of global agricultural land of less than 4 million ha in 2008, 45 million ha worth of large-scale farm deals were announced in 2009. More than 70% of such demand has been in Africa, and poor countries such as Sudan Ethiopia, Mozambique and Madagascar have transferred several millions of hectares to investors in recent years (World Bank, 2010).

Landscapes are defined here as the areas, as perceived by people, whose character is the result of the action and interaction of natural and/or human factors (ELC, 2000). They are the level at which several fundamental ecosystem services are defined and directly implemented (Millenium Ecosystem Assessment, 2005). To procure these services, land-use practices have to integrate landscape dimension and occurring dynamics. It is at this level that the nutrient flows between subsystems can be considered, particularly in African agricultures where a large part of the resources are located out of the households, on collective grassland areas. Reasoning the systems integration at the landscape level may also allow the improvement of production functions and marketing chain performances, due to proximities for producing forage and feed, and for the nutrient recycling capabilities.

The evaluation of the overall ecosystem services on a landscape scale allows us to set up value chains and qualification processes in the socio-ecosystems, an important tool for the sustainable development and dynamic of a green economy (Staal *et al.*, 2008).

Several assessments justify that attention should be paid to the links between land-use systems and to the ways of optimizing the whole landscape system functioning. The ecosystem functions of agriculture are mainly organized at the landscape scale (biodiversity, water cycle, soil characteristics and ground-water reserves). Promoting these functions to improve production systems, under an ecological intensification precept, operates at such scale. The logic is one of 'eco-efficient' landscapes, determined by the relative locations of different land uses. The resources mobilized by agricultural activities are produced and used in differing parts of the landscape, particularly regarding animal nutrition and soil fertility. The system of resource flows operates at this scale. Optimization of these flows, which is an interesting alternative for improving performance in animal production and plant management of natural resources, is also designed at this scale. This is particularly true in systems where mobility is central and operates through livestock movements, crop rotations or shifts. The focus here is on the adequacy of agricultural practices and provision of facilities to manage the flow of matter and resources between the landscape compartments.

28.3 Three Case Studies

The more we focus on long-term developments, the more important it is to reason on a landscape scale, the aim being to characterize the temporal evolution and territorial dynamics of land-use systems. These concepts are illustrated here, using three case studies.

28.3.1 Village systems: 'village land' of the Sudanese zone of Mali

In Mali, the aim is to optimize flow management of organic matter and nutrients between cropping systems and livestock farming systems (Fig. 28.1); specifically:

1. The production of forage and food for cattle, so that they do not solely depend on communal grasslands. Rather than forage itself, it is crop residues or dual-purpose plant (i.e. maize, sorghum and mil) cultivation that provide the most. Co-products are either browsed in the field after harvest or stored and used as fodder during the dry season.

2. The production of organic matter to fertilize the fields. This fertilizer is broadcast on the field, alongside animal trampling and common grazing on cropland residue, or produced and stored in locally adapted facilities, pits or parks, combining a mix of crop, domestic residues and animal dung. These forms of integration address and alleviate two critical points of the system: decreases in soil fertility and cattle feeding (low crop productivity, increasingly longer transhumance or permanent herd relocation outside the territory).

Across the Sudanian region there appears to be a great diversity of integration levels in the livestock farming systems (Rapidel *et al.*, 2006). This depends on resource pressure, conditioned by bush proportions and crop fields in the village territory (less bushes, higher pressure), and the herd size. Trees are also important; their root system will draw nutrients from deeper soil layers and the atmosphere and bring them into the surface circuit: for soil fertility (decaying litter) and feeding use (leaves) for the herds in the dry season. These systems also rely on the facilities and equipment available to the farmers: number of manure pits and parks, number of

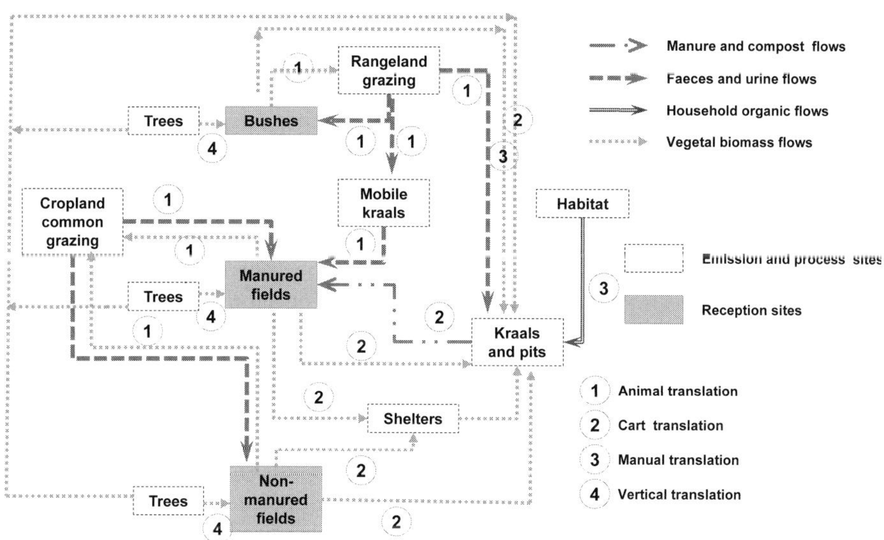

Fig. 28.1. Nutrient flows in a typical agro–silvo–pastoral landscape of South Mali.

manure transport carts, relative distances between fields and ditches, and number of fodder storage places, the latter being usually located close to dwellings.

In South Mali there is currently a transition between a fallow system to maintain soil fertility, associated with savannah grazing to feed livestock, and crop–livestock integration. Sub-humid forest is still largely present, with fertile forest soil, a high biomass production, and a rather limited population. The fallow system has enough space to function; when set aside, individual fields become a collective 'bush' (fallow), where animals of the village graze all year round. Bushes offer a resource for transhumant herds from the north, where forage is seasonally lacking. The only flow practice between farming systems consists of rotating night kraals with large numbers of animals on parcels of land where cereals or cotton will be cultivated the following season.

The situation in the north is the opposite. With high population pressure and low rainfall, open rangelands become rare, land tenure is saturated, and the fallow system has been abandoned. To manage soil fertility and feed herds, technical and organizational innovations have been developed. Mineral fertilizers are used on all permanent fields, according to their low cost and to the credit facilities available. Peasants have produced endogenous key innovations, like production and transportation of organic fertilizers, or the use of crop residues for forage. The new integrated system is based on specific infrastructures and herd management. Thanks to these innovations, herd and field productivities can be stabilized, despite the demographic expansion.

As a consequence of population growth, permanent fields have extended to the detriment of rangelands, including to the summit of the toposequence, where soils are poorer. Under such excessive pressure, the cattle feed supply reduces, the people go on quasi-permanent transhumance, and organic manure production and crop yields regress. At this level, the new system enters a crisis and productivities decrease.

An idea of these two landscapes and integration levels is given in Fig. 28.2. Archetypes describe contrasting situations: on the one hand Dentiola is a landscape under high anthropic pressure and integration is well performed; on the other hand Kanouala is a village where cropping pressure is such that natural biomass production is insufficient to run the system, and cattle have no more physical space.

28.3.2 Agro-industrial cluster systems of Central Brazil

In a similar tropical environment originally marked by poor soils and grassland vegetation, the landscape of Rio Verde de Goiás, Brazil, has now reached a high degree of integration between maize and soybean cropping, grassland and livestock farming. This integration is strongly based on farm nitrogen flow management. Flows integrate the industrial meat transformation chain. Nitrogen flow is based on maize–soybean direct mulch cover (DMC) seeding systems, with two harvests per year. Mixed with maize and turned into animal feed rations, soya is transformed into short-cycle white meat. Nitrogen surpluses of pig and poultry manure are spread on cultivated grasslands that are grazed to produce cattle meat. The forest has virtually disappeared and is limited to strips on the hills. Manured surfaces are considerable and nitrogen concentrations never threaten the aquifers. After a few weeks rest, grazing is allowed on highly productive *Brachiaria* grasslands, rich in nitrogen, that cattle transform into meat or milk, depending on the farms.

The chain is controlled by the slaughtering and processing industry, which has set up a whole arsenal of technologies and industrial and financial resources for poultry and pig mass production, in a basin that was previously relatively poor and only dedicated to soybean production for export. Today, 2.1 million birds are slaughtered each week; 154 farms with 100,000 heads

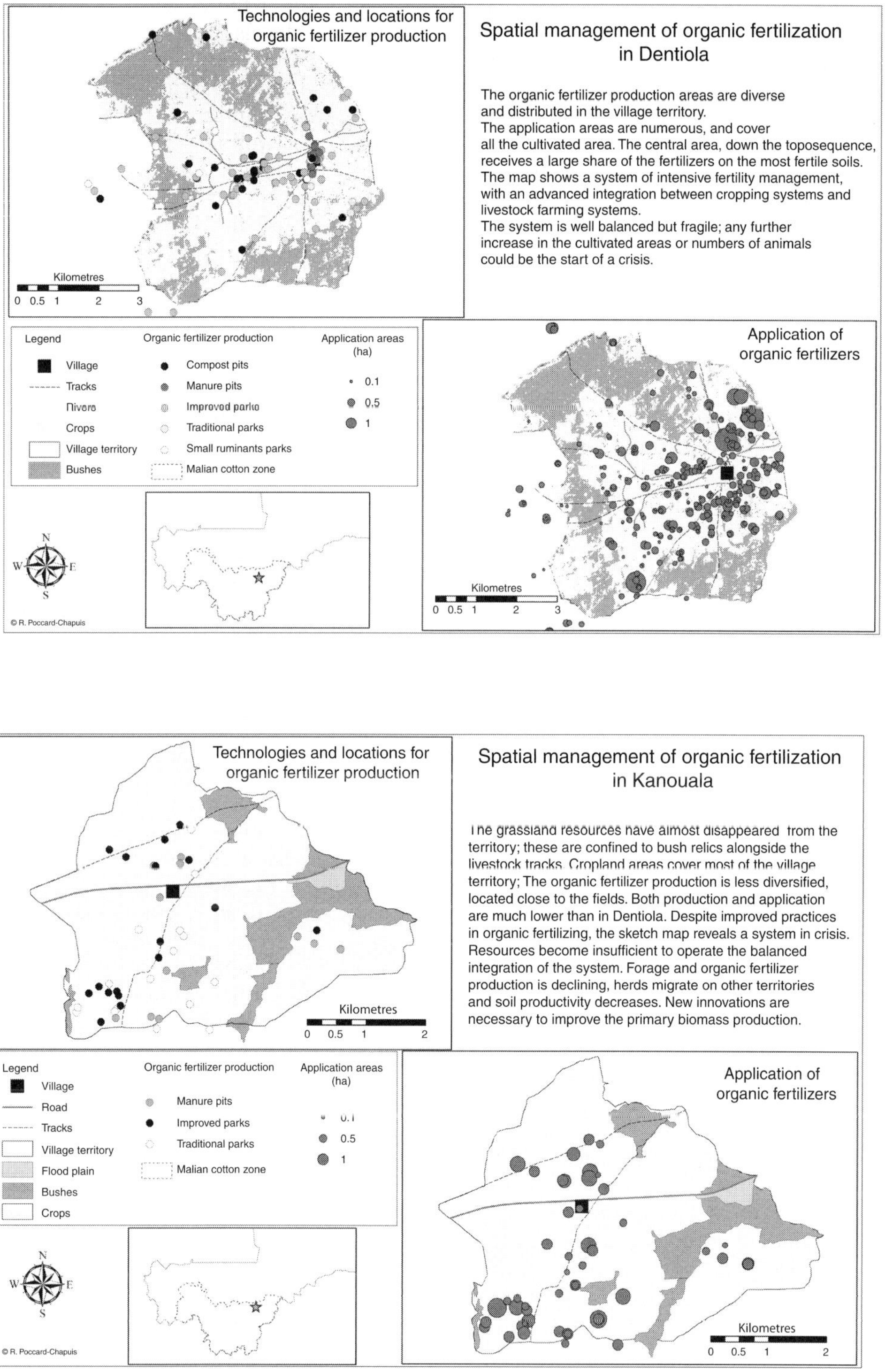

Fig. 28.2. Archetypes of village landscapes in two locations of the Sudanese zone of south Mali.

each produce 133,000 t of white processed meat each year, with more meat frozen and exported without transformation (Coutinho, 2010). Around this gigantic integrated farmers–industry system, most large firms supplying input and agricultural equipment came along. What is fascinating is that integration is piloted by the poultry industry. Recently, the local industrial partner realized that he operated poorly in the nitrogen cycle: pigs and poultry farmers had excellent pasture through manure, but the industry did not benefit, so the partner decided to start a similar system to integrate milk farmers with the rest of the industry.

28.3.3 Pioneering fronts in the Amazonian forest

In this case study, the focus is less on the nutrient flow issues and crop–livestock integration than on deforestation: continuous fragmentation of primary forests for grassland development across the country.

Livestock ranching has been associated for a long time with Amazon deforestation. This negative association has gained increased significance in recent years as the rate of deforestation has increased and the role of livestock in the process has become more pronounced (Mertens *et al.*, 2002). In the Amazon, deforestation accelerated from 1990 to 2000, to about 25,000 km^2/year. Deforestation was estimated at 0.6% per year from 2000 to 2005 (FAO, 2007). Assisted by the advantage of some Brazilian states being free of bovine disease, beef exports soared. Over the past decade, beef total export value increased tenfold, from US$190 million to $1.9 billion, making Brazil the world's largest beef holder (205×10^6 cattle heads) and exporter (Steinfeld *et al.*, 2010). This impressive change in productivity meant that beef-cattle ranching, which at the beginning of the 1990s was only profitable if subsidized and part of a land-use sequence (including cropping), became financially attractive and a major force in continued deforestation.

The integration issue here is that forest formation should keep a functional position in the landscape. In the Amazonian pioneer classical system, forest has no place. The common practice is total clearing for establishing *Brachiaria* grasslands in large blocks (Poccard-Chapuis, 2004). The smallest block is 10 ha, whereas the largest sometimes exceeds 10,000 ha. There is something systematic in these deforestations: they are often geometric, cutting uniform blocks of forest.

Actually, many farmers feel the necessity to reconstruct landscapes. For environmental reasons, there is a high need to manage the water cycle better. Surface water is scarce; all the river and source banks have been deforested and rainfall has decreased. Without forests, the sky is less cloudy and it rains only according to the seasonal intertropical movements of air masses. The dry season has become very hard and, at the end of the season, pastures are like straw: like savannah vegetation in Mali.

For legal reasons, since 2008 the federal government has increased the pressure to 'protect the Amazon' and enforce: (i) the reconstruction of riparian forests around streams; and (ii) the replanting of trees on properties, to reach a 50% proportion of forests (natural or planted), with no deforestation. However, annual crop and livestock farming systems were based on deforestation. Annual crops were produced after slash burning, pasture succession was poorly managed, as fire could control the degradation, and it was simple to open new pastures when the old ones were too degraded. Now, faced with the impossibility of continuing deforestation and extensive systems, silviculture is developing rapidly. It consists of: (i) trees planted in the pastures, to conserve and improve the long-term value of land capital; and (ii) planted forests: orchards or cocoa and other Amazonian fruits for rapidly emerging markets, or tree species for timber or for energy purposes (eucalyptus for local steelmakers, and palm oil for biodiesel). But this silviculture is only just starting; seeds and seedlings are dramatically lacking, and it is mainly the large farms that are informed and able to react in such a new context.

Many farmers are trying to legalize ownership and protect waterways. At the landscape biodiversity level, there are currently no mechanisms to optimize the connections between the forest restorations. Therefore, it is likely that the landscape will remain in a 'moth-eaten' shape, with grazing blocks between forest blocks. As a result, land occupation is relatively loose, settlements are scattered and the entire infrastructure is more expensive to maintain compared with those in completely deforested areas, where housing and infrastructures are concentrated, and those in the fully protected zones. The large blocks of forest can be subject to sustainable forest management, which is impossible in the more fragmented forests.

28.4 Integration: Opportunities and Constraints

In Central Brazil and Amazonia, the increased competition, the need for vertical integration and horizontal concentration are important constraints for industrial chains in their traditional supply basins and markets. Within such a context, survival of an industry necessitates a continuous increase in turnover. The Amazon appears to be a good opportunity for picking up cheap and raw material. The industries from the cerrados – the Brazilian savannahs – have therefore started to move towards the north, taking advantage of opportunities opened up by the improvement and electrification of the main north–south axes. For farmers this is a major opportunity: livestock becomes the sole product whose market access is guaranteed, and stable and predictable prices prevail in all parts of the territory and in all seasons. Grazing and livestock activity and herd management have several interesting functions or features: securing land tenure and social prestige; producer number is increasing, driving the estate market and attracting new farmers and livestock investors; industry is strengthening its presence; and small and middle enterprises are redeemed by large industries, which has the steadying effect of rendering the territory more attractive for farmers, etc. A virtual interaction cycle is taking place, to produce and expand a landscape specialized in cattle production. The industrial firms involved are the largest in the country; they are connected to the global market. Currently, with the deforestation publicity, the image of these firms is suffering, or is likely to become worse. They are playing a leading role in convincing their suppliers to respect the specific environmental legislation in the Amazon region and to improve their practices. These companies compensate, in some ways, for the poorly efficient role of the government on the pioneering front. They wield enough power and pressure to promote practices of deforestation avoidance, ecological intensification, introduction of trees and landscape reconstruction. Indeed, in a win–win strategy, such practices impart an image of environmental sustainability to the product that the companies develop.

Considering the diversity of land-use systems at the landscape and regional level for regulating environmental fluxes and biodiversity, tropical grasslands are a key component. The aims of land-use management, however, differ. In the Brazilian Amazon, the first role of grasslands is to fight against forest regrowth. *Brachiaria* grassland is the vegetation cover that better resists bush and tree recruits, thus 'delaying' forest return. This is important when we consider that, locally, land becomes open, is appropriated and is economically valued only if it is deforested. From an environmental point of view, however, grasslands often have negative impacts, especially for the water cycle, but also for soil and biodiversity, because they are often poorly managed pastures. On the other hand, this mismanagement results in weed encroachment, which evolves into shrub and secondary tree formation, which then supplies ecosystem services, including biodiversity rebuilding and more fixed carbon than seen in a climactic primary forest ecosystem. Secondary forests now cover 20% of the deforested areas in Amazonia, which compensates somewhat for deforestation's negative impact.

In Africa, livestock provide a key mechanism for managing biodiversity and risk, but incoming populations are fragmenting rangelands in many places, making it increasingly difficult for pastoralists to maintain access to feed and water resources. In Mali, grasslands are fundamental to agricultural system functioning at village and landscape levels. Depending on their extent and degradation level of fodder resources, they define the village carrying capacity. Landscapes must provide sufficient space for grazing during the rainy season, when the rest of the cultivated area is not available for cattle grazing, and satisfy some of the herd feeding in the dry season. Too fragmented or too degraded rangelands are indicative of a system in crisis.

Mali grasslands are collective, and therefore it is very often the 'tragedy of the commons'. Everyone exploits the resources as soon as possible, before the neighbour does. And when it comes to implementing grassland-improving practices, nobody agrees to spend their time and effort because it would primarily serve others. So, in these situations, it would be interesting to promote better coordination and collective actions.

In Amazonia, a central problem derived from landscape specialization is *Brachiaria* monoculture, which is the absolute grass. This species suffers from a syndrome that results in the plant death, and is poorly known but spreading to all regions. The Amazon is undergoing a reform in the pasture-management process, to diversify and increase productivity and to offset the area reclaimed from the forest, or to be replanted, while avoiding this syndrome. It is now a somewhat inverse dynamic, with the trend no longer towards simplification but rather towards the diversification of systems. Industrial chains tend to specialize in the territory, but, at the same time and gradually, forestry, fruit growing and mechanized grain production become ever more attractive options. It is particularly the case for dairy farmers, who tend to produce grain for their cows because they have no more forests to cut to improve their feeding resource production.

In other situations, such as Mali and Rio Verde, it is around material flows that interactions are organized. The internal arrangement in the territory has no great importance, whereas, in the Amazon, connections between forest fragments are of high importance. In Mali, transportation is a major constraint: organic matter transformation places, such as pools and parks, have to be located close to the fields to optimize flows.

Herd size is an important factor in the balance and effectiveness of these flows: small cattle numbers mean poor valorization and nutrient recycling of resources and by-products, while too many heads lead to system saturation and inefficiency. In Mali, nobody regulates the sum of individual cattle, goats and sheep herds. Each individual has their own strategy of increasing the herd as much as possible. So, in many situations, landscape capabilities are rapidly saturated. By contrast, in Rio Verde, the poultry and swine industries manage livestock numbers and distribute permissions to open poultry cooperatives and farms. This regulation by the industry is an important element in balancing these chain–territory interactions, between crop and livestock.

All this stresses the necessity for developing spatially explicit models for taking into account transport and flow issues, which become a major constraint in integrated or mixed systems. In Mali, the location of fodder storage and organic matter transformation points is important. In the Amazon, activities need to be located according to environmental potential, notably in relation to water and relief, and the spatial arrangement is important in optimizing the provision of ecosystem services (connections, groups, critical sizes and managing ground-water basins). Nor is it only the agronomic operational functioning that is disrupted if the systems are not spatially organized. Such landscape optimization also manages social problems of dispersion, enclaves, lower access to services and higher infrastructure maintenance costs.

28.5 Conclusions

The consequences of the changing contexts within landscapes differ for the various system uses (extensive, intensifying and mixed to crop grassland), as well as for the different players within these systems. Urbanization and expanding markets provide opportunities for crop and livestock producers and associated market agents to access and increase their basic revenue. However, the increasing importance attached to product quality and to safety by more affluent consumers, particularly those in urban areas, often means that some grassland production systems are relegated to markets for low-value products or squeezed out of the markets altogether.

The three case studies discussed above show the diversity of integration possible between grasslands and other land uses across landscapes. This variability is illustrated by the contrasting balances and trade-offs between animal numbers, cultivated or natural land producing and/or receiving nutrients, and ensuring adequate biodiversity levels and ecological functions at the landscape and watershed scale. In all cases, the balances and flows are controlled by the players. This raises two critical issues: on the one hand, the implementation of innovative practices and, on the other hand, the coordination processes. Innovative practices are necessary, as shown in the three cases, resulting in the long and complex process of building innovations from existing techniques.

The systems and trade-offs can be balanced by coordination on a landscape or territory scale. Again the three case studies show different situations: collective management deficiencies by the villagers in Mali, a complete lack of regulation on the Amazonian pioneer fronts, and full control by the livestock industry in Rio Verde. The latter case seems the most effective but it remains fragile in the long term, subject to the strategies of merging firms faced with global markets. In the two other cases, coordinating the integration flows, through functional institutions, is a priority for agricultural development.

References

Bruinsma, J. (2003) *World Agriculture: towards 2015/2030, an FAO Perspective.* Food and Agriculture Organisation, Rome.

Coutinho, C. (2010) Dinâmicas territoriais e sustentabilidade: impactos dos setores avícola e sucroalcooleiro no município de Rio Verde – Goiás, a partir de 1990. MSc thesis, University of Brasília, Brasília.

ELC (2000) *European Landscape Convention.* Council of Europe, Florence, Italy.

FAO (2006) *World Agriculture: Towards 2030/2050.* Interim report. Prospects for food, nutrition, agriculture and major commodity groups. Global Perspective Studies. Food and Agriculture Organisation, Rome.

FAO (2007) *State of the World's Forests 2007.* Food and Agriculture Organisation, Rome.

Herrero, M., Thornton, P.K., Notenbaert, A.M., Wood, S., Msangi, S., Freeman, H.A., Bossio, D., Dixon, J., Peters, M., van de Steeg, J., Lynam, J., Parthasarathy Rao, P., Macmillan, S., Gerard, B., McDermott, J., Seré, C. and Rosegrant, M. (2010) Smart investments in sustainable food production: revisiting mixed crop-livestock systems. *Science* 327, 822–825.

Mertens, B., Poccard Chapuis, R., Piketty, M.G., Laques, A.E. and Venturieri, A. (2002) Crossing spatial analyses and livestock economics to understand deforestation processes in the Brazilian Amazon: the case of Sao Félix do Xingu in South Para. *Agricultural Economics* 27, 269–294.

Millennium Ecosystem Assessment (2005) *Rapport de Synthèse de l'Évaluation des Écosystèmes pour le Millénaire.* Island Press, Washington, DC.

Poccard-Chapuis, R. (2004) Les réseaux de la conquête. Rôle des filières bovines dans la structuration de l'espace sur les fronts pionniers d'Amazonie Orientale brésilienne. PhD thesis, Paris X- Nanterre University, Paris.

Rapidel, B., Poccard Chapuis, R., Bélières, J.F., Traoré, B.S., Coulibaly, J., Diallo, D. and Lesnoff, M. (2006) Biomass production, transformation and allocation at the village scale in Southern Mali: development of a conceptual model. In: *Changing European Farming Systems for a Better Future. New Visions for Rural Areas.* Wageningen Academic Publishers, Wageningen, The Netherlands, pp. 144–148.

Rosegrant, M.W., Fernandez, M., Sinha, A., Alder, J., de Fraiture, C., Eickhout, B., Fonseca, J., Huang, J., Koyama, O., Pingali, P., Ringler, C., Robinson, S., Thornton, P., van Vuuren, D. and Ahammad, H. (2009) Looking into the future for agriculture and AKST (Agricultural Knowledge Science and Technology). In: McIntyre, B.D., Herren, H.R., Wakhungu, J., Watson, R.T. (eds) *Agriculture at a Crossroads.* Island Press, Washington, DC, pp. 307–376.

Staal, S.J., Nin Pratt, A. and Jabbar, M. (2008) *Dairy Development for the Resource Poor: a Comparison of Dairy Policies and Development in South Asia and East Africa.* PPLPI Working Paper No. 44. (3 part series). Food and Agriculture Organisation, Rome.

Steinfeld, H., Gerber, P., Wassenaar, T., Castel, V., Rosales, M. and de Haan, C. (2006) *Livestock's Long Shadow: Environmental Issues and Options.* Food and Agricultural Organisation, Rome.

Steinfeld, H., Mooney, H.A., Schneider, F. and Neville, L.E. (2010) *Livestock in Changing Landscape. Drivers, Consequences and Responses.* Island Press, Washington, DC.

van Vuuren, D.P., van Vliet, J. and Stehfest, E. (2009) Future bio-energy potential under various natural constraints. *Energy Policy* 37, 4220–4230.

World Bank (2010) *Rising global interest in farmland: Can it Yield Sustainable and Equitable Benefits?* World Bank, Washinton DC.

Index